Earth Rotation: Solved and Unsolved Problems

NATO ASI Series

Advanced Science Institutes Series

A series presenting the results of activities sponsored by the NATO Science Committee, which aims at the dissemination of advanced scientific and technological knowledge, with a view to strengthening links between scientific communities.

The series is published by an international board of publishers in conjunction with the NATO Scientific Affairs Division

A Life Sciences B Physics	Plenum Publishing Corporation London and New York
C Mathematical and Physical Sciences	D. Reidel Publishing Company Dordrecht, Boston, Lancaster and Tokyo
D Behavioural and Social Sciences E Engineering and Materials Sciences	Martinus Nijhoff Publishers The Hague, Boston and Lancaster
F Computer and Systems Sciences G Ecological Sciences	Springer-Verlag Berlin, Heidelberg, New York and Tokyo

Series C: Mathematical and Physical Sciences Vol. 187

Earth Rotation: Solved and Unsolved Problems

edited by

Anny Cazenave

Groupe de Recherche de Géodésie Spatiale,
Centre National d'Etudes Spatiales,
Toulouse, France

D. Reidel Publishing Company

Dordrecht / Boston / Lancaster / Tokyo

Published in cooperation with NATO Scientific Affairs Division

Proceedings of the NATO Advanced Research Workshop on
Earth Rotation: Solved and Unsolved Problems
Château de Bonas, Gers, France
11-13 June 1985

Library of Congress Cataloging in Publication Data

NATO Advanced Research Workshop on Earth Rotation: Solved and Unsolved Problems (1985 : Gers, France) Earth rotation – solved and unsolved problems.

(NATO ASI series. Series C, Mathematical and physical sciences; vol. 187)
"Published in cooperation with NATO Scientific Affairs Division."
"Proceedings of the NATO Advanced Research Workshop on Earth Rotation: Solved and Unsolved Problems, Château de Bonas, Gers, France, 11–13 June 1985"—T.p. verso.
Includes bibliographies.
1. Earth—Rotation—Congresses. I. Cazenave, Anny. II. Title. III. Series: NATO ASI series. Series C, Mathematical and physical sciences; no. 187.
QB633.N38 1985 525′.35 86–17826
ISBN 90–277–2333–8

Published by D. Reidel Publishing Company
P.O. Box 17, 3300 AA Dordrecht, Holland

Sold and distributed in the U.S.A. and Canada
by Kluwer Academic Publishers,
101 Philip Drive, Assinippi Park, Norwell, MA 02061, U.S.A.

In all other countries, sold and distributed
by Kluwer Academic Publishers Group,
P.O. Box 322, 3300 AH Dordrecht, Holland

D. Reidel Publishing Company is a member of the Kluwer Academic Publishers Group

Printed in The Netherlands

TABLE OF CONTENTS

PREFACE

The idea for organizing an Advanced Research Workshop entirely devoted to the Earth rotation was born in 1983 when Professor Raymond Hide suggested this topic to the special NATO panel of global transport mechanism in the Geosciences. Such a specialized meeting did not take place since the GEOP research conference on the rotation of the Earth and polar motion which was held at the Ohio State University (USA) in 1973.

In the last ten years, highly precise measurements of the Earth's rotation parameters and new global geophysical data have become available allowing major advance to be made in the understanding of the various irregularities affecting the Earth's rotation. The aim of the workshop was to bring together scientists who have made important contributions in this field during the last decade both at the observational and geophysical interpretation levels. The conference was divided into four main topics. The first session was dedicated to the definition, implementation and maintenance of the terrestrial and celestial reference systems. A few critical points have been identified as requiring further improvements: (i) appropriate selection of terrestrial sites recognized for their long term stability, (ii) determination of the relationship between terrestrial and celestial references systems as well as between the various terrestrial ones, (iii) improvment of the theory of a rotating elastic earth (the recently adopted theory needs already some corrections). The others papers of this session were devoted to the various methods of observation of the Earth's rotation parameters. For short terms irregularities (years or less) new techniques based on tracking of artificial satellites, Lunar-Laser Ranging and Long Base Interferometry have now supplanted astrometric methods, but as underlined by several speakers, systematic errors affecting each method are not yet fully estimated.

For decade and secular variations, the best approach remains based on observations of solar eclipses (second session). Recomputation of secular changes in the length of day shows a clear evidence for a non tidal component of acceleration in the last millenia, which could be explained by a change in J_2 resulting from post glacial uplift. A review of the spectrum of climatic variations and possible causes in the range 10^4-10^9 years was presented. During the Quaternary ice age, changes in the mean global temperature and subsequent changes in the ice budget and mean sea level could have induced large changes in the Earth rotation.

Changes in the angular momentum of the atmosphere drive most of the fluctuations in the length of day over times of a year or less (third session). The residuals are now known well enough to investigate the role of the oceans in the Earth angular momentum budget. However, while estimates of atmospheric momentum are directly available from the atmospheric models used in weather forecasting, similar oceanic estimates are currently not available. These estimates could be used to explore the interchanges of angular momentum between oceans, atmosphere and Solid Earth. The two main problems of the Chandler wobble (excitation and dissipation) have been discussed in the fourth session. The two most likely candidates for dissipating wobble energy are mantle anelasticity and dynamic response of the oceans to wobble. Concerning the Chandler wobble excitation there is strong evidence that the direct effect of recent earthquakes cannot supply sufficient power. Examination of boundary layer flows at the core-mantle boundary, on the other hand, leads to the conclusion that these may exert sufficient pressure coupling on an elastic mantle to excite a significant part of the Chandler wobble. Models of electromagnetic coupling have been shown to be able to produce fluctuations in the length of day which are quite well correlated with observations over the last century. Presentations of core dynamics models have ended this last session. To conclude the workshop, general recommandations have been formulated. These appear at the end of this book.

Lectures have been given by scientists of international fame who made important contributions in the recent years. Much discussion took place informally during and after the lectures, leading to many fruitfull interactions between lecturers and participants. It was an unanimous impression that the workshop was a success.

We are very grateful to NATO who allowed this workshop to take place offering financial support and travel grants to most participants. We thank also the council of Europe for its co-sponsorship which has been much helpful.

LIST OF ALL PARTICIPANTS

(including the Director (D), Lecturers (L), Students (S) and other Participants/Observers (0))

MELCHIOR P.	Prof.	L	Observatoire Royal de Belgique 3, Av. Circulaire, 1180-BRUXELLES, BELGIQUE
PAQUET P.	Prof.	D	Observatoire Royal de Belgique 3, Av. Circulaire 1180-BRUXELLES, BELGIQUE
BERGER A.	Prof.	L	Institut d'Astronomie et de Géophysique G. Lemaître Université de Louvain 2, Chemin du Cyclotron, BP 1348 LOUVAIN LA NEUVE, BELGIQUE
DEHANT V.		L	Institut d'Astronomie et de Géophysique, G. Lemaître Université de Louvain 2, Chemin du Cyclotron, BP 1348 LOUVAIN LA NEUVE, BELGIQUE
MORRISON L.V.	Prof.	L	Royal Greenwich Observatory, Herstmonceux Castle Hailsham East Sussex, BN271RP, UK
ZSCHAU J.	Prof.	L	Institut für Geophysik, Kiel University, Olshausen Strasse 40-60, D2300 KIEL, RFA
PAULUS M.	Dr.	L	Institut für Sonnen Physik, Schöneckstrasse 6, D-7800 FREIBURG, RFA
HAUSCH W.		0	Institut für Physikalische Geodäsie, 6100 DARMSTADT, RFA
LENHARDT H.		0	Institut für Physikalische Geodäsie, 6100 DARMSTADT, RFA

DJUROVIC G.	Dr.	0	Observatoire de Belgrade, Volgina 7, 11050 Belgrade, Yougoslavie
DICKMAN S.R.	Prof.	L	Dept of Geological Sciences, State University of New York Binghampton, N.Y. 13901, USA
EUBANKS T.M.	Dr.	L	Earth Orientation Measurements Group, J.P.L., 4800 Oak Grove Drive PASADENA, CA 91109, USA
MUELLER I.	Prof.	L	Dept of Geodetic Science, Ohio State University 1958 Neil Av., COLOMBUS OHIO 43210-1247, USA
ROSEN R.D.	Dr.	L	Atmospheric & Environmental Research INC, 840, Memorial Drive, CAMBRIDGE, MAS. 02139, USA
CARTER W.E.	Dr.	L	Geodetic Research & Development Lab., NOAA-NGS Rockville, MY 20852, USA
DICKEY J.O.	Dr.	L	Earth Orientation Measurements Group, J.P.L., 4800 Oak Grove Drive, PASADENA, CA 91109, USA
BABCOCK A.	Dr.	O	USNO, Time Service Dept, Washington DC 20390, USA
SMYLIE D.E.	Prof.	L	Dept of Earth & Atmospheric Science, York University, 4700 Keele Street, DOWNSVIEW, ONTARIO, CANADA
FLODMARK S.	Dr.	L	Stockholm University, 11346 Stockholm, SUEDE
BOUCHER C.	Dr.	L	Institut Géographique National, 2, Avenue Pasteur 94160 ST MANDE, FRANCE
CAPITAINE N.	Dr.	L	Bureau International de l'Heure 61, Av. de l'Observatoire, 75014 PARIS CEDEX, FRANCE

FEISSEL M.	Dr.	L	Bureau International de l'Heure 61, Av. de l'Observatoire 75014 PARIS CEDEX, FRANCE
GAMBIS D.	Dr.	O	Bureau International de l'Heure 61, Av. de l'Observatoire 75014 PARIS CEDEX, FRANCE
GIRE C.	Dr.	L	Institut de Physique du Globe, 4, Place Jussieu 75230 PARIS CEDEX 05, France
GONELLA J.	Prof.	L	Laboratoire d'Océanographie Physique, 5, Rue Cuvier, 75005 PARIS, FRANCE
HINDERER J.	Dr.	L	Institut de Physique du Globe 5, rue Descartes 67084 STRASBOURG, FRANCE
MIGNARD F.	Dr.	L	CERGA, Av. Copernic 06130 GRASSE, FRANCE
BARLIER F.	Dr.	O	CERGA, Av. Copernic 06130 GRASSE, FRANCE
ROZELOT J.P.	Dr.	O	CERGA, Av. Copernic 06130 GRASSE, FRANCE
REMY F.		O	CERGA, Av. Copernic 06130 GRASSE, FRANCE
VEILLET C.	Dr.	O	CERGA, Av. Copernic 06130 GRASSE, FRANCE
SOURIAU M.	Dr.	L	GRGS, 18, Av. E. Belin 31055 TOULOUSE CEDEX, FRANCE
SOURIAU A.	Dr.	L	GRGS, 18, Av. E. Belin 31055 TOULOUSE CEDEX, FRANCE
LEFEBVRE M.	Dr.	O	GRGS, 18, Av. E. Belin 31055 TOULOUSE CEDEX, FRANCE
LAGO B.	Dr.	O	GRGS, 18, Av. E. Belin 31055 TOULOUSE CEDEX, FRANCE
COUTIN S.		O	CNES, Place Maurice Quentin, 75039 PARIS CEDEX 1, FRANCE

CAZENAVE A.	Dr.	D	GRGS, 18, Av. E. Belin 31055 TOULOUSE CEDEX, FRANCE
CALMANT S.		S	GRGS, 18, Av. E. Belin 31055 TOULOUSE CEDEX, FRANCE
GAUDON P.		S	GRGS, 18, Av. E. Belin 31055 TOULOUSE CEDEX, FRANCE
SPAUTE D.		S	GRGS, 18, Av. E. Belin 31055 TOULOUSE CEDEX, FRANCE

LIST OF AUTHORS

BERGER A.	Institut d'Astronomie et de Géophysique G. Lemaître Université de Louvain 2, Chemin du Cyclotron, BP 1348 LOUVAIN LA NEUVE, BELGIQUE
BOUCHER C.	Institut Géographique National, 2, Avenue Pasteur 94160 ST MANDE, FRANCE
CAPITAINE N.	Bureau International de l'Heure 61, Av. de l'Observatoire, 75014 PARIS CEDEX, FRANCE
CARTER W.E.	Geodetic Research & Development Lab., NOAA-NGS Rockville, MY 20852, USA
DEHANT V.	Institut d'Astronomie et de Géophysique, G. Lemaître Université de Louvain 2, Chemin du Cyclotron, BP 1348 LOUVAIN LA NEUVE, BELGIQUE
DICKEY J.	Earth Orientation Measurements Group, J.P.L., 4800 Oak Grove Drive, PASADENA, CA 91109, USA
DICKMAN S.R.	Dept of Geological Sciences, State University of New York Binghampton, N.Y. 13901, USA
DJUROVIC G.	Observatoire de Belgrade, Volgina 7, 11050 Belgrade, Yougoslavie
EUBANKS T.M.	Earth Orientation Measurements Group, J.P.L., 4800 Oak Grove Drive PASADENA, CA 91109, USA

GIRE C. — Institut de Physique du Globe,
4, Place Jussieu
75230 PARIS CEDEX O5, FRANCE

GONELLA J. — Laboratoire d'Océanographie Physique,
5, Rue Cuvier,
75005 PARIS, FRANCE

HINDERER J. — Institut de Physique du Globe
5, Rue Descartes
67084 STRASBOURG, FRANCE

MELCHIOR P. — Observatoire Royal de Belgique
3, Av. Circulaire,
1180-BRUXELLES, BELGIQUE

MIGNARD F. — CERGA, Av. Copernic
06130 GRASSE, FRANCE

MORRISON L.V. — Royal Greenwich Observatory,
Herstmonceux Castle Hailsham
East Sussex, BN271RP, UK

PAQUET P. — Observatoire Royal de Belgique
3, Av. Circulaire
1180-BRUXELLES, BELGIQUE

PAULUS M. — Institut für Sonnen Physik,
Schöneckstrasse 6,
D-7800 FREIBURG, RFA

ROSEN R.D. — Atmospheric & Environmental
Research INC, 840,
Memorial Drive,
CAMBRIDGE, MAS. 02139, USA

SMYLIE D.E. — Dept of Earth & Atmospheric
Science, York University,
4700 Keele Street,
DOWNSVIEW, ONTARIO, CANADA

SOURIAU A. — GRGS, 18, Av. E. Belin
31055 TOULOUSE CEDEX, FRANCE

SOURIAU M. — GRGS, 18, Av. E. Belin
31055 TOULOUSE CEDEX, FRANCE

CELESTIAL REFERENCE SYSTEMS

C. BOUCHER
Institut Géographique National
2 avenue Pasteur
94160 SAINT-MANDE

ABSTRACT

Three concepts are related to the topic of celestial reference system, namely the ideal inertial or celestial systems, their realizations as conventional inertial or celestial systems, implying a set of models and constants, and the materialisation by some objects for which one gives a set of coordinates, forming what is called a conventional inertial or celestial frame.

The ideal systems are defined in the frame of newtonian or better relativistic theories, having an origin, a scale and an orientation. Celestial systems are usually barycentric or geocentric, their orientation is ecliptic or equatorial.

Various realisations are now possible, either cinematical (quasars with VLBI, stars with ground or space optical astrometry) or dynamical (planets with radar and astrometry, Moon with Lunar laser ranging, Earth artificial satellites with various tracking techniques).

Corresponding to these techniques, various frame are available : radio or star catalogues, Lunar, planetary or satellite ephemerides...

Each of these frames have their own interests and limitations. A major achievement is the comparison of these systems through various methods : occultations, VLBI, DVLBI, space astrometry,...

A. Cazenave (ed.), Earth Rotation: Solved and Unsolved Problems, 1–8.

1. CONCEPTS AND REALIZATIONS

The Celestial reference systems can be defined as spatial references which are not rotating with the Earth in its diurnal motion, but rather fixed, i.e. that celestial objects have slow motions, at least in direction from the origin.

To go into more details, and following a terminology which becomes more widely accepted we can distinguish three levels :

a) ideal celestial reference system

This level is the adoption of reference systems which will fulfill the broard idea expressed in the word "celestial". It will depend first of the underlying physical theory which has been selected.

In the frame of the newtonian theory, a reference system will be currently identified as an affine euclidian frame of the 3d-space (O,E) where O, the origin will be any point of the space and $E = (\vec{E1}, \vec{E2}, \vec{E3})$ an basis of the associated vector space.

This basis will be selected as orthogonal, with a common length of basis vector, close to unity (in SI units, i.e. meters).

This choice defines the orientation and the scale of the system.

Having such a system, one can express coordinates of any point of the space, either as cartesian, or as spherical, geographical...

Furthermore, an inertial system is identified in newtonian theory to a galilean system, i.e. where no inertial force occure.

Two such inertial frames are related by a constant rotation and a translation which changes with time linearly.

One can also define a quasi-inertial frame by a frame in which inertial forces are small with regard to a given standard. In particular, one can consider frames with a constant rotation (no Cariolis force) and a time varying translation (with a small acceleration).

In a viable theory of gravitation, such as the General Relativity, one can use the post-Newtonian expansion using the fact that bodies in the Solar System has a slow motion and that the gravitation related fields are weak (Will, 1981).

We can then define a reference system as a local coordinate system where a 3d spatial part is quasi-cartesian.

In particular, PPN coordinates are suitable for celestial systems.

One can also build locally lorentzian systems moving with the Earth but non rotating (Fermi coordinates).

In any case, celestial reference systems will be characterized by :

- their origin

Two choices are currently done :
. barycenter of the Solar System (barycentric system)
. center of mass of the Earth (geocentric frame)

Various slightly different options can also be selected : center of mass of the Sun, barycenter of the Earth - Moon system...

- their scale, close to the SI unit of length (meter).

If one consider a completely different unit (suchas AU) its conversion factor to the SI meter is then a part of the definition of the system.

- their orientation

Several possibilities exist, taking into consideration some particular directions such as the ecliptic or equatorial polar axis. We just can refer to standard textbooks.

b) conventional celestial reference systems (CCRS)

Having selected a particular ideal celestial reference system, it is then necessary to realize it in the modelling of some measurements. All models, constants and algorithms, which identily such a realization, define operationally the corresponding CCRS.

There are two major types of realizations :

1) cinematical realizations where one applies the MACH'S principle to distant objects. Their directions realize such systems.

For instance, compact radiosources (extragalactic like quasars or galactic like radiostars) are used with Very Long Baseline Interferometry (VLBI). The definition must take into account source structure, and proper motion for radiostars.

stars are used by optical (ground or spaceborne) astrometry. The definition here accounts for IAU constants, galactic rotation on proper motions.

2) dynamical realizations use dynamical models for the motion of various objects :

Planets are used with various techniques (optical astrometry, radar, planetary probes...). The definition includes here IAU constants.

The Moon is specifically used with lunar laser ranging. The definition include lunar theory and tidal dissipation.

Artificial Earth's satellites are used by numerous tracking techniques (laser, radio...). Here the definition includes mainly the force model.

c) conventional celestial reference frames (CCRF)

A given CCRS, is materialized by determining a consistent set of coordinates (or directions) of a selected list of objects. This is called a conventional celestial reference frame.

Corresponding to the previously mentionned systems, the currently available frames are :

- catalogues of compact radiosources
- star catalogues
- planetary ephemerides
- lunar ephemerides
- satellite ephemerides

2. REVIEW OF MAJOR CELESTIAL FRAMES

2.1 - Radiosource catalogues

Several catalogues exist, taking as basic object extragalactic compact radiosources (quasars). JPL, NASA/GFSC and NGS has published catalogues containing from a few tens to a few hundreds of source positions at the mas (milliarcsecond) level.

The relative consistency of theese catalogues (on common objects) is also at this level (see for instance Niell et al. 1984).

Theese catalogues are established by VLBI and are the best celestial frame presently achievable. They also correspond to the best realization of an inertial frame based on the Mach's principle as quasars are among the most remote objects of the Universe.

Thanks to VLBI, the underlying celestial system is a pure cinematical one : VLBI ensures a direct geometric connection beetween the celestial frame with quasars and the terrestrial frame with radio-telescopes.

The major present limitation comes from source structures which exist for most objects at the mas level, and their variations with time and frequency.

Present technology can nevertheless ensure to monitor theese structures in order to maintain the frame at the mas level.

2.2 - Optical catalogues

This type of frames is unidely used. The official IAU celestial frame is a member of this type (FK4).

It should be replaced by FK5 sometime in the near future.

As star direction are slowly changing with time due to galactic rotation,..., position and velocities (proper motions) are given in theese catalogues. The current use is also to connect them to dynamical equinox by sun and planet optical observations.

The overall accuracy of optical ground based catalogues does not exceed a few 0"01, ten times worse than VLBI ! (Turon - Lacarrieu 1984).

New space based optical catalogues should be able to compete with the VLBI accuracy level.

This will be the case of the HIPPARCOS catalogue, with 100 000 stars.

2.3 - Planetary/lunar ephemerides

Theese ephemerides are provided by analytical (BDL) or numerical ways (JPL, MIT). Such frames are used in particular for celestial mechanics and spacecraft navigation during planetary missions.

The lunar ephemerides are more specifically used for analysis of lunar laser ranging and Earth Moon System Studies.

The accuracy is depending on the planet.

2.4 - Satellite frames

Satellite ephemerides are currently used for geodetic point positioning (ex. Transit, GPS) or specific reductions (ex. satellite radar altimetry). As they are time functions, they are rather converted from celestial to terrestrial reference system. The accuracy can reach a sub-meter level, and sub-decimeter in a close future, at least for short arcs. Dynamical models are a strong limitation for long term stability in the celestial frame.

3. INTERCONNECTIONS OF CELESTIAL FRAMES

The accurate knowledge of the transformation formula beetween two celestial systems is a key point for unification of systems.

Such intercomparisons also give an estimate of the relative accuracy of the two frames ; repeated intercomparisons also provide ideas about stability.

Many ways has been already done, or are planed, and many others will certainly appear in the future.

We can summarize the present list as follow :

- use of common objects.

This can be done with two frames of the same type :

- optical observation of planets.

This is currently used to build ground based optical catalogues.

For HIPPARCOS, observation of minor planets will tie the HIPPARCOS frame to a planetary frame :

- occultation of stars by planets
- occultation of radiosources by planets
- differential VLBI beetween planetary orbiters/landers and quasars (sue Newhall et al. 1984)
- observation of optical counterparts of quasars
- VLBI measurements an radiostars, already achieved on 8 objects at a few 0"01 (Lestrade et al. 1984, 1985)
- other plans (e.g. Bertotti et al. 1984, Bauersima 1984).

4. CONCLUSIONS

Several types of celestial frames are available and have their own interests.

The major task remains to improve each of them and to connect them together, and particularly to the future primary frame which will likely be a radiosource catalogue.

REFERENCES

NEWHALL X.X.,
PRESTON R.A., ESPOSITO P.B.

Relating the JPL VLBI reference frame and the planetary ephemerides. IAU Symp. 109 "Astrometric techniques" Gainesville 1984.

NIELL A.E. et al.

The JPL/DSN J2000 Radio reference frame.
IAU Symp. 109 "Astrometric techniques" Gainesville 1984

LESTRADE J.F. et al.

Results of VLBI observations of Radio stars and their potential for linking the Hipparcos on Extragalactic reference frames. IAU Symp. 109 "Astrometric techniques" Gainesville 1984.

GUINOT B.

Basic problems in the kinematics of the rotation of the Earth. IAU Symp. "Time and the Earth's rotation" 1979.

BERTOTTI B. et al.

Linking reference systems from space. Astron. Astrophys. 133 (1984) 231-238

LESTRADE J.F. et al.

Milliarcsecond structures, VLBI and optical positions of 8 Hipparcos radio stars. ESA SP-234 pp. 251-253.

KOVALEVSKY J.

Systèmes de reference terrestres et célestes.
preprint 1984.

TURON- LACARRIEU C.

Matérialisation d'un système de référence par un catalogue d'étoiles.
Cours de Technologie du CNES, Toulouse 1984.

BAUERSIMA I

Coupled Quasar, Satellite and Star Positioning (CQSSP)
Mitt. Satelliten-Beobachtungsstation Zimmerwald n° 13
Bern 1984.

FROESCHLE M., KOVALEVSKY J.

The connection of a catalogue of stars with an extragalactic reference frame - Astron. Astrophys. 116-1 (1982) pp. 89-94

WILL C.

Theory and experiment in gravitational physics -Cambridge University Press. 1981.

THE CONCEPTUAL AND CONVENTIONAL DEFINITIONS OF THE EARTH ROTATION PARAMETERS

N. Capitaine
Observatoire de Paris (UA 1125 du CNRS)
61, Avenue de l'Observatoire
75014 - Paris
France

ABSTRACT. In order to interpret and to compare the various results of the Earth Rotation Parameters (ERP) with a real 0.001" level of precision, it is first necessary to define these parameters with the same order of precision. In this purpose, the conceptual definitions (Eichhorn 1983) which should be used for the ERP are given as well as the conventional definitions corresponding to the geometrical or dynamical principles involved in the various methods used for the determination of the ERP. A special emphasis is made on the definitions of the Celestial Ephemeris Pole (CEP) and of the Universal Time UT1 which have both to be improved. The corrections to be applied to the different observed ERP are given in order to refer them either to the CEP (as conventionally defined in this paper) for geometrical or dynamical observations, or to the instantaneous pole of rotation for gravimetric observations.

1. INTRODUCTION

It will be assumed, in the following, that both an Earth-fixed reference frame,denoted by (To), and a space-fixed reference frame, denoted by (Co), are available in order to consider further the definition, with respect to (To) and (Co), of what are commonly called the Earth Rotation Parameters (ERP).

At first sight, this definition does not raise any problem as the common pratice has been to define the ERP, for the two first ones, as the coordinates m_1 and $-m_2$ of the "pole" in the (To) frame with respect to the terrestrial pole of reference and, for the third one, as the value of an angle around the rotation axis wich can be conventionally linked to UT1.

But, in fact, such a definition is not as obvious as it can appear when a 0.001"-precision is required.

Concerning the coordinates of the pole, the considered pole is not always the same pole: the kinematical definition of m_1 and m_2 (Munk & Mc Donald 1960) refers to the instantaneous pole of rotation, the new IAU reference pole, as given when using the 1980 IAU theory of nutation

A. Cazenave (ed.), Earth Rotation: Solved and Unsolved Problems, 9–24.

(Seidelmann 1982) is the so-called Celestial Ephemeris Pole and the poles to which refer the ERP as determined by different methods can be affected by some systematic biases.

Concerning the third ERP, its classical determinations use the conventional relation (Aoki et al. 1982) between UT1 and the sidereal time and refer to the equinox which has itself a rotation around the axis of rotation with respect to the (Co) frame which induces unnecessary complexities in the corresponding definition of UT1.

Until fifteen years ago, the ERP were provided by the only astrometric method and it seemed obvious that they were referred to the instantaneous axis of rotation and that the third ERP was reckoned from the equinox of date, which was the traditional reference of all astrometric reference frames. Since 1970, other methods provide the ERP: dynamical ones (satellite Doppler tracking, satellite laser ranging (SLR) or lunar laser ranging (LLR)) as well as geometrical ones (Very long base interferometry: VLBI) but, until a few years ago (namely ten years), the precision of the results do not allow to distinguish between the slightly different "poles" to which they refer.

But, at the present time, in order to interpret and to compare the polar coordinates with the current 10^{-3} " level of precision, it is necessary that these coordinates be referred to the same pole with a comparable accuracy. Moreover, for modern observations as SLR or VLBI, the equinox of date is an unnecessary intermediate reference which can induce some spurious variations in UT1; it would be better to define UT1 as an angle proportional to the rotation of the Earth in space as suggested by Guinot (1979).

So, with such an improved accuracy of the measurements, have come now the requirements that the definition of both the polar coordinates and UT1 be stated more precisely and especially that the involved concepts and the realization of these concepts be clearly distinguished (Guinot 1984).

In this attempt, Eichhorn's terminology (1983) will be used here concerning the two essentially different kinds of definitions which are the "conceptual definitions" in accordance with fundamental principles, and the "conventional definitions" as given by a numerical model applied in the reductions of the observations. These two kinds of definitions have generally been mixed in the particular case of the ERP, as well as for the pole which is considered, as for the angle linked to the angular rotation of the Earth; the aim of this paper is to give first the conceptual definitions of the parameters related to the Earth's rotation and, then, to see how the observed parameters can realize them by some conventional model.

2. CONCEPTUAL DEFINITIONS

The conceptual definition of the parameters related to the Earth's rotation depends on whether the considered concept is the instantaneous rotation or the celestial orientation of the Earth.

2.1. The conceptual definition of the Earth Rotation Parameters

Let ω_1 , ω_2 , ω_3 be the three components, relative to the (To) frame, of the rotation vector $\vec{\omega}$ of the (To) frame (also denoted $Gx_0y_0z_0$) with respect to the (Co) frame (also denoted $GX_0Y_0Z_0$).

The Earth Rotation Parameters (or ERP), which will be devoted here to the instantaneous rotation of the Earth (and are thus related to the vector $\vec{\omega}$), can then be conceptually defined as the small dimensionless quantities m_1, m_2, m_3 (see Fig 1) such that:

$$\omega_1 = \Omega\, m_1 \quad , \quad \omega_2 = \Omega\, m_2 \quad , \quad \omega_3 = \Omega(1+m_3) \qquad (1)$$

Ω being a constant equal to the mean angular velocity of the Earth.

Due to the proximity of the axes $\vec{\omega}$ and $\vec{Gz_0}$, m_1 and m_2 are the coordinates of the instantaneous pole of rotation, I, with respect to To (intersection of the $\vec{Gz_0}$ axis with the terrestrial sphere) in the tangent plane to the terrestrial sphere at To. m_3 is the relative variation of the Earth's angular velocity around the third axis of the (To) frame.

These parameters m_1, m_2, m_3 also appear as the solutions of Liouville's equations (Munk & Mc Donald 1960) which govern the polar motion $m = m_1 + i\, m_2$ of the axis of rotation with respect to the (To) frame and the lenght of day variation m_3.

The conceptual definition of the ERP is then very clear as soon as the terrestrial reference frame is chosen. In the case considered here the m_1, m_2, m_3 are referred to the so-called "geographic frame" (To) which, after Chao, (1984) can be legitimely identified with the Tisserand's terrestrial frame.

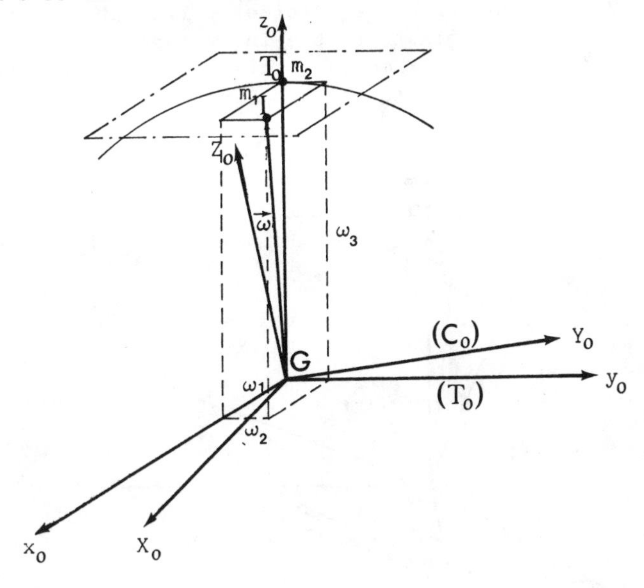

Fig. 1 The Conceptual Definition of the ERP

2.2. The conceptual definition of the Earth Orientation Parameters

The orientation of the (To) frame with respect to the (Co) frame is theoretically given by three parameters which can be chosen in an infinite number of ways. They can be related to the Euler's angles θ,ψ,Φ between these two frames.

The two first angles θ and ψ which give the orientation of the $\vec{GZ_0}$ axis in the (To) frame, can be replaced (because of the proximity of the $\vec{Gz_0}$ and $\vec{GZ_0}$ axes) by the coordinates of the celestial pole of reference, Co , with respect to To, in the tangent plane to the terrestrial sphere at To. The two first Earth Orientation Parameters (or EOP) can then be conceptually defined by these coordinates denoted x and y.

In the real case, it happens that the celestial orientation of the Earth is the sum of a "predictable" component (due to the precession and nutation motions) and an "unpredictable" component (due to the "polar motion" and the variations of the Earth's angular velocity). So the "predictable" component of the coordinates x and y can be conceptually assumed to be perfectly taken into account in order to define a "conceptual point" P. It is the celestial pole of reference, Co, freed from all its forced diurnal motions with respect to the (To) frame.

The terrestrial coordinates u and v of this conceptual pole P of rotation, which is clearly distinct from the instantaneous pole of rotation, can then be chosen (instead of x and y) as the conceptual definition of the two first EOP.

The conceptual definition of the third EOP must involve the specific celestial Earth's angle of rotation (proportional to UT1) around the axis GP. It can be clearly given by the so-called "stellar angle" (Guinot 1979) $\theta_S = \widehat{\varpi\sigma}$ on the equator of P from the non rotating origin σ , as defined by Guinot, to the instantaneous origin of longitude ϖ , such that, when P moves on the celestrial sphere, the triad $\widehat{GP\sigma}$ has no rotation with respect to (Co) around GP.

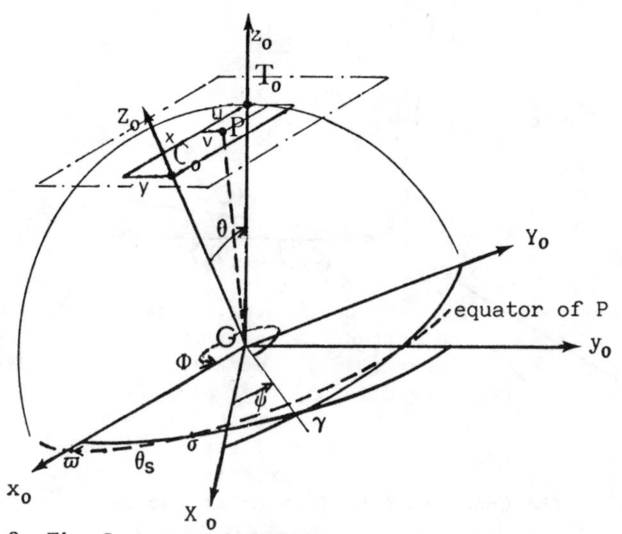

Fig. 2 The Conceptual Definition of the EOP

2.3. The newly adopted definition of UT1: the need for a more convenient conceptual definition.

The classical conceptual definition of the third EOP is, in fact, until now, the true sidereal time around the rotation axis which can be conventionally linked to UT1 (or, for the corresponding ERP, the excess lenght of the mean solar day, which is the time derivative of the rotation angle).

The corresponding definition of UT1 as recommanded by IAU (Aoki et al. 1982) is not a conceptual definition but a conventional one (Xu et al. 1984), which gives a relationship between the mean sidereal time and UT1 to be used with the FK5-based astronomical reference system:

$$\text{GMST1 of 0h UT1} = 241110.54841\text{s} + 8640184.812866\text{s}\ T'u + 0.093104\text{s}\ T'^2u - 6.2\text{s} \times 10\ \ T'^3u \qquad (2)$$

with T'u = d'u/36525, d'u being the number of days of Universal Time since 2000 january 1, 12h UT1.

This does not correspond to a clear concept, as it refers both to the non-rotating origin and to the equinox and as the time argument is a mixture between UT1 and TDT (or Terrestrial Dynamical Time) which would appear in the T'^2u and T'^3u terms.

A more convenient conceptual definition of the third EOP would be, as given in the preceding section, following Guinot (1979,1981), the stellar angle θ_S which itself gives the sidereal rotation of the Earth and its time derivative, directly the angular velocity of the Earth in space.

The corresponding conventional definition of UT1 which was given by Guinot (1979,1981) was:

$$\text{UT1} = k\ (\theta_S - \theta_{S_0}) \qquad (3)$$

the coefficient k being a constant chosen so that a day of UT1 is close to the duration of the mean solar day.

The use of such a parameter θ_S in place of the sidereal time in the rotation matrix from (To) to (Co) implies a corresponding change in the precession and nutation matrixes (Mueller 1981, Aoki & Kinoshita 1983, Guinot et al. 1985).

3. THE CONVENTIONAL DEFINITIONS OF THE PARAMETERS RELATED TO THE EARTH-ROTATION

Concerning the usual polar coordinates, it was considered until recently that they were the terrestrial coordinates of the instantaneous pole of rotation. But Jeffreys (1963), and Atkinson (1973,1975) pointed out that classical astrometry gives in fact the place on the sky of the "Earth's pole of figure". Moreover, Kinoshita et al. (1979) showed that the position of the instantaneous axis of rotation does not specifically occur in the reductions of any of the existing data types. So the new pole as given by the 1980 IAU theory of nutation (i.e. the Celestial

Ephemeris Pole or CEP) was chosen so that it would be the pole to which all observations refer (Leick & Mueller 1979).

However, the definition of the CEP has, itself, to be clarified and moreover each type of method determining the ERP (or EOP) has some characteristics which induce systematic differences in the corresponding computed pole. It can, for instance, exist some kind of observations for which the involved concept is not the celestial Earth's orientation but the Earth's angular velocity.

We will then now consider the conventional definition of the Earth-Rotation (or Orientation) parameters and their realization by the available practical determinations. The expressions of the observed parameters with respect to those as conceptually defined in the preceding section will be given.

Neither the effects of the differences between the (To) frame and the (Co) frame used for each method will be considered here nor the effects of the errors in the modelization of the acting forces, but only the characteristics of each method in its determination of the ERP (or EOP).

3.1. Geometrical observations

3.1 a The general principles

The geometrical observations give the position of the terrestrial frame (To) with respect to the celestial frame (Co).

Following notations of Kinoshita et al. (1979) and of Zhu and Mueller (1983), the orientation matrix between the two frames, can be written:

$$A_T^C(t) = R_2(u_C)\, R_1(v_C)\, R_3(\mathcal{X}_C)\, N_U(t)\, P_R(t) \qquad (4)$$

$R_2(u_C)$, $R_1(v_C)$ and $R_3(\mathcal{X}_C)$ being respectively the rotation matrixes for the "polar coordinates" corresponding to a computed pole P' and to the angle of rotation of the Earth around the GP' axis,
(these three matrixes containing the "unpredictable" part of the Earth orientation in space),

$N_U(t)$ $P_R(t)$ are the rotation matrixes for the precession-nutation of the Earth wich is the "predictable" part of the Earth orientation in space. These two last matrixes derive a computed pole P' (which is a realization of the conceptual pole P) from the celestial pole Co, using a conventional representation of precession (Lieske et al. 1977) and nutation (Wahr 1981, Seidelmann 1982).

The general expression of the orientation matrix $A_T^C(t)$, in addition with the conventional representation of its "predictable" component, give thus the conventional definition of the EOP. This definition is such that the two first EOP, u_C and v_C, are the polar coordinates of the computed pole P' and that the third EOP, $\mathcal{X}_C$, is consequently computed as the angle of rotation around the axis directed to this pole P'. The derived UT1 depends in addition on the conventional relation used between the angle of rotation and UT1, which is presently, as recommanded by IAU, the relationship (2). It should be, as explained in the preceding section, with a slight change in the classical expressions of the matrixes $P_R(t)$ and $N_U(t)$, the conventional relationship between the stellar angle and UT1.

The expressions of the realized parameters with respect to the conceptual ones (as given in the preceding section) can be derived from the involved geometrical and dynamical relations.

Firstly, the Euler's kinematical relations (Woolard 1953, Mc Clure 1973) give for any polar motion $m = m_1 + i\, m_2$ of the rotation axis into the Earth:

$$\dot{\theta} + i\,\dot{\psi}\sin\theta = -\,\Omega\, m\, e^{i\Phi}$$

and then: $$d\theta + i\, d\psi\,\sin\theta = -\Omega\int_{to}^{t} m\, e^{i\Phi}\,dt$$

$d\theta$ and $d\psi$ being respectively the variations in the Euler's angles θ and ψ (between the instants to and t) of the (To) frame with respect to the (Co) frame, corresponding to the terrestrial displacement m of the axis of rotation.

The geometrical relation :

$$u + i\,v = -\,i\,(d\theta + i\,d\psi\,\sin\theta)\,e^{-i\Phi}$$

allows then to write the following expression for the observed EOP u and v as derived by any geometrical method and corresponding to m :

$$u + i\,v = i\Omega\left(\int_{to}^{t} m\, e^{i\Phi}\,dt\right)\, e^{-i\Phi} \tag{5}$$

For a circular polar motion such that:

$$m = \nu e^{i(k\Omega t - \alpha)} \tag{6}$$

The resulting observed coordinates u and v are then such that:

$$u + i\,v = \left[1 / (1 + k)\right] m = \left[1 - \frac{k}{1 + k}\right] m \tag{7}$$

This relation expresses that any observation based on the determination of the celestial Earth's orientation cannot give (due to the geometrical and kinematical principles involved) the terrestrial position of the instantaneous pole of rotation which is moving both into the Earth and in space. It can only give, in fact (Capitaine 1982), the celestial shift $\overrightarrow{ToCo}$ between the terrestrial pole To and the celestial pole Co (which can be replaced by the point P when the "predictable" component of this shift is taken into account). For instance for the chandlerian wobble m of the instantaneous Earth's axis of rotation, the amplitude of the observed motion is the sum of the amplitudes ν of m and $-\left[(C - A)/C\right]\nu$ of the corresponding celestial "sway".

Moreover, the realization P' of the conceptual pole P, is dependent on the used conventional representation of the precession-nutation motion. Thus, the polar coordinates u_C and v_C as derived by any geometrical method countain the unpredictable motion due to geophysical causes as given by (7), but <u>also</u> , (as it appears from (4)) the <u>errors</u> δu , δv due to the imperfections $\delta\theta$, $\delta\psi$ in the precession-nutation model transformed by the rotation of angle Φ , following the relation:

$$\delta u + i\,\delta v = i\,e^{-i\Phi}(\delta\theta + i\,\delta\psi\,\sin\theta)$$

The total observed shift $\overrightarrow{ToP'}$ (including the part $\overrightarrow{ToP}$ due to the kinematical and geometrical principles and the part $\overrightarrow{PP'}$ due to the imperfections in the representation of the precession-nutation motion) appears either as a celestial motion ($d\theta + id\psi \sin\theta$), or as a terrestrial one ($u + i\,v$), according as it is derived before, or after, the rotation of angle χ_c .

3.1 b The CEP: a mixture between concept and realization

The newly defined Celestial Ephemeris Pole (CEP), as adopted in the new IAU 1980 theory of nutation has been chosen in order to be more consistent with the observed one as its forced motion includes the diurnal polar motion of the intantaneous pole of rotation.

But, in fact, its definition is not sufficiently clear. Theoretically it has no diurnal or quasi-diurnal motion with respect to an Earth-fixed or a space-fixed coordinate system and it is thus the pole of the Tisserand's mean axis of the mantle if the free motion is zero (Seidelmann 1981). That gives two conceptual definitions of the CEP.

Practically this CEP is also defined as the pole given by the IAU-1980 theory of nutation which includes the forced diurnal polar motion. This is a conventional definition.

So the definition of the CEP, as given in the 1980 theory of nutation, is a mixture between a conceptual definition and a conventional one.

The considered concept itself is not very convenient because there is no theoretical reason for non observing a diurnal terrestrial polar motion by some kind of observations as, even using a geometrical method, the nearly-diurnal free wobble, if it exists, would be observable. So this concept is not realistic for an Earth-model with fluid core (Xu 1983). There is neither theoretical reason for non observing a diurnal celestial motion such as the "sway" by any kind of method.

Moreover the conventional definition does not specify where belongs the purely diurnal term of 0.0087". In fact, this term is included in the conventional celestial pole and it must be clear that the adoption of the 1980-IAU theory of nutation implies a change from the previous practice in the celestial reference pole of 0.0087" (Capitaine et al. 1985).

The real problem is not that in principle it is not possible to observe a diurnal polar motion either into the Earth or in space, but that the two motions of the axis of rotation in space and the corresponding one into the Earth are not separable by any observation based on the geometrical and kinematical principles considered in 3.1 a.

It would then be better to say that the definition of the CEP is not a conceptual one , but a conventional one : the CEP is the pole as obtained from the celestial reference pole by using the conventional representation of the precession and nutation motion including the so-called forced diurnal polar motion. The CEP is thus in fact the point P' as defined in the section 3.1 a and the convenient conceptual pole corresponding to the CEP should then be the point P as defined in the section 2.2.

Following such a definition, the observations based on the celestial Earth's orientation give, necessarily, the motion of the CEP with respect to the terrestrial reference pole: the observed polar motion is the observed terrestrial motion of the CEP within the Earth.

It should be noted that, when corrected for the conventional precession-nutation motions, the CEP (as defined above) still undergoes retrograde quasi-diurnal motions within the Earth or in space, due firstly to the errors in the conventional representation of this forced motion and secondly to the unpredictable motions of geophysical origin.

The conventional definition of the polar coordinates as derived by observations of the Earth's orientation are then those referred to the CEP (as defined above) which depends on the used conventional representation of the precession-nutation motion. The systematic bias between the CEP and the pole of rotation is the sum of several terms respectively of chandlerian and annual periods (both with an amplitude of the order of 0.001") and of nearly-diurnal periods (with an amplitude of the order of 0.01"), corresponding, through the relation (7), to each free or forced nutation of the Earth.

The third observed EOP is conventionally linked to UT1 and is around the axis directed to the CEP. The resulting systematic difference between the computed UT1 and the corresponding one around the axis of rotation is the sum of several terms respectively of chandlerian, annual and nearly-diurnal periods with amplitudes of the order of 0.0001s.

These conventional definitions of the EOP can be applied, without any correction to all the geometrical observations, as for instance the VLBI observations, but some additional considerations are needed for the classical latitude and time observations.

3.1.c The particular case of classical latitude and time observations.

The classical latitude and time observations (based on the preceding geometrical principles) have two characteristics which can induce additional systematic effects in the determination of the EOP: firstly they are referred to the local vertical and secondly they are approximately made at a constant mean local time.

Concerning the reference to the local vertical (assumed to be corrected for the effect of the luni-solar earth-tides (Melchior 1966)), a systematic effect appears due to the deviation of this vertical produced by the so-called polar tide which is dependent on the ERP. The corresponding variations induced in the observed latitude and Universal Time at a station of latitude φi and East longitude λi can be expressed, using the development of the rotational potential (Munk & Mc Donald 1960) as:

$$\begin{cases} \Delta \varphi_i = (\Lambda\, a\, \Omega^2 / g)\left[(m_1 \cos \lambda_i + m_2 \sin \lambda_i) \cos 2\varphi_i + m_3 \sin 2\varphi_i\right] \\ \Delta UT0_i = (\Lambda\, a\, \Omega^2 / g)\,(- m_1 \sin \lambda_i + m_2 \cos \lambda_i)\, tg\, \varphi_i \end{cases} \quad (8)$$

a being the mean equatorial radius of the Earth, g the acceleration of gravity and ($\Lambda \simeq 1.2$, Melchior 1978), the combination $1 + k - \ell$ of the

Love's and Shida's numbers k and ℓ .

The above variations, dependent on the ERP, induce systematic effects in the classical (BIH or IPMS) global determinations of these parameters which use the well-known relations:

$$\begin{cases} \varphi_i - \varphi_o = u \cos\lambda_i + v \sin\lambda_i + z \\ (UT0-UTC)_i = (UT1-UTC) + (u \sin\lambda_i - v \cos\lambda_i + w) \, tg\varphi_i \end{cases} \quad (9)$$

This effect is then partly removed in latitude when using well distributed stations, but appears in time as a reduction of the observed polar motion by the factor:

$$R = 1 - (\Lambda \, \Omega^2 a/g) \quad (10)$$

which is approximatively equal to 0.9958.

For instance, when u (or v) = 0.5", the resulting systematic error in the computed u (or v) is : du (or dv) = 0.002", corresponding to a spurious displacement of the observed pole which, cannot be neglected when a 0.001" precision is required.

Concerning the constant mean local time of the observations, a systematic effect can appear for any retrograde nearly-diurnal wobble which exhibits itself as a common observed effect for all the stations (Yatskiv 1972). This common effect is thus absorbed in the common non-polar z and w terms when considered in the global determinations (9) of the EOP from classical latitude and time observations. Such computed EOP refer thus to a pole which has no diurnal or quasi-diurnal retrograde motion within the Earth. It must be noticed that this is not the case for the global determinations not including the non-polar terms and that, in both cases, there is no reason for this pole not to undergo diurnal or quasi-diurnal motion in space or such prograde motion within the Earth.

So the pole to which refer the EOP as computed by BIH or IPMS global determinations, is not exactly the CEP as defined above because, in this particular case, any of its nealy-diurnal retrograde motion within the Earth is systematically removed which is is not the case in the general case. The resulting systematic differences in the computed polar coordinates and UT1 are of nearly-diurnal periods with amplitudes respectively equal to the errors in the nutation coefficients and of the amplitude of the celestial free core nutation.

3.2 Dynamical observations

The dynamical observations give the position of the terrestrial frame (To) with respect to a computed orbit which can be the orbit of an artificial satellite or of the Moon. In addition to the geometrical effect as considered in 3.1 a and 3.1 b for the determination of the EOP, the effect due to the dynamical perturbations of the EOP on the observed orbit referred to the (Co) frame must then be considered.

The dynamical effect of the EOP on the orbit is due to the induced variations in the tesseral coefficients C_{ij}, S_{ij} , of the geopotential.

The only significant variations appear in the coefficients C_{21} and S_{21} so that (Gaposchkin 1972, Lambeck 1973): $C_{21} = - u\, C_{20}$, $S_{21} = v\, C_{20}$.

Such modifications in the geopential give rise to corresponding variations in the angular orbital elements (i, ω, Ω) of the satellite (or of the Moon) referred to the (Co) frame, which can be easily expressed, using Lambeck's (1973) developments, when corresponding to a periodic term of the polar motion:

$$m_j = \nu_j \; e^{i(k_j \Omega t + \beta_j)}$$

For the inclination, it gives:

$$\delta i_j = \dot{\Omega} \nu_j \frac{\cos \left[(1+k_j)\Omega t + \beta_j + \Omega \right]}{\left[(1+k_j)\Omega + \dot{\Omega} \right]} \tag{11}$$

The classical geometrical variation Δi_j, (Lambeck 1971) produced by the displacement of the (To) frame with respect to the satellite or the Moon orbit, due to the m_j term of the polar motion, appears with the same argument and with the theoretical amplitude ν_j (which in fact, due to the geometrical considerations in 3 .1 a is not ν_j ,but $\nu_j/(1+k_j)$).

So, the two effects are not separable and the total observed amplitude is :

$$\left[\nu_j / (1+k_j) \right] \left[1 + \frac{\dot{\Omega}}{(1+k_j)\Omega + \dot{\Omega}} \right] \tag{12}$$

The dynamical effect is practically taken into account in the usual orbit computations by using the tesseral coefficients C_{21} and S_{21} corresponding to the extrapolated or interpolated values of u and v , but for short periodic or unpredictable terms, this effect is not computed. In spite of its weak amplitude due to the multiplying factor $\dot{\Omega}$ (the rate of precession of the orbital node), it must not be neglected, especially when a resonance effect occurs ($\dot{\Omega} \simeq -(1+k_j)\Omega$).

So, the polar coordinates, as computed by dynamical methods are referred to the CEP (as defined in 3.1 b) as far as the purely dynamical effect is removed. The determination of the third ERP is itself strongly correlated with the one of the precession of the orbit node of the satellite (or of the Moon) due to the argument $(1+k_j)\Omega + \dot{\Omega}$ appearing in the variations of the orbital elements.

3.3 Gravimetric observations

As it has been shown recently (Richter 1983) that polar motion can be clearly revealed by gravimetric determination of the acceleration of gravity g, the question of the pole to which such kind of observations refer, arises.

The kinematical expression of the absolute gravity in a point M of the Earth of radius vector $\vec{r}\,(r, \varphi, \lambda)$ is, if $\vec{\omega}$ is the Earth's rotation vector:

$$\vec{g}_a = \vec{g}_r - \vec{\Delta g} = \vec{g}_a - \vec{\omega} \wedge (\vec{\omega} \wedge \vec{r}) - \dot{\vec{\omega}} \wedge \vec{r} - 2\, \vec{\omega} \wedge \dot{\vec{r}} \tag{13}$$

The observed variation of the vertical component $\Delta g = \vec{\Delta g}.\vec{n}$ of gravity due to the "polar tide" can then be expressed in function of the ERP m_1, m_2, m_3 as conceptually defined (and of their time derivatives).

The classical expression of this variation (Melchior 1978), only considers the first term, $[\vec{\omega}\wedge(\vec{\omega}\wedge\vec{r})].\vec{n}$, which neglects the velocity of the displacement of the axis of rotation within the Earth and the velocity of the displacement of the point M. This first term being the gradient of a potential, the classical Love's representation can be applied with the multiplying factor $\delta = 1 + h - \frac{3}{2}k$ (h and k being the Love's numbers of degree 2) which represents the effect of the tidal deformation (Melchior 1978) of the Earth on the gravimetric observations.

The variation of the intensity of gravity due to the polar tide can then be written:

$$\Delta g(a) = \delta(a)\left[\ 2\,\Omega^2 a m_3 \cos^2\varphi - \Omega^2 a \sin 2\varphi(m_1 \cos\lambda + m_2 \sin\lambda)\right] \quad (14)$$

Considering the second term of $\vec{\Delta g}$ (in which appear the time derivatives $\dot{m}_1$ and $\dot{m}_2$) does not modifies the expression of $\Delta g(a)$ as the effect of this additional term is removed in the scalar product $\vec{\Delta g}.\vec{n}$. Considering the third term of $\vec{\Delta g}$ (in which appear the time derivatives $\dot{x}_1$, $\dot{x}_2$ and $\dot{x}_3$ of the terrestrial coordinates of the point M) gives rise to an additional term in the vertical component of gravity at M (due to the local non-radial deformation) which can be written:

$$dg(a) = \Omega^2 a \sin 2\varphi\left(\left[-\dot{x}_2/a\Omega\sin\varphi\right]\cos\lambda + \left[\dot{x}_1/a\Omega\sin\varphi\right]\sin\lambda\right) \quad (15)$$

In the case of any <u>free wobble</u> $m = m_1 + i\,m_2$, $dg(a)$ is only due to the deformation corresponding to the polar tide and is then such that the term between the parentheses can be written:

$$\Lambda(a\,\Omega^2/g)\left[-(\dot{m}_1/\Omega)\sin\lambda + (\dot{m}_2/\Omega)\cos\lambda\right].$$

It is clear that for non quasi-diurnal components, such as the chandlerian and annual ones (for which $a\,\Omega\,\dot{m}/g = O(10^{-6} m)$), dg can be neglected and the chandlerian and annual terms of the polar motion computed by gravimetric observations are then referred to the rotation axis. The reduction introduced for the amplitude of the observed free quasi-diurnal component (for which $\dot{m}/\Omega = O(m)$), (such as the free-core nutation) can be neglected due to the order of magnitude of the amplitude itself (< 0.001").

In the case of any luni-solar <u>diurnal forced wobble</u> m (such that $\dot{m}/\Omega = O(m)$), the additional term (function of $\dot{m}$) appearing in d g and due to the displacement of the point produced by the corresponding luni-solar tidal deviation of the vertical, cannot be separated from the corresponding term in m. It prevents the observation of such a wobble to be referred to the axis of rotation.

So, the polar motion computed by gravimetric observations is nearly referred to the axis of rotation as it is the case for its principal components (chandlerian + annual); there is however a systematic bias in its observed quasi-diurnal component due to the principle of these observations and to their reference to the local vertical.

4. CONCLUSION

As shown in the preceding sections, the available observed polar motion is not referred to the instantaneous pole of rotation (as it can be conceptually defined), nor to a pole which has no diurnal or quasi-diurnal motion with respect to an Earth-fixed or a space-fixed reference frame (as conceptually defined in the 1980-IAU theory of nutation). This observed motion is, in fact, referred to a conventional pole which depends, at a 0.001" level of precision, on the systematic biases due to the principles and particular effects of the method of observation. In addition, the third ERP (or EOP), which is conventionally linked to UT1, suffers presently from a lack of a clear corresponding concept.

A clarification of these notions are thus necessary in order to intercompare and interprate the observed available ERP with a 0.001" level of precision.

In order to clarify the notion of the pole, it has been shown that the CEP must first be considered not as a conceptual pole (as it was in the 1980-IAU theory of nutation), but as a conventional pole obtained (from the celestial pole of reference) by using the conventional representation of the Earth's precession-nutation motion. It corresponds to a conceptual pole which can be defined as "the celestial pole of reference freed from all its terrestrial forced diurnal motions". With such a definition, the observed polar coordinates by any of the presently classical method (astrometric, LLR, SLR, VLBI), for which the only critical element is the Earth orientation in space, can be referred (after some particular corrections for the classical astronomical determinations and for the dynamical determinations) to the CEP.

It results of that principle that the observed available polar motion is in fact the sum of the terrestrial motion of the instantaneous pole of rotation and of its corresponding celestial motion. For instance, the observed chandlerian motion is not, as usually considered, the terrestrial motion of the pole of rotation but is, in fact, the sum of the chandlerian motion of the pole of rotation and of the so-called "sway" of now non-negligible amplitude (0.001"). A similar effect appears for the annual wobble. This pole also undergoes, except for the global astrometric determinations, the quasi-diurnal terms (with respect to the Earth-fixed reference frame) due to the errors in the conventional representation of nutation and to the neglected celestial free core nutation. The access to the instantaneous pole of rotation can be obtained for each wobble of known frequency, following the corresponding relation given in this paper.

It must also be noticed that the polar motion, as it will be determined in a near future by gravimetric observations using a supraconducting gravimeter (for which the critical element is the velocity of the observer), does not refer to the same pole. It can be, with the corrections due to the luni-solar tidal deformations, referred to the pole of rotation.

In order to clarify the notion of the third ERP (or EOP) it will be better to use a more convenient parameter or "stellar angle" (Guinot 1979) which can be clearly conceptually defined and then conventionally expressed when using a conventional representation of precession and nutation. The determination of this third EOP must always be understood with respect to the axis directed to the used conventional pole and not with respect to the axis of rotation.

REFERENCES

Aoki, S., Guinot, B., Kaplan, G.H., Kinoshita, H., Mc Carthy, D.D., Seidelmann, P.K., 1982, 'The new definition of Universal Time', Astron.Astrophys. 105, 359.

Aoki, S. and Kinoshita, H., 1983, 'Note on the relation between the equinox and Guinot's non-rotating origin', Celestial Mechanics 29, 335-360.

Atkinson, R. d'E., 1975, 'On the "dynamical variations" of latitude and time', Astron. J. 78, 147.

Atkinson, R. d'E., 1975, 'On the earth's axes of rotation and figure', Mon.Not.Roy.Astron.Soc. 171,381.

Capitaine, N., 1982,'Effets de la non-rigidité de la Terre sur son mouvement de rotation: étude théorique et utilisation d'observations', Thèse de Doctorat d'Etat, Université de Paris VI.

Capitaine, N., Williams, J.G. and Seidelmann, P.K., 1985, 'Clarifi cations concerning the definition and determination of the Celestial ephemeris pole, Astron.Astrophys. 146, 381-383.

Chao, B.F., 1984, 'On the excitation of the Earth's free wobble and reference frames', Geophys.J.R.astr.Soc. ,79, 555-563.

Eichhorn, H., 1983, 'Conceptual and conventional definitions', Philosophia Naturalis.

Gaposchkin, E.,M., 1972, 'Pole position studied with artificial Earth satellites' in Rotation of the Earth ,ed. by P. Melchior and S.Yumi, D.Reidel Publishing Company.

Guinot, B., 1979, 'Basic problems in the kinematics of the rotation of the Earth', in Time and the Earth's rotation, 7-18, D.D. Mc Carthy and J.D. Pilkington eds.

Guinot, B., 1981, 'Comments on the terrestrial pole of reference, the origin of longitudes and on the definition of UT1',in Reference coordinate systems for Earth Dynamics, 125-134, D. Reidel Publishing Company.

Guinot, B., 1986, 'Concepts of reference systems', published in the Proc. of the IAU Symp 109.

Guinot, B., Capitaine, N. and Souchay, J., 1985, in preparation.

Jeffreys, H., 1963, in Nutation and Forced motion of the Earth's pole, foreword, p vii-xix.

Kinoshita, H., Nakajima, Y.K, Nakagawa, I., Sasao, T. and Yokoyama, K., 1979, 'Note on nutation in ephemerides', Publ.Int.Latit.Obs. Mizusawa 12, 71-108.

Lambeck, K., 1971, 'Determination of the Earth's pole of rotation from laser range to satellites', Bull Geod. 101, 263-281..

Lambeck, K., 1973, 'Precession, nutation and the choice of reference system for close earth satellite orbits', Celestial Mechanics 7, 139-155.

Leick, A. and Mueller, I.I., 1979, 'Defining the Celestial Pole', Manuscripta Geodaetica 4, 149-183.

Lieske, J., Lederle, T., Fricke, W., Morando, B., 1977, 'Expressions for precession quantities based upon the IAU (1976) system of Astronomical constants', Astron.Astrophys. 58, 1-16

Mc Clure, P., 1973, 'Diurnal polar motion', Goddard Space Flight Center Doc. No. X - 592-73-259.

Melchior, P., 1966, The Earth Tides , Pergamon Press.

Melchior, P., 1978, The tides of the Planet-Earth , Pergamon Press.

Munk, W.H. and Mac Donald, G.J.F., 1960, The Rotation of the Earth , Cambridge University Press, London.

Mueller, I.I., 1981, 'Reference coordinate systems for Earth Dynamics: a preview ' in Reference coordinate systems for Earth Dynamics , ed. by E. M. Gaposchkin and B. Kolaczek, D. Reidel Publishing company, p 1-22.

Richter, B., 1983, 'The long periodic tides in the Earth Tide spectrum', Proc. of the XVIII IUGG General Assembly, Vol 1, IAG Symposium a, 204-216.

Seidelmann, P.K., 1982,'1980 Theory of nutation: the final report of IAU working group on nutation', Celestial Mechanics 27, 79-106.

Wahr, J.M., 1981, 'The forced nutations of an elliptical, rotating, elastic and oceanless earth', Geophys. J. R. astr. Soc. 64, 705-727.

Xu B.X., 1983, 'Definition and Realization of the celestial pole', Proc. of the 9th International Symposium on Earth Tides, Schweizerbart'sche Verlagsbuchhandlung publisher.

Xu B.X., Ahu, S.Y., Zhang H., 1986 'Discussion of meaning and definition of UT', published in the Proc. of the IAU Symp 109.

Yatskiv, Ya. S., 1972, 'On the comparison of diurnal nutation derived from separate series of latitude and time observations' in Rotation of the Earth, ed. by P.Melchior and S.Yumi, D.Reidel Publishing Company.

Zhu S.Y. and Mueller,I.I., 1983, 'Effects of adopting new precession, nutation and equinox corrections on the terrestrial reference frame', Bull. Geod 57, 29.

HOW TO MEASURE THE EARTH ROTATION

P. Pâquet
Observatoire Royal de Belgique
Av. Circulaire, 3
1180 Brussels
Belgium

ABSTRACT. The methods to measure the Earth Rotation Parameters (ERP) are reviewed together with their advantages and desadvantages. Results obtained during the Merit campaign allow to estimate the std. dev. of each techniques while their intercomparisons shows clearly the high quality of VLBI and SLR. However the other methods as Doppler and classical astronomy are still very valuable respectively for Polar Motion and UT_1 observations.

1. INTRODUCTION

The spectrum of fluctuations in the Earth rotation is very wide and for their measurements the approaches are totally different according the frequency changes considered.

The eclipse recomputations are the best support to determine the secular fluctuations occurred during the last 3000 years (Morrison, 1985) while since 1963, the palaeontologists with the analysis of growth rhythms of corals, shells of mollusc bivalves and stramadolites allow to contribute not only to the understanding of the dynamics of the Earth-Moon system but also to confirm the secular and variable amplitude of the slowing down of the Earth rotation till period as far as 2000 MY in the past, (Wells, 1963; Lambeck, 1980).

On a more intermediate scale of a few thousands to 10^6 years, palaeoclimatologists, analyzing deep sea sediments, succeeded to trace the evolution of the inclination ε of the equator with respect to the ecliptic (Berger, 1980); their results are becoming so confident that their estimations of ε and of the excentricity of the Earth orbit could be considered as an input for the theories of celestial mechanics which attempt to design models of the evolution of the Sun-Earth-Moon system.

Since the beginning of this century, detection of short term variations, such as Polar Motion and seasonal variations, became technically possible. However it has been necessary to wait the years 1950-1960, when quartz clocks were extensively used in the observatories, to notice a greater interest for Earth rotation observations and their modelling. For evidence, just notice that :

A. Cazenave (ed.), Earth Rotation: Solved and Unsolved Problems, 25–44.

. the first recognition of the exchange of energy between ER and atmosphere were published in 1950 (W.H. Munck and al, 1950), (F.H. van den Dungen and al, 1952).
. the needs for a better time scale conducted to account for the Polar motion in 1950 (IAU Trans., Vol. VII) and the seasonal variations in 1955 (IAU Trans., Vol. IX).

From this epoch the precision of the measurements of the three ERP (x, y, UT) has been improved rapidly with the design of new astronomical instrument (Danjon's astrolabe; PZT), new star catalogue (FK4, 1964), new method of computations (B. Guinot and al, 1971) which lead, at the end of the sixties, to one determination, every 5 days, with a std. dev. of 0"02.

Beginning with the zeventies several new techniques started to be deployed and contributed very successfully to the progress of ERP measurements :

- in 1970, the Lunar Laser Ranging (LLR) station of Mc Donald started operations and provided regular UT informations.
- in 1970, the TRANSIT system determined the pole position with performances better than those of astronomy; actually it conducts to one position every two days with a std. dev. of 0"006 (P. Pâquet and al, 1973).
- in 1979, the three ERP were deduced simultaneously from the LASER observations of the LAGEOS satellite. This network is presently providing ERP every 5 days with a std. dev. of 0"003 (M. Feissel, 1985).

Last but not least, during the last decade and with occasional experiments, the radio observations of quasars demonstrate the capabilities to measure ERP with a std. dev. of few milliarcseconds.

Since 1983 a VLBI world network is operational and provides regular ERP on a 5 days basis (Carter and al, 1984). Short Base Interferometers (SBI) are also providing UT information on a regular basis, (D. Mc. Carthy and al, 1984).

All astronomical and geodetic methods for ERP measurements have a common basis which is the pre-knowledge of the dynamical processes responsible for precession and nutations. They are based on new Earth model (Wahr, 1979).

Over this pre-requirement all methods have different level of complexity concerned with their different approaches :

- classical astronomy is doing geometry by measuring angles, but also kinematics to correct the observations of the proper motion of stars;
- space geodesy (DOPPLER and LASER) outputs the ERP, together with the satellite orbit determination, from a pure dynamical process;
- Lunar Laser Ranging is a geometrical method with a strong support of dynamics to provide the Lunar ephemeris;
- Interferometric methods are measuring Earth position with respect to fixed radio-sources; it is a pure geometric method.

In the next paragraphs, we review the implied basic measurement schemes together with their advantages and disadvantages.

2. THE CLASSICAL ASTRONOMICAL APPROACH

The observations measure two quantities :
- the instantaneous latitude
- the difference between two time scales which are respectively the Universal Time (UT_0) (deduced from the direct observation of stars) and the UTC.

The difference $UT_0 - UTC$ is a function depending from the Earth Rotation (ER) fluctuations and Polar Motion (PM); the contribution of PM is noted $UT_0 - UT_1$, and the other variations $UT_1 - UTC$.

If we call :

x, y the instantaneous coordinates,
Δt the fluctuations of the ER,
ϕ_i the latitude observed by the observatory i,
$\phi_{o,i}$ the adopted latitude of the observatory i,
$UT_{o,i}$ the UT deduced by the observatory i,
$\lambda_{o,i}$ the adopted longitude of the observatory i,
UTC the uniform time scale,

the differential effects of PM on latitude and longitude are given by

$$x \cos \lambda_{o,i} + y \sin \lambda_{o,i} = \phi_i - \phi_{o,i}$$

$$-x \operatorname{tg} \phi_{o,i} \sin \lambda_{o,i} + y \operatorname{tg} \phi_{o,i} \cos \lambda_{o,i} + \Delta t = UT_{o,i} - UTC \qquad (1)$$

It is important to note that the set of adopted values $\phi_{o,i}$ and $\lambda_{o,i}$ determine the reference system with respect to which x, y, Δt are expressed. The Earth reference system is composed of :
- a fixed origin CIO near the instantaneous pole; latitudes are defined with respect to CIO.
- a fixed point in the equator which is the origin of the longitudes.

The parameters measured are angles related to latitude variations and the time of transit of stars accross the local meridian or some other particular astronomical reference which is related to longitude variations.

Each station having at its disposal a star catalogue, which is a long term stable reference, computes immediately on site the local variations of latitude and longitude.

The local results of about 50 stations are then gathered by a central Bureau (BIH, IPMS) and combined in the system (1) which is solved for x, y, Δt, once every five days.

From this, it is easy to understand the difficulty to keep always the same Earth system of reference. Indeed all stations defining the reference are not operating with the same continuity, do not produce observations of the same quality, do not use the same instruments. For each particular 5 days solution, the set of stations could be different and theoretically the reference is also different. Practically, adequate method of computations allows to keep as well as possible the homogeneity of the solution.

The astronomical methods presently in use do not allow to determine the pole position with an accuracy better than 0"01, 30 cm at the Earth surface.

The main sources of inaccuracy are induced by :

. the error and inhomogeneity of the star catalogues,
. the instrumental and observation errors,
. the modelisation of the atmospheric refraction.

During the forthcoming 10 years and if enough instruments of high level are still in operation, we could expect an improvement by a factor 2 or 3 with the support of :

. the new star catalogue FK5,
. the set up of automated instruments that will reduce the personal error.

At a longer term (1995 ...) better catalogues will be available from satellite observations. The european project HIPPARCOS is well situated for such an evolution. Indeed it will deliver star positions with an accuracy of 0"002 while the present accuracy is near 0"05.

We must also keep in mind that classical astronomical methods have a very good long term stability and could be adequate for the estimation of decade fluctuations of ERP.

Recently they have also allowed to identify ER fluctuations of the order of few 0^s001 with periods of 122 days (D. Djurovic, 1974) and 50 days (M. Feissel and al, 1980).

3. ERP FROM NEAR EARTH SATELLITE OBSERVATIONS

3.1. Principle of measurements

If all the forces acting on a satellite were perfectly known it would not be difficult to predict the satellite position and as the satellite motion is independant of the ER it would be also possible to detect the various motions of the Earth with respect to the orbit as a reference. The satellite orbit would replace the stars of the astronomers.

However as it is still impossible to extrapolate the satellite position with the required precision, the determination of the ERP is associated with an orbit adjustment.

In the present technology the parameters supposed to be known are :

- the Earth reference system represented by the station coordinates and gravity field;
- the newtonian forces created by the Sun and the Moon;
- the directions of the perturbating forces generated by the atmospheric drag and solar pressure;
- a first approximation of
  . the position $\vec{p}_0$ and velocity $\vec{v}_0$ of the satellite at a fixed epoch, (initial state vector)
  . the modulus F_D and F_P of the drag and solar pressure forces.

The unknowns to be determined are :

- the correction $\Delta\vec{p}_0$ and $\Delta\vec{v}_0$ of the initial state vector;
- two scaling factors λ_D, λ_F readjusting the modulus of the drag and solar pressure factors;
- the geophysical unknowns x, y, Δt
- local parameters depending of the type of observations, such as clock

error, atmospheric refraction.

The observational material being collected by a world network, the orbit and ERP determinations result from the adjustment of a set of equations having the following form.

Let W_o the observed quantity and W_c the equivalent quantity computed with the first estimation of all the parameters to be determined :

$$W \equiv W\,(\Delta\vec{p},\ \Delta\vec{v},\ \phi_i,\ \lambda_i,\ \ldots),$$

ϕ_i, λ_i being the instantaneous coordinates of the ground stations. Between the observed and computed quantities we have :

$$W_o - W_c = \frac{\partial\vec{W}}{\partial p}\,\Delta\vec{p} + \frac{\partial\vec{W}}{\partial v}\,\Delta\vec{v} + \frac{\partial W}{r_i\,\partial\phi_i}\,\Delta\phi_i + \frac{\partial W}{r_i \sin\theta_i\,\partial\lambda_i}\,\Delta\lambda_i + \ldots,$$

remembering equation (1), the set of $\Delta\phi_i$, $\Delta\lambda_i$ can be replaced by the unique unknowns x, y, Δt, (PM and UT). The equation of observation become :

$$W_o - W_c = \left[\begin{array}{l} \frac{\partial\vec{W}}{\partial p}\,\Delta\vec{p} + \frac{\partial\vec{W}}{\partial v}\,\Delta\vec{v} \\ + x\left[\cos\lambda_i \frac{\partial W}{r_i\,\partial\phi_i} - \mathrm{tg}\,\phi_i \cos\lambda_i \frac{\partial W}{r_i \cos\phi_i\,\partial\lambda_i}\right] \\ + y\left[\sin\lambda_i \frac{\partial W}{r_i\,\partial\phi_i} + \mathrm{tg}\,\phi_i \sin\lambda_i \frac{\partial W}{r_i \cos\phi_i\,\partial\lambda_i}\right] \\ + \Delta t\left[\frac{\partial W_i}{r_i \cos\phi_i\,\partial\lambda_i}\right] \\ + \text{other parameters (scaling drag, offset, ...)} \end{array}\right] \qquad (2)$$

Presently two types of observations are providing observations to solve system (2) :

- *radiolectric observations* of TRANSIT satellites, the observables being the integrated DOPPLER effect or difference of distances between the ground station and two consecutive satellite positions at time t_i, t_{i+1}.
  The satellites of the TRANSIT system are radiating two frequencies at 150 and 400 Mhz, orbiting on a low altitude orbit (1000 km). They are very sensitive to the high frequency terms of the gravity field and the atmospheric drag.
  Most of the satellites were designed in the year 1964-1972. Aiming to correct for the non-gravitational forces acting along track, a drag free satellite is currently used since 1983; the orbit improvement with respect to the old TRANSIT series is of the order of 60 %.

  However a gravity field dedicated for such a satellite has not yet been used and as a consequence of a lack of precision in the determination of the orbits node, the TRANSIT system does not solve the

system (2) for the third Earth rotation parameter Δt.
Using the drag free satellite data, each coordinate of the pole position is given each two days with a precision of about 0"006; the improvement with respect to the old TRANSIT satellite is of the order of 50 %.
New improvements can still be expected as soon as a dedicated gravity field will be determined.

- *LASER ranging on satellite LAGEOS*, the observables being range measurements.
The spherical shape of the satellite LAGEOS, launched in 1975 and its high altitude orbit (a = 13000 km) are well dedicated for geophysical studies. Both concur to reduce the effect of non-gravitational forces and of the higher harmonics of the gravity field.
With the range data, the system (2) is solved for the three ERP. It provides one solution every five days and the precision is of the order of 0"003 for each coordinate of the pole position and 0^s0005 for UT_1.
For short term measurements at 5 days intervals, the VLBI is presently the most and accurate complete method, as the three ERP are given. However and because the observations are weather dependent, it appears difficult to get result on a time basis shorter than 5 days.

3.2. Atmospheric perturbations of the measurements

For radio signals the main perturbations concern the signal propagation through the ionosphere and troposphere. LASER ranging is only concerned with the dry component of the troposphere.

The first order of the ionospheric perturbations is removed by mixing two received signals; however it has been shown that for a set of low frequencies (150 and 400 Mhz) the second order effects generate apparent periodic changes of the measured distance between station and satellite. The total amount could reach one meter during high solar activity (Dehant V. and al, 1983).

To overcome this perturbation a definitive solution is to use a pair of higher frequencies. For example in the Global Positioning System (GPS), by using the pair (1227, 1575) Mhz, the ionospheric perturbation is reduced at the level of 2 or 3 centimeters (J. Saint-Etienne, 1981). On the other hand, radio signals are perturbated by the refraction of the lower part of the atmosphere. The effect is practically independant from the frequency ($<$ 15 Ghz) and for different elevations the amplitudes of the dry and wet contributions are given in Table 1.

TABLE 1
Magnitude of the tropospheric refraction

Elevation	Total effect in meters	Contribution (m)		Error (cm)	
		dry	wet	dry	wet
5	24.4	22.0	2.4	5.	49.
10	13.3	12.0	1.3	3.	27.
20	6.9	6.2	.7	1.	14.
30	4.7	4.2	.5	1.	9.
90	2.4	2.2	.2	.5	5.

Between the zenith and an elevation of 5 degrees, the wet component is about 10 % of the total tropospheric effect. Below 5 degrees it can reach 20 or 25 %. From Table 1 it is also interesting to note that the models are excellent for the dry component, the error indeed remains below 0,2 %; unfortunately the wet component leaves error of the order of 1 or 2 % of the total refraction effect (H.S. Hopfield, 1971).

This is the reason why, during the processing, the measurements below 5 or 10 degrees of elevation are systematically rejected. For the future we must also note that to reach the subdecimetric level the tropospheric refraction should be considerably improved. The most promising approaches is based on radiometer techniques which allow to deduce the profile of the water vapour distribution (Elgered, 1982). From such experiments it seems possible to monitor the tropospheric refraction with residuals less than 1 or 2 centimeters.

An other approach is proposed by an ESA project of geophysical studies, with a satellite transmitting radio frequencies (8 and 2 Mhz), with a spherical shape and on a high altitude orbit (13000 km). With a classical method of refraction correction, simulation demonstrated the possibility to reject systematically the observations below 30 degrees of elevation, while the level of precision of 0"002 is maintained (Ch. Reigber and al, 1983).

4. RADIO INTERFEROMETRIC METHODS

The radio signals generated by a distant quasar or radio-source is recorded by at least two stations. A local oscillator provides the reference frequency and time marks which will be super-imposed to the registered phase of the received radio signals.

If the distance between the two antennas, defining the baseline, is small enough to use the same local oscillator, the system is called Short Base Interferometry (SBI).

Since 1967 experiments were conducted by far away stations, each using their own local oscillators; the system is known as Very Long Base Interferometry (VLBI). From the phase recorded at each station of the network, computer process determines the maximum cross-correlation between the signals of each pair of stations aiming to estimate the time delay and/or its time derivative.

The time delay τ represents the projection of the baseline $\vec{b}$ on the direction of the radio source $\vec{d}$; if σ is the angle between $\vec{b}$ and $\vec{d}$, then

$$\tau = \frac{b}{c} \cos \sigma$$

As for the near Earth satellite approach, the modeling of τ depends of parameters of different origins :

. astronomical constants (precession, nutations),
. long term geophysical parameters (crustal deformations),
. short term geophysical parameters (the 3 ERP),
. local conditions (epoch and frequency bias, troposphere, ionosphere),
but with the advantage that all sets of unknowns, including precession and nutations, could be included in a dedicated solution.

The greatest interest of VLBI is to refer to a celestial reference frame which for the time being is considered as ideal because the radio sources are far enough to be considered as fixed points. The long term stability of the system is thus preserved and major problems, as the refinement of precession and nutations, could be addressed in a processing of a long term series of observations (Walter H.G., 1977; O.J. Sovers and al, 1984). Over a short period of time, τ is particularly sensitive to the local conditions (clock offset) while the time evolution of τ is more sensitive to the Earth rotation parameters (Alan E.E. Rogers, 1970).

Since 1984 a network composed of 5 stations is performing regular sessions of 24 hours of observations, repeated every 5 days. The processing delivers the three ERP with a formal error of 0"002 on each component of the PM and $0^{s}0002$ on UT1 (W.E. Carter and al, 1983).

Presently and taking into account the reliable long term stability, the VLBI is the best system for the ERP measurements.

The desadvantage is the heavy task of data processing that allows, only exceptionally, a continuous operation aiming to monitor the very short term variations of the ERP one day and less, (D.S. Robertson and al, 1985). On the other hand and as for the satellite radio observations, to improve the precision the main limitation is the perturbation induced by the troposphere; its modeling is a key step for a new improvement of VLBI observations.

5. LUNAR LASER RANGING (LLR)

In 1970 the variations of the ER were determined with range observations of the Lunar Laser in a single station (Mc Donald). As only one station was operating, PM values were taken from other sources (BIH).

Since a few years other stations succeeded to range to the Moon :
- CERGA (France) started regular operation in 1982-1983,
- Haleakala (Hawaii) and a second station on the site of Mc Donald obtained valuable measurements since 1984.

Altough the full power of the LLR is not yet proved, the partial results are very encouraging. Figure 1 displays the precision of LLR against the duration, in hour angles, of an experiment; the original data of this figure were taken from (J.O. Dickey and al, 1985).

As for, VLBI, LLR accesses an extended set of unknowns (J.O. Dickey and al, 1983) :
- tidal acceleration of the Moon,
- astronomical constants (precession, nutation)
- long term geophysical parameters
- short term geophysical parameters.

Altough the experiment is technically difficult, one of the advantages of the LLR is on the other hand that the data analysis is easier and does not require a large amount of computations. However the prediction of the orbit of the Moon and the librations remains a difficult task because the numerical integration is the only technique that gives the required decimeter level accuracy.

Analytical orbit theories are less good by at least one order of magnitude and their improvement will require, most probably, new theoretical approaches.

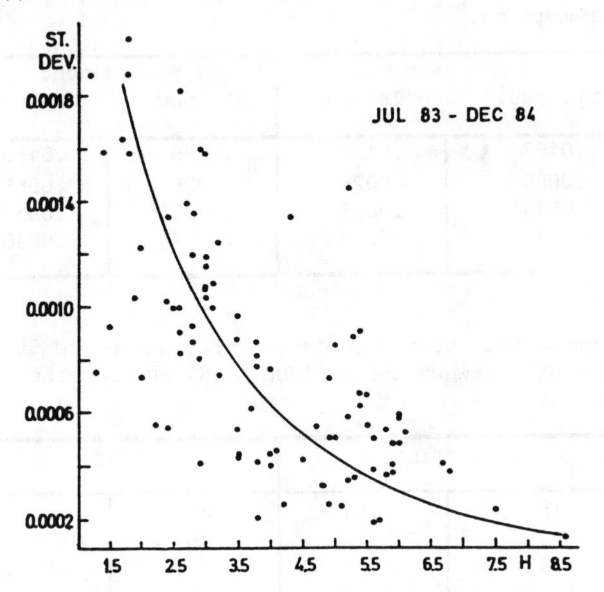

Fig. 1. In sec. and for LLR determination of UT_1, std. dev. versus the duration, in Hour angles, of the experiment.

6. COMPARISONS OF FEW PM AND UT MEASUREMENTS

Aiming to intercompare all the techniques a special observation campaign "MERIT" has been organized from 1 September 1983 to 31 October 1984. Data from all the techniques, excepted LLR, were collected on a regular basis. Details on preliminary results are given in (M. Feissel, 1985) and (W. Carter, 1985). However to complete this paper and to have a quick idea of the present possibilities, figures 2 and 3 present the residuals in the (X,Y) components of the Polar Motion (PM). They are obtained after having removed the annual and chandler components. Each series has been fitted with the Vondrak's method ($\varepsilon = 10^{-6}$) and although the trends are very comparable, the scattering clearly shows the respective precision of each series. Numerical values are given in Table 2; however for IPMS the mean square error is not given because the original data were already smoothed.

It is also interesting to note, between each series, the high correlations and the associated Student significant tests which are given in Tables 3 and 4.

TABLE 2. In arc sec., standard deviation and Allan variance of the (X, Y) components of the PM, after having removed the annual and chandler motions.

Series	X component		Y component	
	St. dev.	Allan var.	St. dev.	Allan var.
BIH	.0167	.0171	.0138	.0153
DOPPLER	.0086	.0026	.0068	.0068
LASER	.0044	.0085	.0036	.0037
VLBI	.0043	.0039	.0034	.0040

TABLE 3. For the X component, coefficients of correlation and Student significant tests between the residuals (R) and smoothed curves (F) of Figure 2.

Series	Correlation		t Student	
	(R)	(F)	(R)	(F)
BIH - DOPPLER	0.270	0.637	2.5	6.8
BIH - LASER	0.454	0.857	4.4	11.6
BIH - VLBI	0.573	0.909	5.0	11.7
DOPPLER - LASER	0.613	0.840	6.5	11.0
DOPPLER - VLBI	0.444	0.800	3.7	8.4
LASER - VLBI	0.764	0.943	7.7	13.5

TABLE 4. For the Y component, coefficients of correlation and Student significant tests between the residuals (R) and smoothed curves (F) of Figure 3.

Series	Correlation		t Student	
	(R)	(F)	(R)	(F)
BIH - DOPPLER	0.229	0.716	2.1	8.2
BIH - LASER	0.297	0.728	2.8	8.4
BIH - VLBI	-0.217	0.315	1.4	2.0
DOPPLER - LASER	0.596	0.938	6.3	15.7
DOPPLER - VLBI	0.211	0.349	1.4	2.3
LASER - VLBI	0.314	0.563	2.1	4.0

The Table 2 reflects clearly the high quality level of the LASER and VLBI methods which provide each component of the pole position with a precision of the order of 12 cm, while the precision of the DOPPLER and astrometry are respectively at the level of 23 and 45 cm.

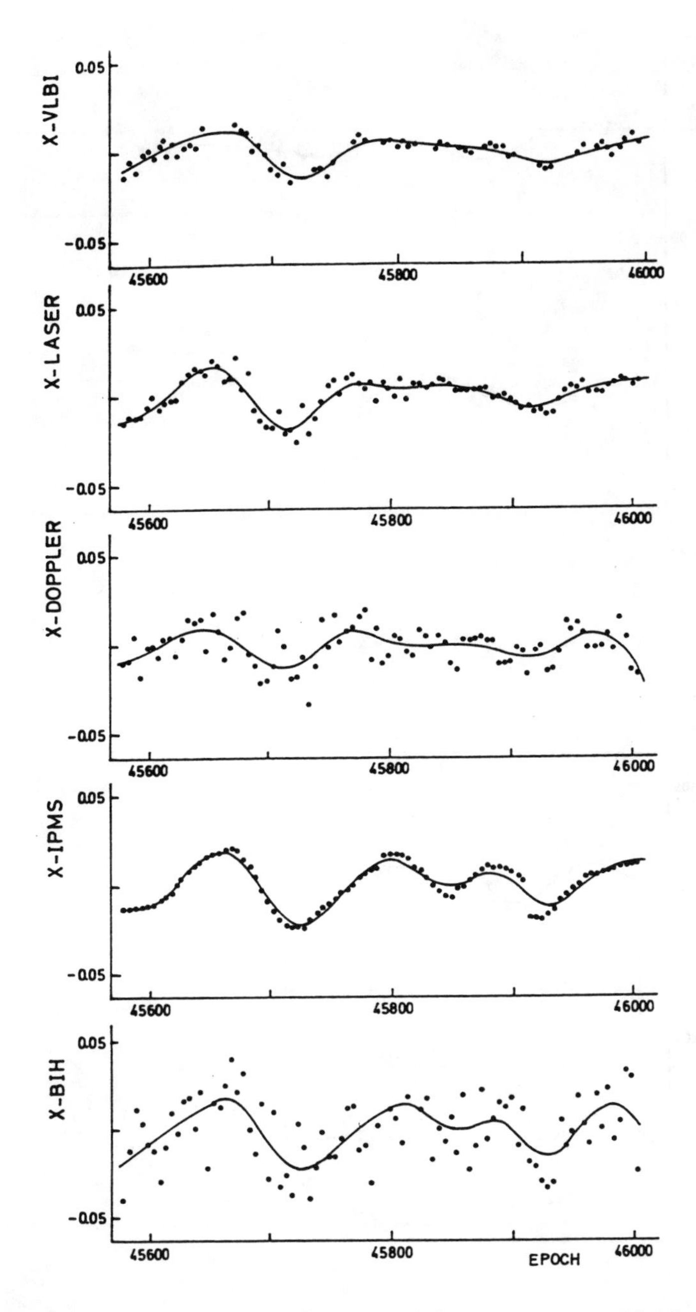

Figure 2. For different techniques and in arc sec residuals of the X component of the PM after elimination of the annual and chandler components.
The continuous curves are a Vondrak smoothing with $\varepsilon = 10^{-6}$.
The epoch is given in days.

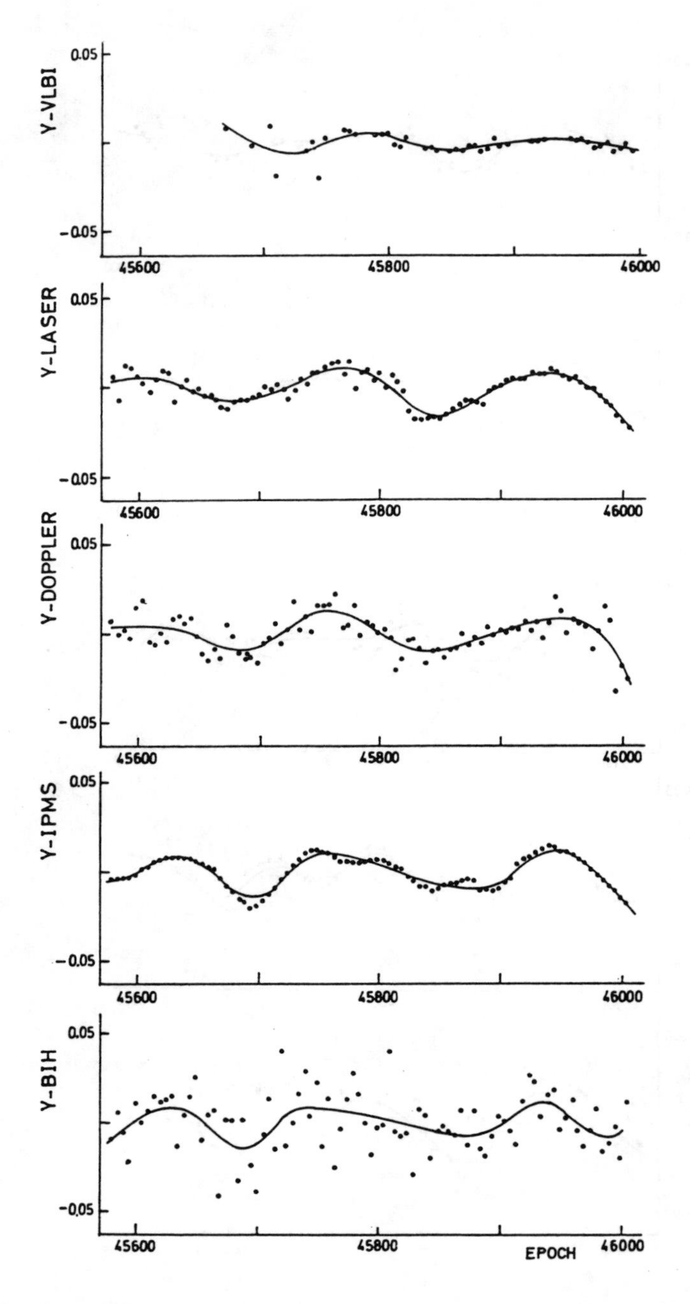

Figure 3. For different techniques and in arc sec residuals of the Y component of the PM after elimination of the annual and chandler components.
The continuous curves are a Vondrak smoothing with $\varepsilon = 10^{-6}$.
The epoch is given in days.

The UT1 informations obtained from classical astronomy and processed by BIH and IPMS are compared to the VLBI results.

In the three cases the original data were corrected, for a linear drift. The resulting residuals (R1) and associated smoothed curves, by the Vondrak's method with ($\varepsilon = 10^{-6}$), are given in figure 4(a).

A set of residuals of second order R2 has been also deduced and again smoothed by the Vondrak's method ($\varepsilon = 10^{-4}$). They are given in figure 4(b). It must be noticed that the 50 days period appears very clearly in all the cases but, very unexpectively, the curve deduced from astronomical data and processed by IPMS is of the best quality. This is confirmed by the st. dev. given in Table 5.

Table 5. For BIH, IPMS, VLBI and in seconds, std. dev. of the residuals R1, R2, R3 :
R1 = original UT1 data - linear drift
R2 = R1 - smoothed
R3 = R2 - smoothed

Series	R1	R2	R3
BIH	.0044	.0016	.0016
IPMS	.0039	.0008	.0003
VLBI	.0040	.0008	.0005

It confirms the very good performance of the VLBI. It demonstrates also the high quality of the new IPMS data processing and it reinforces the evidence that the UT1 data deduced from classical astronomy contain valuable and competetive informations, at least for components whose periods are ranging from few weeks till few years.

It appears clearly that the agreement between VLBI and classical astronomy is quite better for UT1 measurements than for PM. However for the PM components greater than 30 days, which corresponds to the filtering capability of the Vondrak's method with $\varepsilon = 10^{-6}$, all the techniques contain the same informations at a very high degree of significance (> 99 %). The 5 days residuals deduced from LASER and DOPPLER have the higher correlation which is also very significant (Table 4).

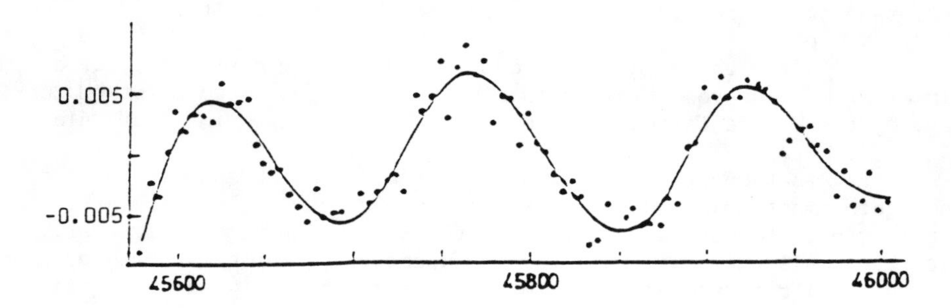

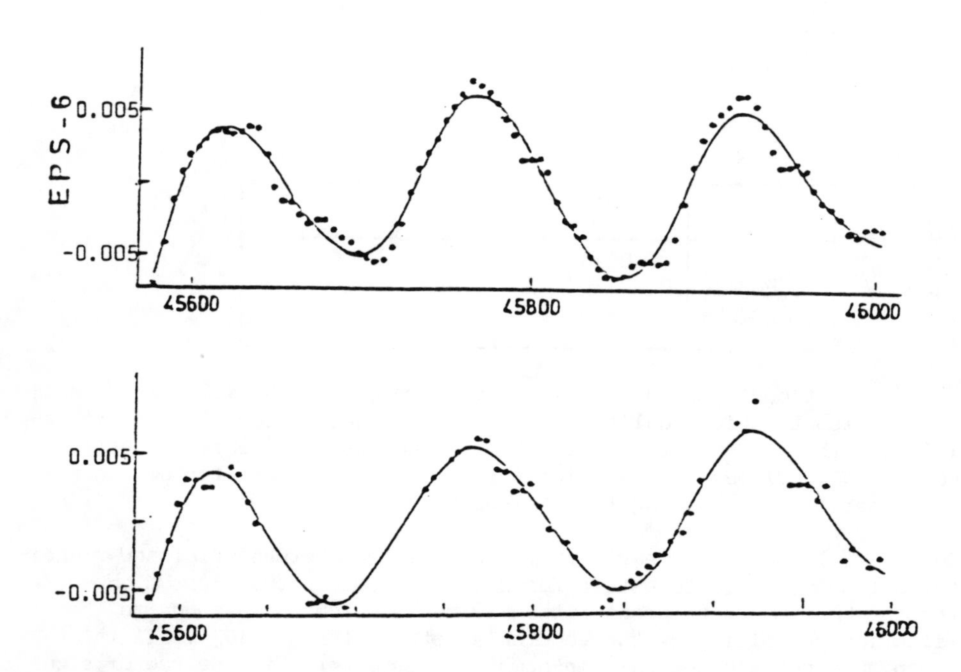

Figure 4(a). For BIH (upper), IPMS (middle), VLBI (lower), UT1 residuals after elimination of a linear drift of the original data. The continuous curves are a Vondrak smoothing with EPS=10-6. The epoch is given in days.

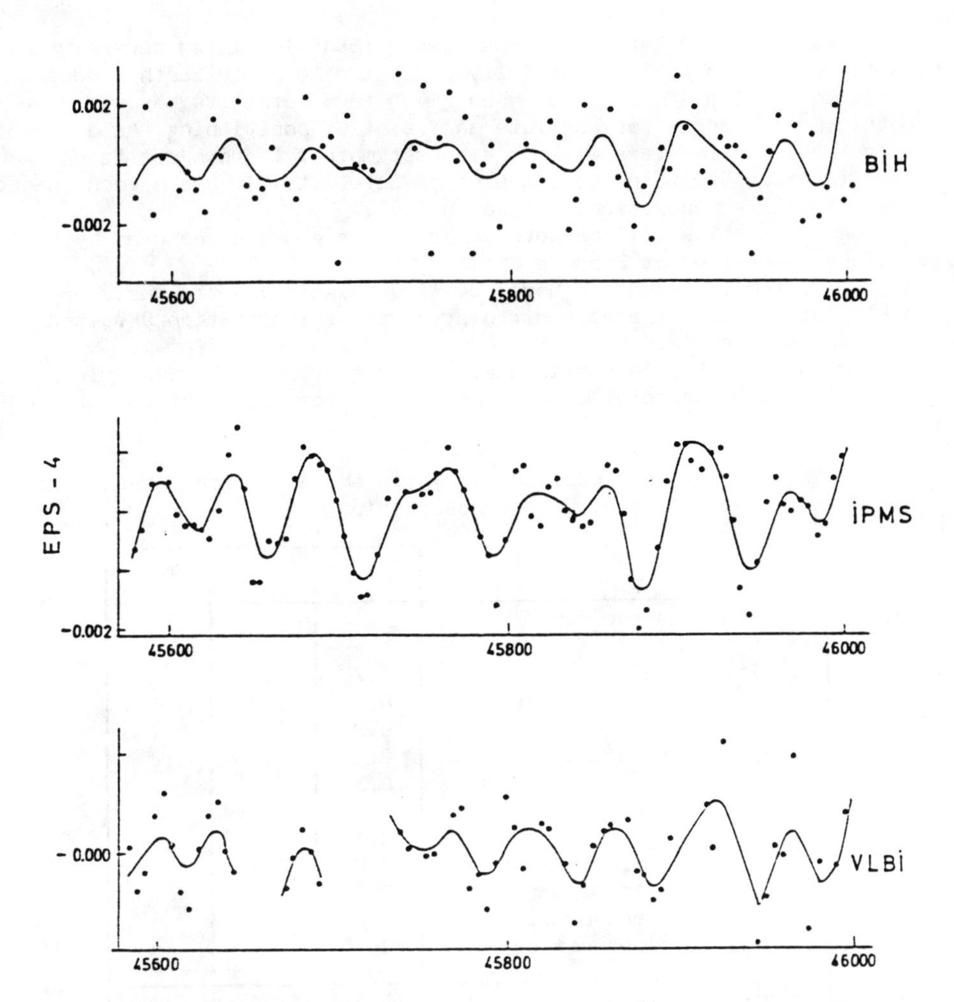

Figure 4(b). For BIH, IPMS, VLBI, UT1 residuals after elimination, from the original data, of a linear drift and a smoothing by the Vondrak' method with $\varepsilon = 10^{-6}$.
The continuous curves are a Vondrak smoothing with $\varepsilon = 10^{-4}$.
The 50 days component of UT1 appears very clearly.

6. EUROPEAN PROJECT OF A DEDICATED GEOPHYSICAL SATELLITE

Since four years the European Space Agency (ESA) initiated several studies to define a satellite specially dedicated to Solid Earth studies. In this scope, the priority has been given to a satellite having subdecimetric performances for absolute and relative positioning. As a consequence the ERP parameters will be also estimated. The project is called POPSAT (Precise Orbit and Positioning of Satellite) it has a good chance to mature till an operational phase in 1992.

The observables will be both range and range rate measured by radio techniques operating at 2 and 8 Ghz.

The system performances have been investigated on the basis of 16 tracking stations, for a near circular orbit at inclination 98°.6 and a semi-major axis of 13.370 km.

The following table published in C. Reigber and al (1983) gives the budget error and the total error expected for the daily determination of the three ERP.

TABLE 6. Accuracy of pole position x, y (in cm) and Earth rotation rate ΔUT_1 (msec) from 1 day of observation.

	x	y	ΔUT_1
Individual error contribution			
Gravity field model	4.7	0.5	
Gravitational parameter	0.0	0.0	
Albedo	0.3	0.9	
Solar radiation	0.1	0.1	
Earth tides	0.3	0.2	
MEX stations ϕ, λ, H	3.4	3.2	
uplift	0.1	0.1	
troposphere	0.2	0.1	
ionosphere	0.0	0.0	
R/RR measurement bias	—	—	
R/RR measurement noise	0.4	0.4	
Total error (RSS)	6.0	3.4	0.15

Those results are very encouraging since the pole position and ΔUT_1 can be derived respectively with a precision of 6 cm and 0.15 msec from one day of tracking data. Such radio system is very appropriate for the short term determinations of the ERP, from few hours to few months.

Table 7 summarizes the main characteristics of the different techniques presently in use or expected to be available in a near future.

TABLE 7. Main characteristics of the different method of ERP measurements.

	Classical Astronomy	Space Geodesy TRANSIT	SLR	VLBI	LLR	Future project POPSAT
Solutions for						
UT	×	—	×	×	×	×
X	×	×	×	×	—	×
Y	×	×	×	×	—	×
Solution every ... days	5	2	5	5	(*)	1
Accuracy of the solution PM	0"015	0"006	0"003	0"003	—	0"002
UT	0ˢ001	—	0ˢ0002	0ˢ0002	0ˢ0005	0ˢ0002
Long term stability	×	—	—	×	×	—
Short term	×	×	×	×	×	×
Access to very short term (≤ 1 day)	—	—	—	×	×	×
All weather capabilities	—	×	—	×	—	×
Day time capabilities	—	×	—	×	—	×

(*) irregular.

7. NEW MEASUREMENTS TECHNIQUE

In the last few years the new technology of superconducting gravimeters seems to be a possible candidate to detect the ERP variations. Two such instruments have recorded data during a period of three years in Bad Homburg (RFA) from April 1981 to May 1984 and Brussels from April 1982 to May 1985 (B. Richter, 1985(a); B. Ducarme and al, 1985(b)). Both show the signature of the PM, but the registration of Bad Homburg is particularly spectacular as displayed in Fig. 5, reproduced with the permission of the author (B. Richter, 1985b).

Altough the site of Brussels is noisier than the site of Bad Homburg, the PM is also clearly identified (B. Ducarme and al, 1985b).

For those sites the amplitude of the gravity variations due to the PM is of the order of 4 microgals but the st. dev. depends of the frequency; it ranges from few nanogals to 0.5 microgals for phenomena whose the period ranges from few hours to one week (P. Melchior and al, 1985).

In the future it can be expected that this instrumentation be a valuable contributor to the measurement of ERP but many improvements are still necessary both in equipment and in modeling of the external perturbation like the effects of atmospheric pressure, the underground water, instrumental drift whom induce residuals that can reach 1 microgal.

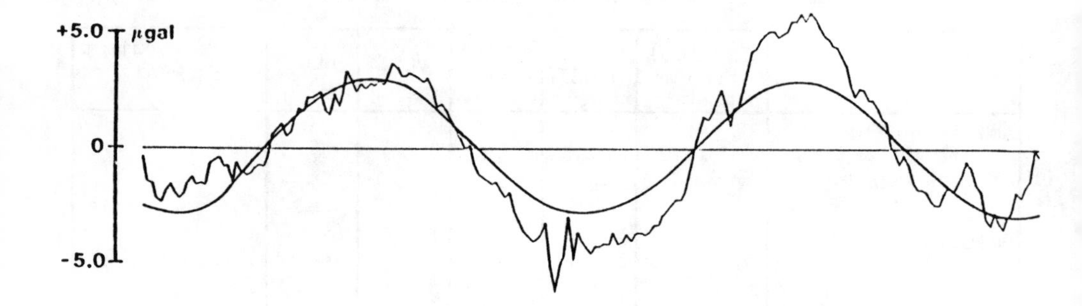

Figure 5. Residuals of a complete Earth tide analysis compared with the 432 days chandler wobble. (Reproduce with permission of the author B. Richter, 1985b).

REFERENCES

A. Berger, 1980. 'The Milankovitch astronomical theory of paleoclimates : a modern review', *Vistas in Astronomy*, 24.

W.E. Carter, D.S. Robertson, J.R. MacKay, 1983. 'Polaris Earth Rotation Time Series'. *Proceed. of the Int. Association of Geodesy Symposia*, Vol. 2, IUGG General Assembly Hamburg. Publ. by Dept. of Geodetic Science and Surveying, Ohio State University, Columbus.

W.E. Carter, D.S. Robertson, 1984. 'IRIS earth rotation and Polar motion results'. *Proceed. of Int. Symp. on Space Techniques for Geodynamics*. Vol. I. Ed. J. Somogyi, Sopron (Hungary).

W.E. Carter, 1986. 'Accurate Polar Motion and UT_1 series from VLBI observations'. *Proceed. of the NATO Workshop, Earth's Rotation : Solved and Unsolved Problems*. Ed. A. Cazenave - Reidel Publ.

V. Dehant, P. Pâquet, 1983. 'Modeling of the apparent height variations of a TRANET station'. *Bull. Géodésique*, 57.

J.O. Dickey, J.G. Williams, X.X. Newhall, C.F. Yoder, 1983. 'Geophysical Applications of Lunar Laser Ranging'. *Proceed. of the Int. Association of Geodesy Symposia*, Vol. 2, IUGG Gen. Assembly, Hamburg. Publ. by Dept. of Geodetic Science and Surveying, Ohio State University Columbus.

J.O. Dickey, X.X. Newhall, J.G. Williams, 1985. 'Earth Rotation (UT_0) from Lunar Laser Ranging'. *JPL Geodesy and Geophysics*, preprint N° 124.

D. Djurovic, 1974. 'Sur un terme harmonique de 122 jours dans la rotation de la Terre et dans le mouvement du pôle'. *Obs. Roy. de Belgique. Com. Ser.* B, n° 84; *Série Géophysique* N° 119.

B. Ducarme, M. Van Ruymbeke, C. Poitevin, 1985a. 'The Superconducting gravimeter of the Royal Observatory of Belgium'. *Bulletin d'informations des Marées Terrestres*, N° 94. Ed. Prof. P. Melchior, Obs. Roy. de Belg., Bruxelles.

B. Ducarme, M. Van Ruymbeke, C. Poitevin, 1985b. 'Three years of registration with a Superconducting gravimeter at the Royal Observatory of Belgium'. *Proceed. of the X Int. Symp. on Earth Tides*, Madrid 1985. (In Press).

G.K. Elgered, 1982. 'Tropospheric wet path. Delay measurements'. *IEEE Trans.*, Vol. AP-30, N° 3.

M. Feissel, D. Gambis, 1980. 'La mise en évidence des variations rapides de la durée du jour'. *Comptes Rendus Acad. des Sciences de Paris.* Vol. 291, B 271 - B 276.

M. Feissel, 1986. 'Measurements of the Earth's rotation in the 80's'. *Proceed. of the NATO Workshop, Earth's rotation : Solved and Unsolved Problems.* Ed. A. Cazenave. Reidel Publ.

B. Guinot, M. Feissel, M. Granveaud, 1971. *Bureau International de l'Heure. Annual Report for 1970.*

H.S. Hopfield, 1971. 'Tropospheric Effect on Electromagnetically measured range : Prediction from surface weather data'. *Radio Science*, Vol. 6, N° 3.

IAU, 1950 and 1955. 'Report of Commission 31'. *IAU Transactions*, Vol. VII and IX.

D. Mc Carthy, A.K. Babcock, D.N. Matsakis, 1984. 'A comparison of CEI, VLBI, BIH and USNO Earth orientation information'. *Proceed. of Int. Symp. on Space Techniques for Geodynamics.* Vol. I. Ed. J. Somogyi, Sopron (Hungary).

K. Lambeck, 1980. 'The Earth's Variable Rotation'. *Cambridge University Press.*

P. Melchior, B. Ducarme, 1985. 'Detection of Inertial Gravity Oscillations in the Earth's Core with a Superconducting Gravimeter at Brussels'. Letter to the Editors to *Physics of the Earth and Planetary Interiors.* (In Press).

L.V. Morrison, 1986. 'Historical and observations of the Earth's rotation secular variations'. *Proceed. of the NATO Workshop, Earth's Rotation : Solved and Unsolved Problems.* Ed. A. Cazenave. Reidel Publ.

W.H. Munck and R.L. Miller, 1950. 'Variation in the Earth's angular velocity resulting from fluctuations in atmospheric and oceanic circulations'. *Tellus 2*, 93-101.

P. Pâquet, R. Dejaiffe, 1973. 'Analysis of the first DOPPLER observations performed at the Royal Observatory of Belgium'. *Proceed. of Symposium on Earth's Gravitational Field and Secular Variations in Position*. Ed. R. Mather and P. Angus-Leppan, School of Surveying, Sydney.

Ch. Reigber, S. Hieber, E. Achterman, 1983. 'POPSAT - An Active Solid Earth Monitoring System'. *Proceed. of the Int. Association of Geodesy Symposia*, Vol. 2, IUGG General Assembly, Hamburg. Publ. by Dept. of Geodetic Science and Surveying, Ohio State University, Columbus.

B. Richter, 1985a. 'Three years of registration with the Superconducting gravimeter'. *Bulletin d'Informations des Marées Terrestres*, N° 94. Ed. Prof. P. Melchior, Obs. Roy. de Belg., Bruxelles.

B. Richter, 1985b. 'The Spectrum of a Registration with a Superconducting Gravimeter'. *Proceed. of the X Int. Symp. on Earth Tides*, Madrid 1985. (In Press).

D.S. Robertson, W.E. Carter, J. Campbell, H. Schub, 1985. 'Daily Earth rotation determinations from IRIS very long baseline interferometry'. *Nature*, Vol. 316.

Alan E.E. Rogers, 1970. 'Very long baseline interferometry with large effective bandwith for phase delay measurements'. *Radio Science*, Vol. 5, N° 10.

J. Saint-Etienne, 1981. 'Erreur ionosphérique résiduelle dans les systèmes de radiolocalisation spatiale bifréquences'. *An. Géophysique*, t. 37, fasc. 1.

O.J. Sovers, J.B. Thomas, J.L. Fanselow, E.J. Cohen, G.H. Purcell, D.H. Rogstad, L.J. Skjerve and D.J. Spitzmesser, 1984. 'Radio Interferometric Determination of Intercontinental Baselines and Earth Orientation Utilizing Deep Space Network Antennas : 1971 to 1980'. *J. G. R.* Vol. 89, N° B9.

F.H. van den Dungen, J.F. Cox et J. Van Mieghem, 1952. 'Les fluctuations saisonnières de la rotation du globe terrestre et la circulation atmosphérique générale'. *Tellus* 4, 1952.

J.M. Wahr, 1979. 'The tidal motions of a rotating, elliptical, elastical and oceanless Earth'. *Ph. D. Thesis*, University of Colorado.

H.G. Walter, 1977. 'Precision Estimation of Precession and Nutation from Radio Interferometric Observations'. *Astron. and Astrophys*. 59, 197.

J.W. Wells, 1963. 'Coral growth and geochronometry. *Nature*, 197.

POLAR MOTION AND SIGNAL PROCESSING

M. Souriau and C. Rosemberg-Borot
Centre National d'Etudes Spatiales
Groupe de Recherche de Géodésie Spatiale
18, Avenue Edouard Belin
31055 TOULOUSE CEDEX - FRANCE

ABSTRACT

The source(s) of the polar motion are still a contoversial topic so that many techniques of signal processing have been applied but with elusive results. After recall of the basic equation of motion and its solution, the most usual methods are surveyed with emphasis on the frequently implicit hypotheses. The 1900-1977 ILS series is used to apply a simple variable spectral analysis. In addition the Fourier transform restricted to the 1967-1984 BIH series is computed in order to obtain more reliable high frequencies. Some significant peaks in the frequency ranges [1, 2] and [-2, -1] cycles per year are observed.

The small perturbations of the Earth rotation are defined by 2 independant components, namely the length of day, i.e. the rotation velocity and the polar motion, i.e. the location of the polar axis. It has been shown recently (Lambeck and Hopgood, 1981) with increasing evidence that the first component is governed by atmospheric driving forces. On the contrary the driving forces related to the polar motion are still to be ascertained in spite of numerous attempted detections dealing with data of increasing accuracy. This has given rise to many competing techniques of signal processing. The following short survey will emphasize the basic assumptions and capabilities of the most popular procedures in order to understand their controversial results.

We will first refer to the basic equation of the polar motion explicitly used in many procedures.

1. THE POLAR MOTION IN THE FREQUENCY DOMAIN

Without damping the Earth wobble is governed by the following equation (Munk and McDonald, 1960) :

A. Cazenave (ed.), Earth Rotation: Solved and Unsolved Problems, 45–59.

$$\dot{m}_1 + \omega_0 m_2 = \omega_0 \psi_2,$$

$$\dot{m}_2 - \omega_0 m_1 = - \omega_0 \psi_1,$$

where dot means time derivative, $_0$ the Chandler frequency, (m_1,m_2) the angular displacement according to direct cartesian coordinates (e.g. Greenwich meridian and 90° East longitude) and (ψ_1,ψ_2) the excitation function. In complex notation identifying the plane of motion with the complex plane, i.e. $m = (m_1 + im_2)$, $\psi = (\psi_1 + i\psi_2)$, one obtains the following equation :

$$\dot{m} - i\,\omega_0\, m = - i\,\omega_0\, \psi \tag{1}$$

Then using the Fourier transform of complex time variables, with upper case letters denoting the frequency domain, and ω frequency, the solution is :

$$i(\omega - \omega_0) = i\,\omega_0\, \Psi \tag{2}$$

The transfer function $H = M/\Psi = \omega_0/(\omega - \omega_0)$ is real. $\Psi(\omega > 0) = \Psi^+$ refers to a uniform prograde rotation with angular velocity ω while $\Psi(\omega > 0) = \Psi^-$ is a reverse spin. In the vicinity of ω_0 $|H^+| \to +\infty$. This behaviour must be restricted by some internal friction which can be accounted for by the complex eigenfrequency $w_0 = \omega_0 + i\alpha$ or $\omega_0 (1 + (i/2Q))$, where α^{-1} is a relaxation time or Q the dimensionless quality factor.

With a weak damping, i.e. $\alpha << \omega_0$, one obtains :

$$|H| \sim (((\omega/\omega_0) - 1)^2 + (1/2Q)^2)^{-1/2} \tag{3}$$

Instead of complex variables, one can start from the two-dimensional vectors ψ_j, m_j, but the transfer function $\Psi_j \to M_j$ is a matrix ; it is consequently more convenient to refer to the previous analysis.

Since the input ψ is not constrained by some observed strong correlation, there is a fundamental ambiguity in the analysis of the polar motion. One has to assume either that the input is given in order to retrieve the oscillator parameters or to start from known parameters to compute the input. Frequently when dealing with computed excitation functions, an order of magnitude of the related motion is estimated by invoking results from the first approach. This is confusing because it is obviously a circular argument. We will first deal with these two approaches before reviewing spectral methods avoiding explicit references to the transfer function.

2. PARAMETER ADJUSTMENT

In order to compute the oscillator parameters, namely w_0 or (ω_0,Q) without known input, one must invoke a class of excitation functions avoiding any explicit use of the input. This is the true justification for assuming a white noise. An additional but weak argument is given by the persistence of the Chandler motion. Then the output is proportional to the transfer function. Its spectrum H (see formula (3)) shows a single peak located at ω_0 all the more narrow that the damping is weak ; this allows to compute Q without knowing the amplitude.

A straightforward spectral analysis shows that this asumption is invalid because at least 2 additional peaks are present, one located at the annual period the other one at the zero frequency. The zero frequency pattern can be removed by adjusting a linear drift ; the second peak is close to the Chandler frequency ; however with a sufficiently long time window, say at least 15 years, these 2 peaks are well separated. After removing the annual peak, a pseudo-white noise may be assumed as an input.

Then the parameters ω_0 and Q of the theoretical spectrum are adjusted to fit the observed spectrum. Early attempts (see Munk and McDonald, 1960) give $\sigma_0 = \omega_0/2\pi$ ranging from 0.83 to 0.85 cycle per year (cpy) and Q from 8 to 80. The Chandler frequency is well constrained but the uncertainty relevant to Q denotes either a discrepancy between the input and the assumed white noise or an additional instrumental noise. In this case the noise can be reduced by 2 procedures. The first one consists in applying a strong tapering window to the data before the frequency analysis in order to cancel the frequency leakage due to the side lobes, but the resolving power of the Fourier analysis is decreased and the Q estimate derived from the frequency band is questionable. The second procedure proposed by Jeffreys (1940) is a statistical adjustment using the maximum likelihood principle. This method consistent with the hypotheses is unbiased.

One can expect better performances with a long time window and cleaner data. The "homogeneous ILS series" covering the time interval 1900-1977 has been analysed according to this procedure by Wilson and Vicente (1980). They have obtained the following values with the 90 % confidence level in brackets, $\sigma_0 = 0.843$ (0.837-0.849) and Q = 170 ($47 < Q < +\infty$). The higher Q values show a reduced noise but its large uncertainty argues for a physical discrepancy.

In fact with this large time window (see Fig. 1), there are at least 2 close peaks so that the single peak model cannot be sucessful. Complexity has to be introduced in the input and/or in the oscillator.

3. DECONVOLUTION

With known oscillator parameters, it is possible to compute the excitation function ψ from the observed motion m by a deconvolution procedure. In complex notation, the convolution is :

$$m(t) = e^{iw_0 t} \left(m_0 - i\, w_0 \int_0^t \psi(\tau)\, e^{-iw_0}\, d\tau\right) \tag{4}$$

With the recorded wobble there is no obvious initial time, so that the value m_0 given by the beginning of the time window is not justified ; however because of its damping, the oscillator has a fading memory and the discrepancy due to ignoring the data preceding the time window is negligible after a time lag equal to α^{-1}. With the parameters given by Wilson and Vicente, α^{-1} = 404 years ! Namely, the initial value may have a drastic influence on the recorded singal even with a long time window. This is a major drawback when computing m(t) from $\psi(t)$, a problem discussed at length by Chao (1985), although using a different approach.

A simple deconvolution procedure is derived from the time sampling of (4) with rate Δt, namely :

$$\psi(t) = i(w_0 \,.\, \Delta t)^{-1}\, (m(t) - e^{iw_0 \Delta t}\, m(t - \Delta t)) \tag{5}$$

This is a differential procedure so that the problem arising from the integration constant is cancelled but at the expense of the signal/noise ratio outside the short frequency window covering the resonance (see e.g. Wahr, 1983, fig. 1). This noise is able to overwhelm the time signal ; but a significant coherence can be expected in a narrow frequency band between the observed and computed spectra of ψ.

4. SOME USUAL SPECTRAL ANALYSES

4a. Discrete spectra

The splitting of the Chandler wobble into several peaks by extending the time window suggests a complex discrete spectrum. Many procedures have been used to localize these peaks. We have selected 3 works, not at random, but because procedures, time windows, dates of completion are totally different. By chronological order they are :

a) tuning by numerical filters (Labrouste and Labrouste, 1946) ;

b) F.F.T. algorithm (Gaposhkin, 1972) ;

c) autoregression, namely Prony's method (Chao, 1983).

Results are given in Table 1 with in addition data from the analysis shown in Fig. 1. The autoregressive method provides harmonic components with varying amplitude while the other methods define peaks with stable amplitude. In spite of the close peaks able to generate side lobes, one observes stable frequencies.

The complexity has been "resolved" into 2 large peaks of equal amplitude surrounded by 2 smaller ones but its physical meaning is unclear : "is it located in the input? in the oscillator?" is still an unsolved problem.

Over a short time window the 2 main peaks ω_1 and ω_2 are too close to be distinguished ; the resulting spectrum is obtained by adding two

Prograde Spectrum				
σ (cpy)	A (milliarcs.)	ϕ(1900.0) deg.	Window (years)	References
1.0000	85.7	118	1900-1977	This paper
0.8984	23.2	-144		(ILS Series)
0.8567	106.8	-79		FFT
0.8341	99.1	-13		
0.8076	43.6	132		
1.0000	86.0	104	1900-1979	Chao (1983)
0.8986	21.8	-142		(ILS series)
0.8574	119.7	-86		Autoregression
0.8349	(57.2)	-17		Complex ampl.
0.8068	(62.7)	147		
1.0000	91.0	117	1900-1940	Labrouste &
0.9025	22.2	-176		Labrouste (1946)
0.8547	102.0	-72		(mixed data)
0.8278	89.0	23		tuning filters
0.8000	36.0	168		
1.0000	84.2	117	1900-1977	Wilson & V. (1980) (ILS) least-squ.
1.0000	78.5	119	1891-1970	Gaposhkin (1972)
0.8985	28.3	-169		(mixed data)
0.8665	22.7	143		FFT
0.8534	119.2	-45		
0.8349	92.1	21		
0.8072	36.1	140		

Table 1. Discrete spectra of the polar motion as detected by several authors. Whenever necessary the prograde component has been computed from the spectra of the (x,y) components and the phases shifted to the date 1900.00. For the amplitudes within parentheses, the damping given by the complex amplitude is too large for significant comparisons with the amplitudes of the other analyses.

vectors, with the relevant amplitudes, rotating with a relative velocity $\Delta\omega\ t = (\omega_1 - \omega_2)\ t$. For 2 peaks of equal amplitude one obtains a period with a rapid change in phase without there being a significant change in amplitude followed by a period with amplitude fluctuations without there being any noticeable shift in phase, this pattern being periodic with a period $T = 2\pi/\Delta\omega$. Then, the time-varying features derived from specific techniques are not a priori contradictory with a discrete spectrum.

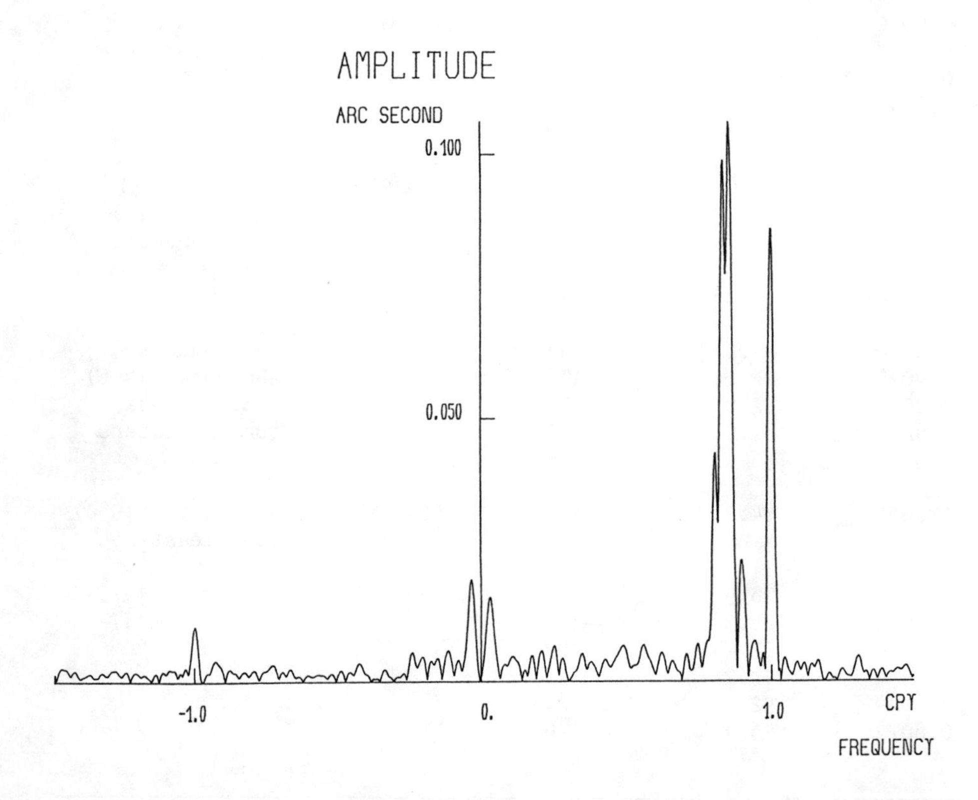

Figure 1. Amplitude spectrum (assuming steady spectral line) of the ILS series between 1900 and 1977, hanning-windowed FFT ; drift rejected by an adjustment according to a 3rd order polynomial prior to spectral analysis.

4b. Time-variable spectra

These techniques are well developped in seismology where they are used for obtaining the surface wave dispersion curves. For this purpose a systematic frequency-time spanning is carried out either by computing successive spectra along a moving window or by defining, over successive small frequency intervals, time-dependant spectral parameters (see Dziewonski and Hales, 1972).

As the 2 polar wobbles are well seperated and each one may be restricted to a small frequency interval, the first method with only one central frequency ω_0 for each wobble is a well fitted time analysis ; this is the method intuitively chosen by many analysts since Wahl (1939).

A signal defined by a spectral line is obviously a stable frequency. On the contrary, a time-varying frequency implies a spectrum spread over a frequency band B. In many analyses, the location of B which is ignored leads to confusion. For example, when analysing the Chandler wobble, the annual wobble must be removed. If the annual component is time-varying, a frequency band has to be removed ; consequently when only a single line is adjusted, some features pertaining to the annual component are included in the Chandler signal. Frequently the high frequency content is not removed (see below) ; this part is consequently included in the Chandler wobble and/or in the annual wobble. To avoid confusion, non overlapping frequency windows suggested by the global spectrum should be used systematically. However the selection of boundaries may be debatable.

This type of analysis implies fluctuations around a mean frequency $\langle\omega\rangle$, otherwise the analysis is meaningless. Starting with a complex variable, the Fourier transform for $\omega_0 = \langle\omega\rangle$ is :

$$Z(t) = (2L)^{-1} \int_{t-L}^{t+L} z(t= \exp(i\omega^0 t)\, dt \qquad (6)$$

where z and Z are complex varaibles, ω_0 the central frequency and (t - L, t + L) the moving window ; the instantaneous amplitude and phase are $|Z|$ and arg(Z). This is the procedure followed by Dickman (1980) and Chao (1983). A similar procedure can be applied to real components but the prograde and retrograde components are mixed. In the vicinity of the Chandler wobble, the retrograde component is small ; consequently the two procedures are nearly equivalent.

Obviously (6) is computed with a box-car window ; with a short L, the Gibbs' phenomenon becomes prevailing, but a tapering window cannot be used without cancelling useful information. This can be overcome by an extrapolation procedure according to the following adjustment usually carried out with real components, either :

$$X(t) = A(t) \cos (\omega_0 t + \phi(t)) \qquad (7)$$

or :

$$X(t) = A(t) \cos (\omega(t) . t + \psi) \qquad (8)$$

(7) is easily identified with (8) by putting :

$$\omega(t) = d\phi/dt + \omega_0,\ \psi = \phi(0) \qquad (9)$$

Starting from (7), then selecting a central frequency ω_0 and a moving window T, where the spectrum is assumed to be stable, the retrieval of A and ϕ is a linear problem easily solved by a least-square adjustment. With respect to $<\omega>$, the adjustment will be successful if $\omega_0 = <\omega>$; however with a small bias $\omega_0 = <\omega> + \Delta\omega$, a linear phase drift (see formula (9)) will be generated. This can be observed in some recent analyses, e.g. Dickmann (1980) and Chao (1983), while Guinot's analyses (1972, 1982) are trendless. Apparently, the least-square method is more accurate in the time-domain than the simple Fourier transform. In fact, one can detect short term fluctuations belonging to the high frequency content of the global spectrum. The corresponding frequency sampling is a patchwork which is not explicit and can lead to the above-mentionned confusions.

An adjustment to formula (8) is a non-linear procedure, however with smooth fluctuations, it can be linearized so that (7) and (8) are equivalent.

Guinot's last analysis shows a stable frequency ($\sigma_0 = 0.840$) in the time intervals 1900-1923 and 1945-1978 with significant amplitude fluctuations and a phase shift $> 180°$ in the time interval 1923-1940 with a steady amplitude level. According to the analysis concerning discrete spectra, this pattern corresponds to a global spectrum with two close peaks of equal amplitude. However a stable frequency pattern implies a new phase shift starting in the 80 s. It is still too early to be able to check the steadiness of these 2 spectral-lines.

The methods used so far are restricted to analyses with a moving time window. As mentioned above it is possible to work directly in the frequency domain. The envelope of a real oscillation s(t) and its instantaneous angular frequency can be obtained from the polar coordinates of a two-dimensional motion defined by the two orthogonal coordinates s(t) and q(t), where q(t) is the signal derived from the quadrature spectrum of s(t) ; namely if $S(\omega) = TF\ (s\ (t)) = A(\omega) + iB\ (\omega)$, then $Q(\omega) = A(\omega) - iB(\omega)$. In fact an explicit computation of q(t) is not necessary because s(t) + iq(t) is directly obtained by cancelling the negative frequencies of $S(\omega)$ before applying the inverse Fourier transform. One can check that the resulting time function is complex and corresponds to s(t) + iq(t). In addition, the preliminary filtering in order to obtain s(t) is easily done in the frequency domain. Nevertheless the polar motion is defined at the very beginning by two components. This has led us to a simplified procedure in agreement with the initial complex notation.

The complex notation is used direclty to define a time-varying frequency line according to the following scheme :

$$m(t) \xrightarrow{\text{FFT}} M(\omega) \times W(\omega) \xrightarrow{(\text{FFT})^{-1}} m_F(t) \qquad (10)$$

$\downarrow$ Spectral window

The window is restricted to a positive (resp. negative) frequency band around the main frequency $<\omega>$. It is essential to cancel the negative (resp. positivbe) side. The complex value m_F (t) yields :

$$A(t) = |m_F(t)| \ , \ \omega(t) = (\arg (m_F(t) - \arg (m_F (t - \Delta t)) / \Delta t \quad (11)$$

being respectively the instantaneous amplitude and angular frequency of the filtered signal.

This is obviously a very simple algorithm with an explicit location of the frequency band and without adjustment to an a priori central frequency. Curiously it has not been used so far as we are aware. Frequency bands of 0.77177-0.91699 cpy and 0.91699-1.0752 cpy have been chosen for the Chandler and annual components. Since these boundaries have nearly zero power, box-car windows have been used for $W(\omega)$. However the boundary between Chandler and annual wobbles is to some extent arbitrary. The results given in Fig. 2 are not far from the previous analyses as it could be expected, but here the frequency band is explicit. The artifacts observed at the time boundaries are easily explained. These boundaries are cut-offs in a persistent signal, implying in the frequency domain Gibb's phenomena, i.e. high frequencies which are rejected by the procedure.

5. SOME EXPECTED OUTPUTS FROM EXPECTED INPUTS

5a. Low frequencies

For $\omega = 0$, $|H| = 1$ and $dH/d\sigma = 1/\sigma_0 \sim 1.2$ cpy^{-1}, consequently the input is practically unbiased in the vicinity of the zero frequency. The recorded data show an obvious low frequency drift which is essentially linear ; its instrumental or physical origin has been a matter of controversy ; but the recent accurate data favour a physical input.

Accordingly we now analyse the low frequency content of 3 basic inputs following 3 different statistical distributions :

a-a white noise ; its energy spectrum is flat, consequently the mean value which is fixed does not fit the data ;

b-a Brownian motion ; it is a diffusion process generating a drift $\propto t^{1/2}$; in the vicinity of the zero frequency, the spectrum $\propto \omega^{-2}$;

c-a self-similar process, i.e. a distribution with identical properties whatever the time-scale. Then the spectrum $\propto \omega^{-1}$

Mandelbrot and McCamy (1970) have shown that the observed low frequency band favours a self-similar behaviour in spite of the restricted time window. It has been shown (e.g. Nakiboglu and Lambeck,

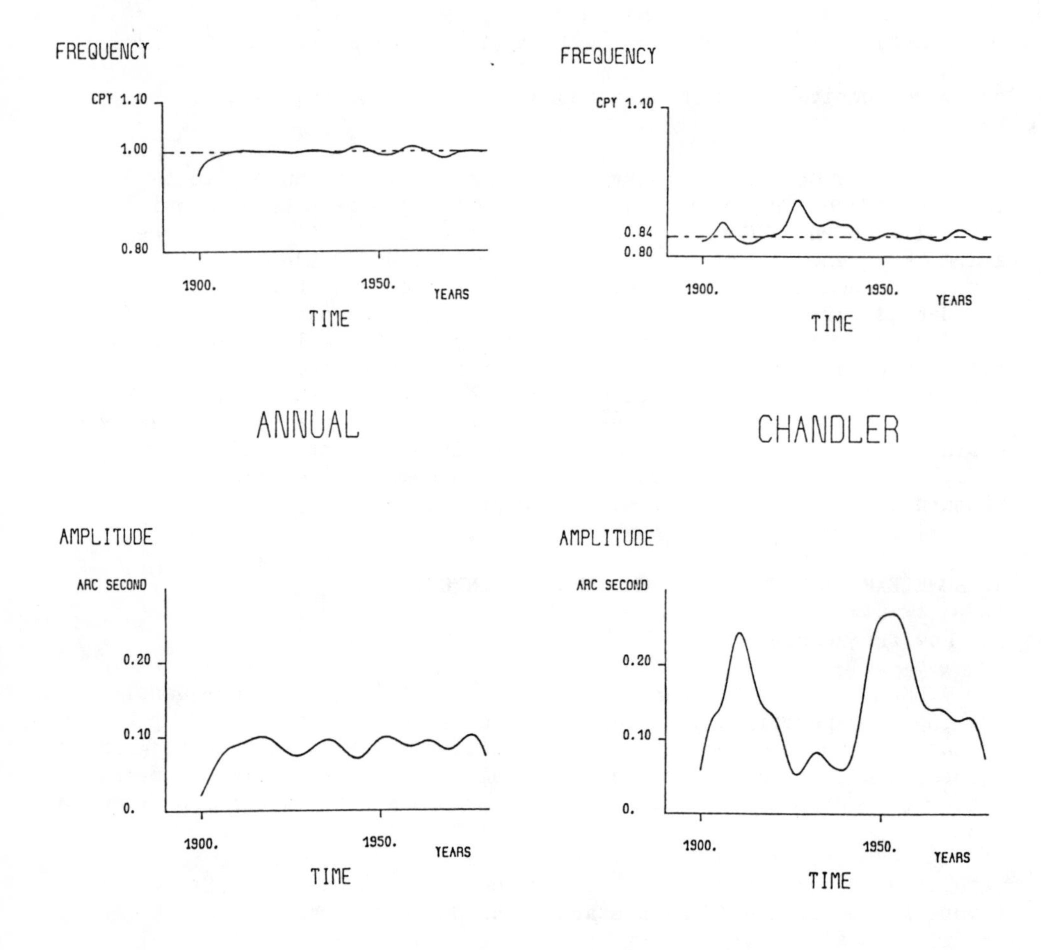

Figure 2. Instantaneous amplitude and instantaneous frequency of the ILS series (1900-1977) according to the frequency domain analysis (see text) for the Chandler frequency band and the annual frequency band.

1980) that the last deglaciation can induce a linear drift in the observed direction. Then assuming a drift $d = kt$, its derivative is $d' = k$ and the corresponding spectrum $D' = k\,\delta$ (δ : Dirac function) so that $D = k\delta/(i\omega)$. As $S(\omega) = D(\omega) * W(\omega)$ where * denotes convolution. Assuming a box-car window of width L, the spectrum $|S|$ is $\sim (\omega^{-1} - L^2\,\omega/6)$. For vanishing ω, $|S| \sim \omega^{-1}$. This could have been expected because a linear curve is the simplest self-similar process.

To sum up, there are some arguments for a linear drift, but additional features exist close to zero, see e.g. Fig. 1 where 2 peaks surround $\omega = 0$ with period around 31 years, the so-called Markowitz wobble, but the time window is not large enough to check its stability.

5b Chandler frequency

Assuming the motion is governed by the simple oscillator defined in § 2, one can test some hypothetical inputs :

1-a white noise with its equal amplitude spectrum does not fit the data (see § 2).

2-a discrete spectrum, an assumption which has not been dealt with so far, is now tested. Without a priori information, we assume inputs with equal amplitude. Referring to (3), amplitude A, eigenfrequency σ_0 and quality factor Q can be computed with 3 spectral lines. Namely let $p = \sigma_1 + \sigma_2$, $q = \sigma_1 - \sigma_2$, and $r = |M_1^+|^{-2} - |M_2^+|^{-2}$, then :

$$\sigma_0 = (r_2\, p_1\, q_1 - r_2\, q_1 - r_1\, q_2)),$$
$$1/A^2 = r\,\sigma_0^2\,/(q\,(p - 2\sigma_0)),$$
$$1/(2Q^2) = ((\sigma_1\,/\,\sigma_0) - 1)^2 + A^2\,(M_1^+)^{-2} \qquad (11)$$

It is possible with additional lines to check the solution stability. Application to data given in Table 1 shows that the values given by Labrouste & Labrouste and those corresponding to our analysis are self-consistent, yielding $\sigma_0 = 0.840$, $A = 1.7 \times 10^{-3}$ arcsecond and $7 < Q < 20$. These results which are sensitive to fluctuations in the input magnitudes must be taken with caution ; they suggest nevertheless that a discrete spectrum is possible.

3 - according to Okubo's simulations (1982), small frequency fluctuations may generate double peaks in agreement with the observed time-scale if $50 < Q < 100$. As mentioned above a transient spectrum is possible but cannot be ascertained before many years.

An alternative is to assume a complex oscillator. For example, part of the value of the eigenfrequency is controlled by the socalled pole tide, i.e. the ocean response to polar motion (see Dickmann, 1979). It is possible to assume a frequency modulation due to a positive feedback between solid earth and ocean (see e.g. Carter, 1981).

A knowledge of the physical input(s) should certainly help to resolve the previous ambiguities ; but all the candidates proposed so far poorly match the data possibly because many of them are still to be quantified with sufficient accuracy.

6. HIGH FREQUENCIES

The ILS series relies almost completely on astrogeodetic data. These are linked to the local vertical of each station. At the required level of sensitivity, the orientation at each station is highly dependent on the local mass motions and atmospheric perturbations. Fortunately, this noise is uncorrelated from station to station so that a global motion can be measured. The noise level is again lowered by reducing the high frequency band. But significant information can be lost. Data from space techniques are much more accurate roughly by an order of magnitude ; they are included in the BIH data since 1972. Consequently for appraising the high frequency content, the unsmoothed BIH series with a sampling rate of 5 days and a time window covering 1967-1984 has been analysed seperately. Its spectrum is given in Fig. 3. Above 1 cpy, 3 small peaks are observed at $\pm$ 1.33, + 1.70 and $\pm$ 2.0 cpy.

The frequency 1.33, corresponding to a period of 274 days, is not related to an obvious physical process. One can invoke an artifact due to a side lobe ; an upper boundary for this effect has been estimated by computing $k.\,|G(\sigma - 1.)| + |G(\sigma - 0.84)|$, where $G(\sigma)$ is the spectrum of the hanning window with a length of 17 years and k is the amplitude ratio of the annual/Chandler peaks ; for $1.2 < \sigma < 1.4$, the amplitude spectrum ratio $|S(\sigma)| / |S(1)| \leqslant 2 \times 10^{-3}$. The observed amplitude ratio is 7×10^{-2}. Furthermore the peak located at -1.33 is observed in a frequency range without large peaks. Consequently a frequency leakage is ruled out. However an aliasing due to a space technique is possible ; nevertheless a similar positive peak can be seen in the ILS series although surrounded by a higher noise. Thus an unknown physical process (so far as we are aware) may be invoked.

The peak located at 1.70 cpy may be the first harmonic of the Chandler wobble. It has no negative counterpart suggesting it is a consequence of the wobble. For example, the drastic enhancement of the pole tide in a shallow sea such as the North Sea is able to induce harmonics.

The semi-annual peaks at $\pm$ 2.0 cpy are expected since they are detected in the atmospheric driving signal and a semi-annual peak is recorded in the l.o.d. perturbations, but it is hardly seen in the ILS spectrum because of the noise level.

The amplitudes of the main peaks are given in Table 2.

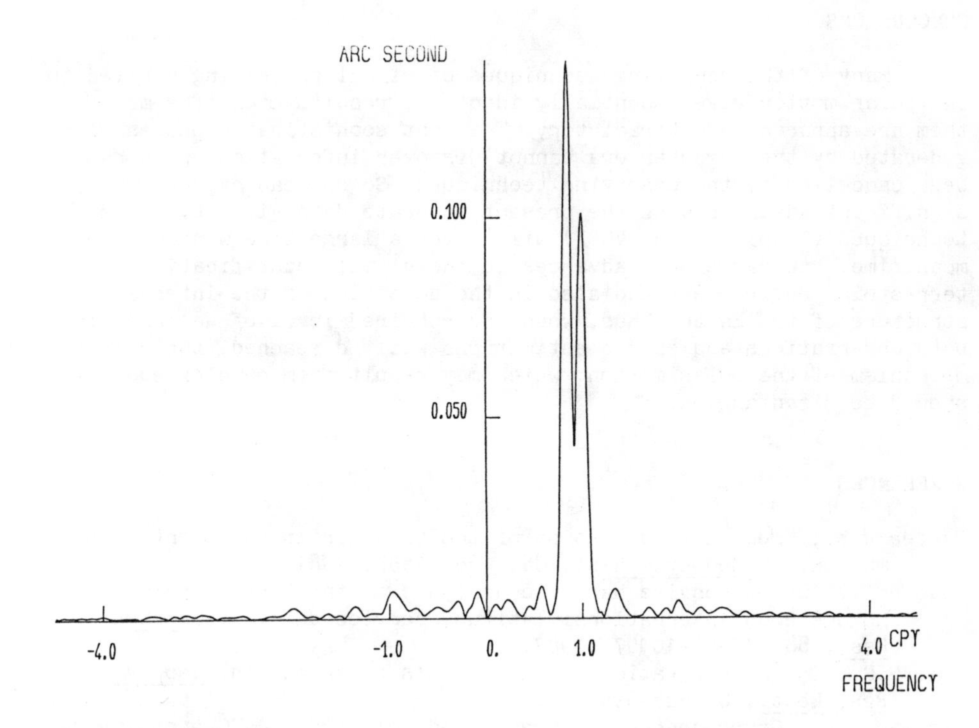

Figure 3. Amplitude spectrum for steady spectral lines of the unsmoothed BIH series (1967-1984), hanning-windowed FFT ; drift rejected by an adjustment according to a 3rd order polynomial prior to spectral analysis.

σ (cpy)	A (positive) (milliarcs.)	A (negative)
0.58	7.9	1.8
0.84	139.8	2.9
1.00	101.1	6.6
1.33	7.3	3.0
1.70	2.5	0.5
2.00	4.3	2.8

Table 2. Locations of the significant peaks in the spectrum of the unsmoothed BIH series (1967-1984). Prograde (positive) and retrograde (negative) components, hanning-windowed FFT.

CONCLUSIONS

Many of the competing techniques of signal processing applied to the polar motion give essentially identical results even if some of them are apparently contradictory. The most sophisticated processes generated by the computer era cannot discover information which has been cancelled by the observing techniques. So one can expect significant advances when the present accurate data given by space techniques (Doppler, SLR, VLBI) will cover a large time window ; in the mean time, one can expect advances in the global quantification of the terrestrial environment and also in the detection of the internal structure of the Earth. Thus, when the required level of accuracy in both observations and source estimations will be reached, the driving mechanism of the polar motion, which may result from complex sources, should be disentangled.

REFERENCES

Carter W.E., Frequency modulation of the Chandlerian component of polar motion, J. Geophys. Res., **86**, 1653-1658, 1981.

Chao B.F., Autoregressive harmonic analysis of the Earth's polar motion homogeneous international latitude service data, J. Geophys. Res., **88**, 10299-10307, 1983.

Chao B.F., On the excitation of the Earth's polar motion, Geophys. Res. Res. Lett., **8**, 526-529, 1985.

Dickman S.R., Consequences of an enhanced pole tide, J. Geophys. Res., **84**, 5447-5456, 1979.

Dickman S.R., Investigation of controversial polar motion features using homogeneous international latitude service data, J. Geophys. Res., **86**, 4904-4912, 1981.

Gaposhkin E.M., Analysis of pole position from 1846 to 1970, in Rotation of the Earth, Edit. Melchior and Yumi, 19-32, Reidel, Dordrecht-Holland, 244 pp., 1972.

Guinot B., The Chandler wobble from 1900 to 1970, Astron. & Astrophys., **19**, 207-214, 1972.

Guinot B., The Chandler nutation from 1900 to 1980, Geophys. J. R. Astr. Soc., **71**, 295-301, 1982.

Jeffreys H., The variation of latitude, Mon. Not. R. Astr. Soc., **100**, 139, 1940.

Labrouste H. and Labrouste Y., Composantes périodiques de la variation des latitudes, Ann. Geophys., **1**, 115-130, 1946.

Lambeck K. and Hopgood P., The Earth's rotation and atmospheric circulation from 1963 to 1972, Geophys. J. R. Astr. Soc., **64**, 67-89, 1981.

McCarthy D.D., The variation of latitude based on the U.S. Naval Observatory Photographic Zenith Tube Observations, J. Geophys. Res., **79**, 3343-3346.

Mandelbrot B.B. and McCamy K., On the secular polar motion and the Chandler wobble, Geophys. J. R. Astr. Soc., **21**, 217-232, 1970.

Munk W. and McDonald G., The rotation of the Earth, a Geophysical Discussion, Cambridge University Press, 323 pp., 1960.

Nakiboglu S.H. and Lambeck K., Deglaciation effects on the rotation of the Earth, Geophys. J. R. Astr. Soc., **62**, 49-58, 1980.

Okubo S., Is the Chandler period variable?, Geophys. J. R. Astr. Soc., **71**, 629-646, 1982.

Wahl E., Ueber die periodischen Eigenshaften der Polbahn, Zentralblatt fur Geophysik, Meteorologie und Geodasie, **4**, 1-48, 1939.

Wahr J.M., The effects of the atmosphere and oceans on the Earth's wobble and on the seasonal variations in the length of day-II Results, Geophys. J. R. Astr. Soc., **74**, 451-487, 1983.

Wilson C.R. and Vicente R.O., An analysis of the homogeneous ILS polar motion series, Geophys. J. R. Astr. Soc., **62**, 605-616, 1980.

ACCURATE EARTH ORIENTATION TIME SERIES FROM VLBI OBSERVATIONS

W. E. Carter and D. S. Robertson
National Geodetic Survey Division
Charting and Geodetic Services
National Ocean Service, NOAA
Rockville, Maryland 20852

ABSTRACT. Since January 1984, the IRIS Earth orientation monitoring service has regularly operated a VLBI network to determine polar motion, UT1, nutation, and precession. The IRIS polar motion time series comprises x and y values at 5-day intervals with accuracies, verified by intercomparisons with SLR values, of better than 2 milliseconds of arc. The UT1 series includes values at 5-day intervals with estimated accuracies of 0.05 to 0.10 millisecond, and daily values during the periods April through June 1984 and 1985 having estimated accuracies from 0.05 to 0.20 millisecond. The IRIS observations have also been used to estimate corrections to the IAU 1984 nutation series, yielding an annual term with an amplitude of approximately 2 milliseconds of arc.

1. INTRODUCTION

January 1984 marks a milestone in the history of the study of the rotation of the Earth. At that time the International Radio Interferometric Surveying (IRIS) Earth orientation monitoring service (EOMS), began regular operations using a network of North American and European very long baseline interferometry (VLBI) observatories equipped with the MARK III instrumentation system [Carter _et al_., 1985]. The IRIS EOMS immediately began producing estimates of the x and y components of polar motion accurate to ± 1 to 2 millisecond of arc and UT1 accurate to ± 0.05 to 0.10 millisecond, averaged over 24-hour periods, at 5-day intervals. In addition, during the MERIT intensive campaign [Wilkins, 1983], April through June 1984, the IRIS EOMS successfully tested a method of obtaining daily estimates of UT1 accurate to ± 0.10 millisecond averaged over a period of about 1 hour using only a single interferometer. Since April 1985, these daily UT1 values have been a regular product of the IRIS EOMS, and improved procedures and instrumentation introduced after the 1984 test have made it possible to improve the accuracy to the 0.05 to 0.10 millisecond level averaged over a period of only 40 minutes.

The IRIS Earth orientation series produced since January 1984 have the highest accuracy and temporal resolution ever obtained. And the operating cost of the IRIS EOMS, about $2 million per year, is a fraction of the operating cost of other services based on less powerful methods

A. Cazenave (ed.), Earth Rotation: Solved and Unsolved Problems, 61–67.

of observation including even the classical optical astrometry based services, which have been operating for several decades.

In the following sections we will present the IRIS time series which demonstrate the capabilities quoted above and briefly describe some early results using IRIS data to probe the structure of the Earth.

2. POLAR MOTION

Figure 1 is a plot showing the path of the pole determined from IRIS observations during the period January 1984 through June 1985. The solid line simply connects the x and y values estimated from the 24-hour observing sessions conducted at 5 day intervals. The lengths of the lines forming the crosses within the diamonds are proportional to the formal standard errors of x and y, but at the scale of figure 1, are generally too small to be discerned. Also plotted in figure 1 are x and y values produced by the University of Texas Center for Space Research from satellite laser ranging (SLR) observations of the laser geodynamics satellite (LAGEOS) collected from a global network of stations cooperating with the National Aeronautics and Space Administration (NASA) Crustal Dynamics Project (CDP) [Tapley et al., 1985]. Offsets in x and y have been removed from the SLR values to align the IRIS and SLR coordinate systems. The IRIS and SLR pole positions are not tabulated at identical dates, and intercomparing the two series requires interpolation, which may increase the apparent discrepancies between the series. We obtained a weighted root-mean-square (RMS) difference of approximately 2 milliseconds of arc in both x and y. Given the very different natures of the two methods of observation, we believe that this intercomparison places a reliable upper bound on the errors of both methods over the 18 months compared. For additional discussion of this subject see Robertson et al. [1985a]. It is certainly possible, indeed expected, that the two series will show a significant divergence as the period of intercomparison is extended, if for no other reason than that the stations in the IRIS and CDP networks are not coincident, and crustal deformations and plate motions will affect the two networks differently.

3. UT1

Figure 2 is a plot of the UT1 values determined from IRIS observations during the period January 1984 through June 1985, differenced from the BIH UT1R series. Unfortunately, there are no UT1 series available from other methods of observation comparable to the IRIS series over periods longer than, at most, a few months. Carter et al. [1984] compared a modified SLR UT1 time series, produced by high pass filtering the differences between VLBI and SLR values, with VLBI series produced prior to the advent of the IRIS EOMS, and obtained an RMS difference of 0.7 millisecond. This is several times the estimated accuracy of the IRIS UT1 series based on error analysis, simulations, and the confirmed accuracy of the polar motion series.

Figures 3 and 4 are enlarged plots of the April to June 1984, and April to June 1985, portions of figure 2. The UT1 values determined from the multistation 24-hour, 5-day interval sessions are plotted as diamonds, while those from the daily single baselines sessions are plotted as simple

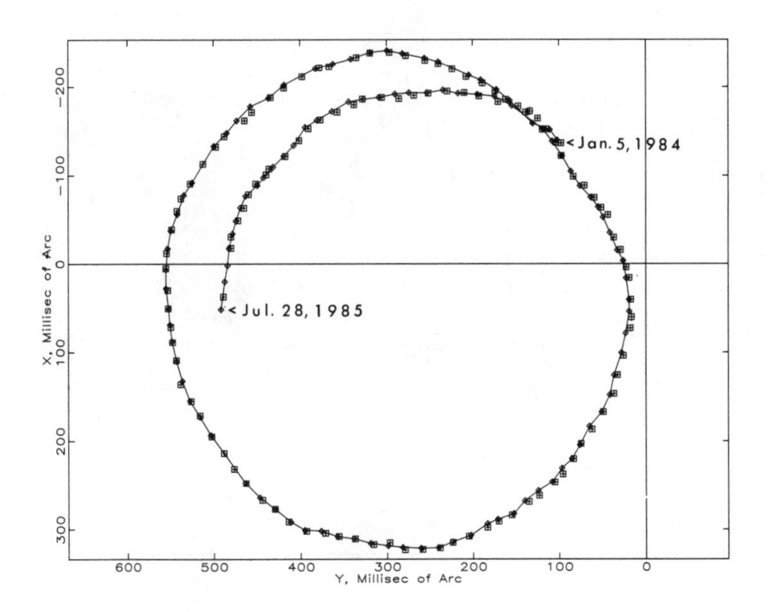

Figure 1. Plot of the IRIS (diamonds) and SLR (squares) pole positions at approximately 5-day intervals during the period January, 1984 to July, 1985.

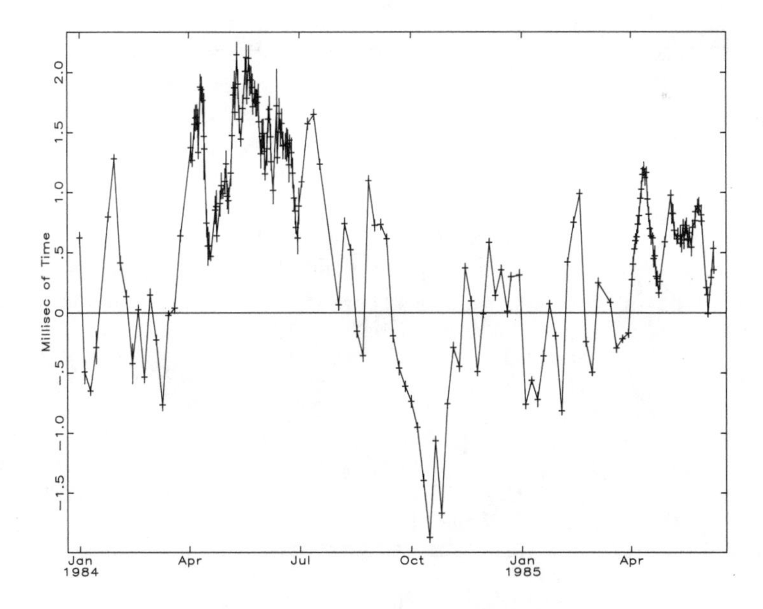

Figure 2. Plot of the IRIS UT1 values, differenced from the BIH UT1R series.

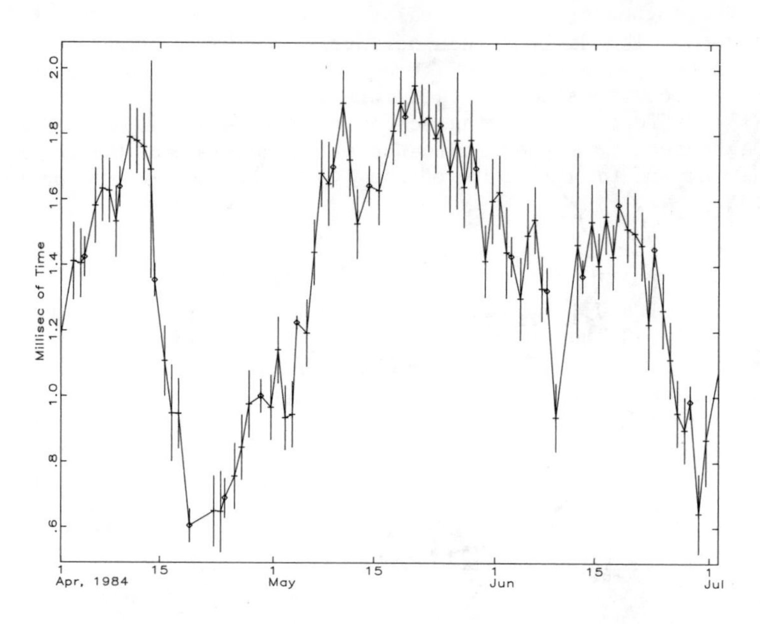

Figure 3. Expanded plot of the April through June 1984 portion of figure 2, allowing the IRIS daily UT1 values determined during the MERIT intensive campaign to be better discerned.

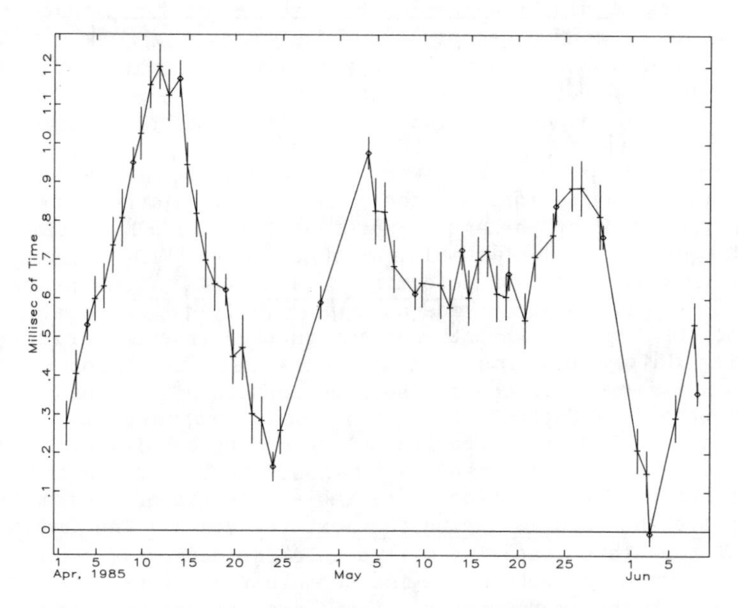

Figure 4. Expanded plot of the April through June 1985 portion of figure 2, allowing the IRIS daily UT1 values to be better discerned.

crosses. The lengths of the vertical bars are proportional to the formal standard errors. The daily values obviously track smoothly at the 1-sigma level between the 5-day values. The systematic errors should be different for the two series, considering the differing number of observatories and radically different observing schedules. Thus we believe that the excellent agreement is a good indicator of the accuracy of the two series. Detailed analysis and interpretation of the 1984 daily series are presented in Robertson _et_ _al_. [1985b].

4. NUTATION

The polar motion and UT1 time series presented above were computed using the standard IAU 1984 nutation series [Wahr, 1981]. The IRIS observations can be used to estimate corrections to the calculated nutation. Figures 5 and 6 show plots of corrections to the nutation parameters ψ and ϵ estimated from the IRIS observations. There are clear systematic variations having a major component at a period of 1 year. Independent analysis of the observations and interpretation of the errors detected in the nutation series in terms of the physics of the core and core-mantle interface are reported in a pair of companion papers by Herring _et_ _al_. [1985] and Gwinn _et_ _al_. [1985].

5. CONCLUDING REMARKS

The time series currently being produced by the IRIS EOMS are already beginning to yield improved knowledge of the physics of the Earth. The application of the IRIS nutation values to probe the interior structure of the Earth is particularly exciting, because it indicates that one of the primary goals of modern geodesy, i.e., to "monitor the time variations of the orientation, shape, and topography of the Earth with sufficient accuracy to refine and discriminate among earth-models and geodynamic theories" [Carter and Robertson, 1985] is beginning to be realized.

As remarkable as the current IRIS results are, there is still much room for improvement. The IRIS network is far from optimal. Improvements in both the number of stations and the geometry of the network could provide an improvement of perhaps a factor of 2 in the polar motion, UT1, and nutation values. The establishment of a new VLBI network with sufficient geometric extent to determine Earth rotation independently from the IRIS network would provide an opportunity to place accurate bounds on the total errors in both systems through intercomparisons of the resulting determinations. It would be especially important to establish such a network in the southern hemisphere so that many of the seasonal atmospheric effects would be reversed. Improvements in the instrumentation, including increasing the recording density of the MARK III tape transports, wider receiver bandwidths to improve the delay resolution function, and addition of reliable water vapor radiometers, will all improve the results during the next few years. The development of additional correlator capacity will enable the density of observations to be increased even to the subdaily regime, should that prove interesting. We look forward to the improvements yet to come, recognizing that they will be possible only because of the pioneering efforts of the many people that have made the IRIS EOMS a success, beginning with the very first observations in January, 1984.

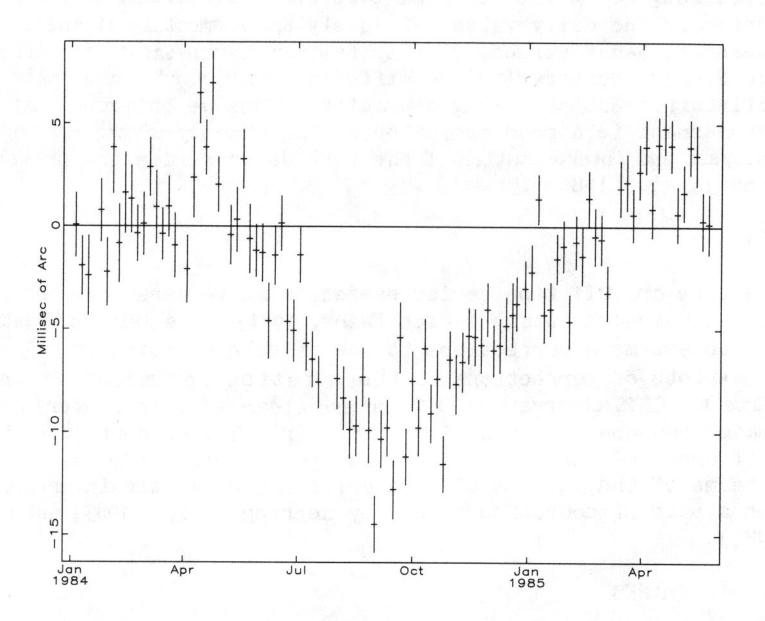

Figure 5. Plot of the corrections to the nutation parameter ψ estimated from the IRIS observations.

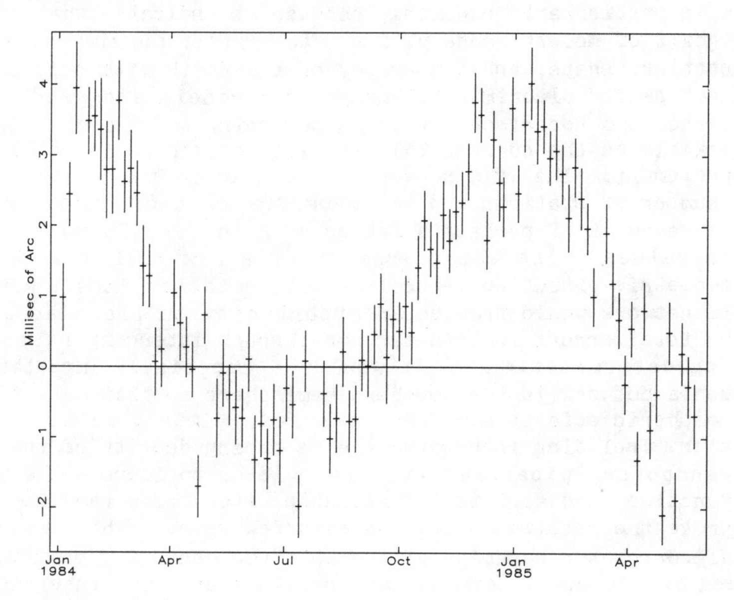

Figure 6. Plot of the corrections to the nutation parameter ϵ estimated from the IRIS observations.

6. ACKNOWLEDGMENTS

VLBI is unique in the degree of cooperation required to schedule, collect, reduce, and analyze each and every observation. The entire process is made possible only by the highly automated and nearly flawless performance of the MARK III VLBI system, developed by a team of talented scientists, engineers, and technicians at the Massachusetts Institute of Technology, Haystack Observatory, NASA Goddard Space Flight Center, and the National Radio Astronomy Observatory. Utilizing VLBI for an operational Earth orientation monitoring service requires that an extraordinary degree of cooperation be maintained day-by-day among participants of different nationalities, on different continents, separated by thousands of kilometers. The IRIS EOMS could not function without the personal commitments and skills of the staff of the Wettzell Observatory, and Max Planck Institute correlator center, FRG; Onsala Space Observatory, Sweden; Westford Observatory, Richmond Observatory, George R. Agassiz Station, Haystack correlator center, and the IRIS operational center at the National Geodetic Survey, USA.

7. REFERENCES

Carter, W.E., D.S. Robertson, J.E. Pettey, B.D. Tapley, B.E. Schutz, R.J. Eanes, and Miao Lufeng, Variations in the Rotation of the Earth, Science, 224, 957-961, 1984.

Carter, W.E., D. S. Robertson and J. R. MacKay, Geodetic Radio Interferometric Surveying: Application and Results, J. Geophys. Res., 90, no. B6, 4577-4587, 1985.

Carter, W.E., and D.S. Robertson, A Modern Earth Orientation Monitoring Service: Functions, Goals and Methods of Observation, in Proceedings of International Conference on Earth Rotation and the Terrestrial Reference Frame, Ohio State University, Columbus, Ohio, 1985.

Gwinn, C. R., T. A. Herring and I. I. Shapiro, Geodesy by Radio Interferometry: Study of the Earth's Nutation, Part II: Interpretation, submitted to J. Geophys. Res., 1985.

Herring, T. A., C. R. Gwinn, and I. I. Shapiro, Geodesy by Radio Interferometry: Studies of the Forced Nutation of the Earth Part I: Data Analysis, submitted to J. Geophys. Res., 1985.

Robertson D.S., W.E. Carter, B.D. Tapley, B.E. Schutz, R.J. Eanes, Polar Motion Measurements: Sub-Decimeter Accuracy Verified by Intercomparison, Science, 229, 1259-1261, 1985a.

Robertson, D.S., W.E. Carter, J.A. Campbell, and H. Schuh, Daily UT1 Determinations from IRIS Very Long Baseline Interferometry, Nature, 316, 424-427, 1985b.

B.D. Tapley, R.J. Eanes, and B.E. Schutz, Center for Space Research Analysis Center Results using LAGEOS SLR data, in Proceedings of International Conference on Earth Rotation and the Terrestrial Reference Frame, Ohio State University, Columbus, Ohio, 1985.

Wahr, J.M., The Forced Nutations of an Elliptical, Rotating, Elastic and Oceanless Earth, Geophys. J. R. Astron. Soc., 64, 705-728, 1981b.

Wilkins, G.A., (ed.), Project MERIT: A Report on the Second MERIT Workshop held at the Royal Greenwich Observatory, Herstmonceux, U.K., 1983.

OBSERVATIONS OF SECULAR AND DECADE CHANGES IN THE EARTH'S ROTATION

L V Morrison
Royal Greenwich Observatory
Herstmonceux Castle
Hailsham
East Sussex BN27 1RP

F R Stephenson
Department of Physics
University of Durham
Durham
DH1 3LE

ABSTRACT. Changes in the length of the day over the past 2700 years are derived from observations of occultations of stars by the Moon and observations of solar and lunar eclipses. The data since AD 1600 show that the length of the day fluctuates by about 4 milliseconds on a timescale of decades. The more ancient data indicate that there are two long-term trends. The first, between 700 BC and AD 1000, gives an increase in the length of the day of 2.4 milliseconds per century: the second, between AD 1000 and the present, gives an average increase of 1.4 milliseconds per century. By comparison, the tidal torques produce an increase of 2.4 milliseconds per century. Hence, there is a non-tidal component in the past millenium which acts to decrease the length of the day by 1 millisecond per century.

1. INTRODUCTION

If $\omega(t)$ is the instantaneous sidereal rate of rotation of the Earth, then astronomical observations measure $\int \omega(t)dt$, where the interval of integration ranges from 1 hour to 10^6 days (2700 years). Meaningful results are obtained over 1 hour using VLBI (Carter et al. 1985) This paper, however, will be concerned with the range 1 year to 2700 years. The standard against which $\int \omega(t)dt$ is compared today is the timescale derived from caesium-beam clocks. These were introduced in 1955. Before then no man-made clocks were stable enough to act as long-term standards because their variation in rate was greater than the variations of $\omega(t)$.

Before 1955 the only available standard of comparison is the timescale constructed from observations of the regular and predictable motions in the solar system, such as those of the Galilean satellites of Jupiter, the planets Mercury and Venus, and the Moon. The Moon's motion affords the best standard because its large angular motion and continuous sequence of observations provide the finest resolution. The Moon has, however, one major drawback - its tidal acceleration. Unless this is known accurately, the timescale constructed from observations of

A. Cazenave (ed.), Earth Rotation: Solved and Unsolved Problems, 69–78.

its position will drift in rate and be misconstrued as a variation of $\omega(t)$. This problem has plagued the analyses of lunar observations in the past; but in the last 10 years a number of independent measurements of the Moon's tidal acceleration have been made and these are in tolerably good agreement. The main problem associated with the use of lunar observations for constructing a timescale against which to compare the Earth's rotation is thus removed.

In this paper we discuss the astronomical observations in the period 700 BC to AD 1984 which lead to the measurement of variations in $\omega(t)$, or, equivalently, fluctuations in the length of the mean solar day. The fluctuations in the length of the day are found to lie between -3 ms and +4 ms on a timescale of decades, with long-term trends of +2.4 ms per century between 500 BC and AD 1000 and +1.4 ms per century between 1000 AD and the present. When compared with the tidal increase of +2.4 ms per century, this leaves a non-tidal component of -1.0 ms per century in the last millenium. A fuller version of this lecture can be found in the paper by Stephenson & Morrison (1984).

2. TIMESCALES, LOD AND SIDEREAL ROTATION

The directly observed quantity discussed in this paper is ΔT - the difference between a uniform timescale and UT1, the timescale based on the period of rotation of the Earth;

$$\Delta T(t) = \text{TAI} - \text{UT1} + 32^{s}.184 \quad \text{after } 1955.5,$$
$$\Delta T(t) = \text{TDT} - \text{UT(1)} \quad \text{before } 1955.5.$$

In terms of the length of the mean solar day (lod),

$$\Delta T(t) = \int (\text{lod}-\text{lod}')dt + \Delta T(t_o),$$

where the reference lod' is exactly 86400 SI seconds, and the epoch t_o is 1977 January 1.0 0^h TAI (IAU, 1977). The value of $\Delta T(t_o)$ is $+47^{s}.52$ (see Morrison, 1979).

The sidereal rate of rotation, ω, is related to the lod, measured in ms, by

$$\omega-\omega' = -0.843 \times 10^{-12} (\text{lod}-\text{lod}') \text{ rad/s},$$

where $\omega' = 1.00273\ 78119\ 06\ (2\pi/86400)$ rad/s,

and the sidereal acceleration, $\dot{\omega}$, is given by

$$\dot{\omega} = -2.67 \times 10^{-22}\ d(\text{lod})/dt \quad \text{rad/s}^2.$$

3. MOON'S TIDAL ACCELERATION

The value of the Moon's tidal acceleration ($\dot{n}$) is crucial in our analysis. A change, $\Delta\dot{n}$, in the adopted value will produce a change, $-\Delta\dot{n}/2n\ (t-19.555)^2$, in the value of ΔT, where n is the Moon's mean motion (0".549/s) and t is the year divided by 100. Recent results for $\dot{n}$ are displayed in Table 1.

Table 1. Values of the Moon's tidal acceleration, $\dot{n}$

method	reference	$\dot{n}$ ("/cy^2)
transits of Mercury	Morrison & Ward (1975)	-26 ± 2
numerical tidal model	Lambeck (1980, p.335)	-29.6 ± 3.1
artificial satellites	Cazenave (1982)	-26.1 ± 2.9
lunar laser ranging	Dickey & Williams (1982)	-25.1 ± 1.2

The value of $-26''/cy^2$ is adopted in this analysis because it was available at the outset of this work. Subsequent results support our choice, but perhaps the correct value is nearer $-25''/cy^2$. In which case, the values of ΔT in this analysis should be altered by $-0^s.91(t-19.555)^2$, and the rate of change of lod by -0.05 ms/cy.

Lambeck (1980, equation 10.5.2) has calculated the acceleration of the Earth's spin in terms of the lunar orbital acceleration on the basis of conservation of angular momentum in the Earth-Moon system. He allowed for both lunar and solar tides in the solid body of the Earth, oceans and atmosphere. His result is

$$\dot{\omega}_T = 51\,\dot{n} \qquad ''/cy^2.$$

Substituting $-26''/cy^2$ for $\dot{n}$ and converting the units to rad/s^2,

$$\dot{\omega}_T = -6.4\times10^{-22} \qquad rad/s^2,$$

and

$$d(\Delta\,lod)/dt = +2.4 \qquad ms/cy.$$

This is equivalent to a quadratic difference in time given by

$$\Delta T_T = +44t^2 \qquad s.$$

The values of ΔT_T at representative epochs are given in Table 2, where the time interval, t, is measured in centuries from the year AD 1800. It will be explained later why this epoch has been chosen. Historical observations which have an accuracy smaller than the values of ΔT in Table 2 can contribute materially to the investigation of changes in the lod.

Table 2. Expected values of ΔT resulting from tidal deceleration.

Epoch	t	$\Delta T_T = 44t^2$
−500	−23	~ 6^h
0	−18	4^h
+500	−13	2^h
+1000	− 8	45^m
+1500	− 3	7^m
+2000	+ 2	3^m

4. OBSERVATIONAL RESULTS FOR ΔT

4.1 Lunar occultations 1620-1955.

Occultations of stars by the limb of the Moon are - in the present context - virtually instantaneous. The accuracy with which ΔT can be obtained from occultations is not limited, therefore, by the nature of the phenomenon itself, but by the methods of timing employed by the observer and the approximations introduced in correcting the observations for the effect of the lunar profile. Observations of occultations have been catalogued and published by Morrison (1978) and Morrison et al. (1981).

These have been analysed to give values of ΔT. The standard deviation (σ) and number of observations lying within 3σ are displayed in Table 3.

Table 3. Standard deviation of ΔT and number of observations within 3σ

Period	σ	N < 3σ	Period	σ	N < 3σ
1620-1669	1m	94	1820-1860	1.5s	1265
1670-1699	15s	65	1861-1942	1.3s	24800
1700-1759	5s	169	1943-1954	1.0s	~10000
1760-1819	2s	313			

The results for the earliest period, 1620-1669, have been supplemented by values of ΔT derived from timings of the 4th contacts of solar eclipses. In this period, the use of star and Sun altitudes as the principal method of timing occultations and eclipses contributed the dominant error to ΔT. This common source of error was removed after the introduction of clocks; then the occultations became superior because of the sharpness of the events.

In the period 1620-1860 the values of ΔT were smoothed by cubic splines. From 1861 onwards annual means were calculated and smoothed using a 5-point convolute (see Morrison, 1979). The resultant smoothed curve for ΔT is shown in Fig. 1. In the period 1955 to 1984, annual values of ΔT were calculated from TAI-UT1+32^{s}.184, where TAI-UT1 is published by the Bureau International de l'Heure. The dashed curve in Fig. 1 is the parabola $\Delta T = 25.5t^2$ which is derived from Arabian observations circa AD 950 (see section 4.4). This is seen to be a reasonable fit if the vertex is placed near AD 1800. Therefore, in the absence of decade fluctuations, the lod would be equal to 86400 SI seconds around AD 1800. This epoch is adopted as the origin of time, t_o, for the calculation of long-term trends in the lod.

4.2 Solar and lunar eclipses AD 500-1600

The path width of a total solar eclipse is less than ~ 200 km at the equator. The projected width along a parallel of latitude is generally

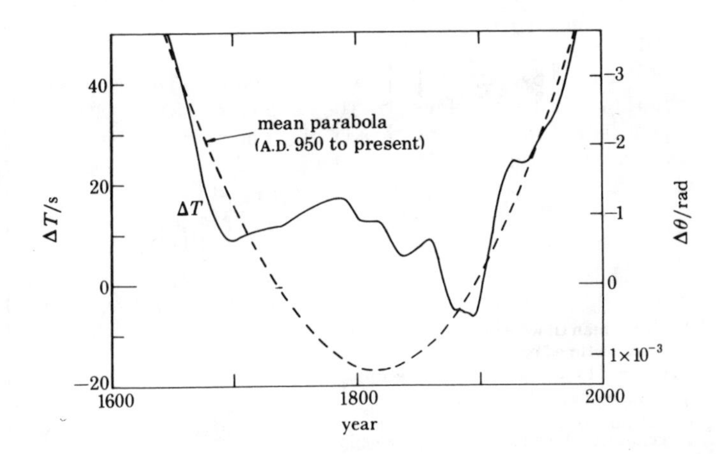

Figure 1. ΔT curve for telescopic period. The mean parabola is derived from Arabian data circa AD 950.

several times greater than this, say ~ 500 km, which is about 18 min in longitude. Therefore, given the place and date of a reported total solar eclipse, ΔT can be calculated to within a range of about 18 min. However, the accuracy can be better than this if the path is very narrow, as in the limiting case of an eclipse which is almost annular. The eclipse of AD 1567 is a case in point: here the uncertainty in ΔT is about 2 min. If an eclipse is recorded as being nearly total, then sometimes an upper or lower bound can be placed on ΔT depending on which side of the belt of totality the observation was made.

Besides 18 untimed eclipse observations scattered over the period AD 500-1600, there are 25 timed observations of lunar and solar eclipse contacts made by Arab astronomers in Cairo and Baghdad between AD 800-1000. They were timed using altitudes which imply a timing error of about 5 min.

The results for ΔT in this period are shown in Fig. 2.

4.3 Babylonian and Chinese observations 700 BC - AD 500.

A series of timings of the first contacts of lunar eclipses in the period 700 BC to 100 BC which were made by Babylonian astronomers has survived on clay tablets. The timings were made relative to the time of sunset or sunrise, whichever was nearer to the eclipse. The times were recorded to the nearest 4 min. The shorter the duration between sunset/sunrise and the eclipse, the less will be the error due to clock drift (probably a water clock). Results for ΔT derived from observations where the time interval was less than 80 min are plotted as filled circles in Fig. 2. There are also some Babylonian observations of the Moon rising or setting while eclipsed. These give one boundary condition for ΔT. They are represented by dotted arrows in Fig. 2.

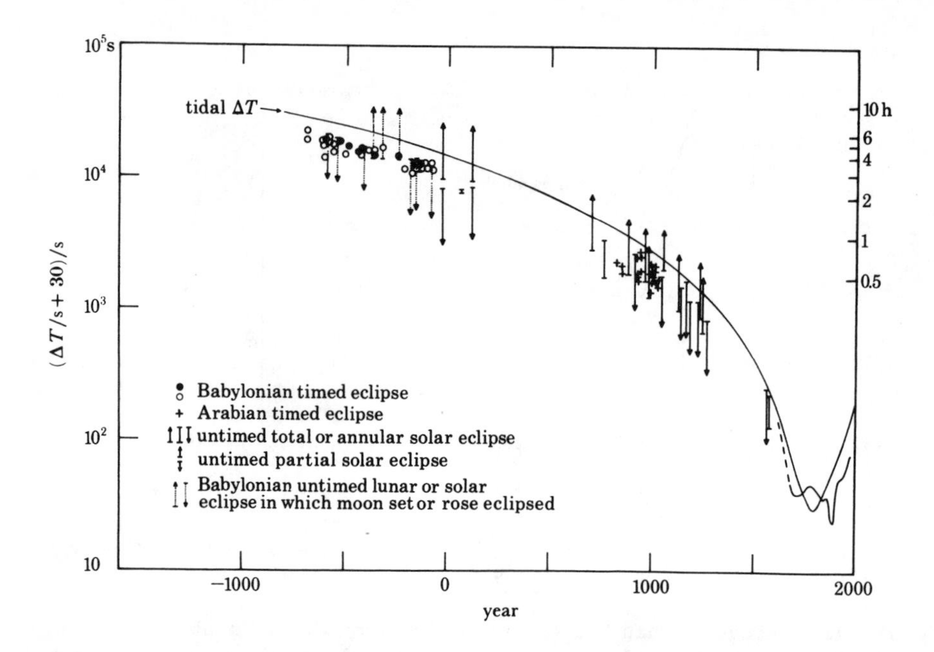

Figure 2. ΔT plotted on a logarithmic scale for the period 700 BC to AD 1984. 30s has been added to make all the values positive. The continuous line shows the tidal deceleration.

Six reliable untimed observations of total or near-total solar eclipses are available from China and Babylon. These have been added to Fig. 2.

4.4 Collected results for ΔT; 700 BC - AD 1984.

Fig. 2 contains all the reliable results for ΔT that we have obtained from historical records. As shown in section 3, the expected shape of the data due to tidal friction is given by the parabola $\Delta T_T = 44t^2$, where t is measured in centuries from AD 1800. Nearly all the data lie below the tidal curve before AD 1600, so we can conclude immediately that there must be an accelerative component in the Earth's spin working against the tidal deceleration. No one parabola will pass satisfactorily through all the data plotted in Fig. 2; therefore, this accelerative component must be variable on a timescale of millennia. The data fall naturally into three groups; so we assume that one acceleration applies between 700 BC and AD 1000 and another between AD 1000 and the present. This is a simplification of the actual situation where the acceleration is probably changing continuously. However, the data do not permit this degree of resolution.

The timed Arabian observations are used to derive the acceleration between 1000 AD and the present. An acceleration coefficient, c, is estimated for each result of ΔT derived from Arabian timings:

$$c = \Delta T/t^2,$$

where t is in centuries from AD 1800. The mean value of c is +25.5 ± 1.1 s/cy^2, which leads to a mean result of +1850 ± 80 s for ΔT at epoch AD 948.

5. DERIVED VALUES OF LOD AND RATE OF CHANGE OF LOD

5.1 Telescopic period AD 1620-1984

The first time-derivative of $\Delta T(t)$ gives the change in the lod compared to the standard of 86400 SI seconds. Before 1861 the derivative of the spline curves was evaluated at yearly intervals. From 1861 onwards the derivatives were calculated annually by applying a five-point quadratic convolute to the annual mean values of ΔT. The resultant values have been joined with a smooth curve in Fig. 3. The dashed line shows the

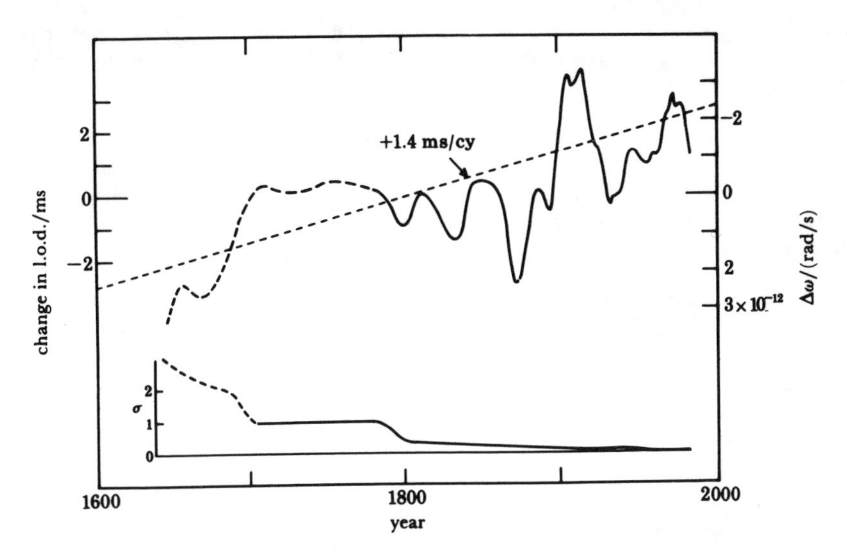

Figure 3. Changes in lod for the period AD 1620 to 1984. The dotted line represents the average rate of change in lod since medieval times. The standard deviation (σ) is shown underneath.

rate of change of +1.4 ms/cy which is equivalent to the mean value of +25.5 s/cy^2 derived for c in section 4.4.

5.2 Medieval period AD 500-1600.

The change in the lod is derived from each observed ΔT as follows;

$$\text{lod}-\text{lod}' = 2\Delta T/36.525(t-18.00) \quad \text{ms}.$$

This assumes a constant acceleration of the Earth's spin between the epoch of each observation and AD 1800. The result for the mean value of +1850 ± 80s for ΔT obtained from the timed Arabian eclipses gives

$$\text{lod}-\text{lod}' = -11.9 \pm 0.5 \text{ ms at AD 948.}$$

The average rate of lengthening of the day between AD 948 and AD 1800 is thus

$$d(\text{lod})/dt = +\ 1.40 \pm 0.04 \text{ ms/cy.}$$

The changes in the lod derived from all the medieval observations are plotted in Fig. 4. It will be seen that only one untimed solar eclipse observation violates the assumption of a constant acceleration between AD 948 and the present.

5.3 Ancient period 700 BC – AD 500.

A constant acceleration is assumed to hold between 700 BC and the mean epoch of the timed Arabian observations (AD 948). The change in the lod is given by

$$\text{lod}-\text{lod}' = 2(\Delta T - 1850)/36.525(t-9.48) + 11.9 \text{ ms.}$$

The average epoch for the forty or so Babylonian timed observations is 390 BC. The mean value of ΔT for these observations is 15600 ± 350 s, which gives the result

$$\text{lod}-\text{lod}' = -44.4 \pm 1.0 \text{ ms at 390 BC.}$$

The average rate of lengthening of the day between 390 BC and AD 948 is thus,

$$d(\text{lod})/dt = +\ 2.43 \pm 0.07 \text{ ms/cy.}$$

6. CONCLUSION

6.1 Tidal deceleration of Earth's spin.

Adopting the value $\dot{n} = -26''/\text{cy}^2$ for the lunar tidal acceleration, we find the following changes in the Earth's spin due to the lunar and solar couples:

$$d(\text{lod})_T/dt = +2.4 \text{ ms/cy}$$

$$\equiv \quad \text{spin acceleration, } \dot{\omega}_T = -6.4 \times 10^{-22} \text{ rad/s}^2$$

$$\text{torque, } C\dot{\omega}_T = -4.6 \times 10^{16} \text{ Nm.}$$

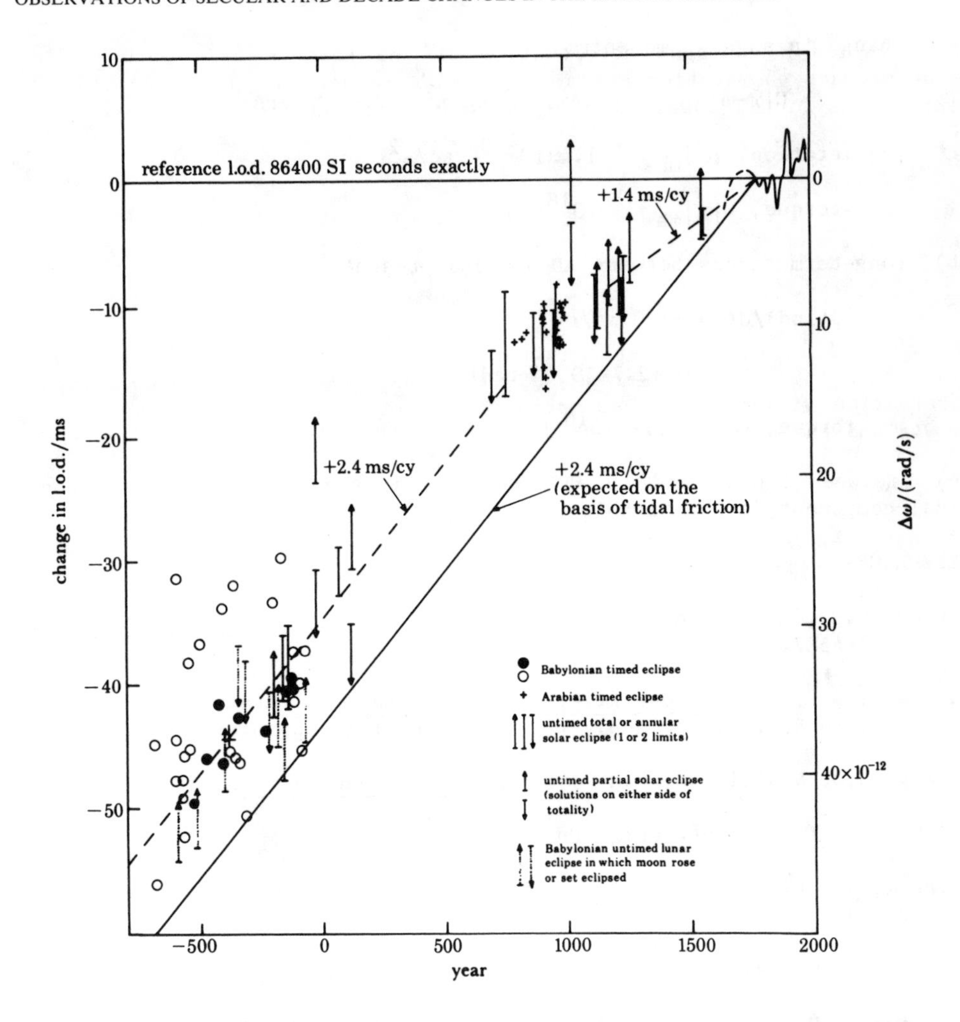

Figure 4. Changes in lod for the period 700 BC to AD 1984. The expected rate of change due to tidal braking is shown as a continuous line. The mean observed rates are represented by dashed lines.

6.2 Non-tidal changes in Earth's spin

(a) Decade fluctuations since AD 1780:

$$|\mathrm{lod}-\mathrm{lod}'|_{max} \sim 4 \text{ ms in 7 years}$$

$$\equiv |\omega-\omega'|_{max} \sim 3\times10^{-12} \text{ rad/s in 7 years.}$$

Change in angular momentum,

$$C|\omega-\omega'|_{max} \sim 2.4\times10^{26} \text{ kg m}^2\text{/s in 7 years}$$

acceleration, $|\dot{\omega}|_{max} \sim 1.4\times10^{-20}$ rad/s^2

torque, $C|\dot{\omega}|_{max} \sim 10^{18}$ Nm.

(b) Long-term trends between AD 950 and present:

$$d(lod)/dt = -1.0 \text{ ms/cy}$$

$$\equiv \dot{\omega} = +2.7\times10^{-22} \text{ rad/s}^2$$

torque, $C\dot{\omega} = +1.9\times10^{16}$ Nm.

(c) Between 700 BC and AD 950 there is no significant secular, non-tidal component in the Earth's spin.

REFERENCES

Carter, W.E., Robertson, D.S. and Mackay, J.R. 1985 J. Geophys. Res. **90**, 4577-4587.

Cazenave, A. 1982 In Tidal friction and the Earth's rotation, II (ed. P. Brosche & J. Sundermann), pp. 4-18. Berlin: Springer.

Dickey, J.O. & Williams, J.G. 1982 Trans. Am. geophys. Un. **163**, 301.

I.A.U. 1977 Trans. Int. astr. Un. **16** B, 56.

Lambeck, K. 1980 The Earth's variable rotation. Cambridge University Press.

Morrison, L.V. 1978 Royal Greenwich Observatory Bull. 183.

Morrison, L.V. 1979 Geophys J.R. astr. Soc. **58**, 349-360.

Morrison, L.V., Lukac, M.R. and Stephenson, F.R. 1981 Royal Greenwich Observatory Bull. 186.

Morrison, L.V. and Ward, C.G. 1975 Mon. Not. R. astr. Soc. **173**, 183-206.

Stephenson, F.R. and Morrison, L.V. 1984 Phil. Trans. R. Soc. Lond. **A313**, 47-70.

EARTH ROTATION AND SOLID TIDES

Paul MELCHIOR
Director, Observatoire Royal de Belgique
Avenue Circulaire, 3
1180 Bruxelles
Belgium

ABSTRACT. The relations between earth tides and the earth's rotation will be investigated here along two ways :

1. *The effects of the earth's rotation upon the earth tide parameters and measurements :*

ROTATION OF THE EARTH		EARTH TIDES
Coriolis Force and ellipticity	→	Latitude Dependence of Love numbers
Polar motion centrifugal force variation	→	Gravity variation ∿ 5 microgal

2. *The effects of the tidal potential and resulting deformations upon the earth's rotation :*

TIDAL POTENTIAL		ROTATION OF THE EARTH	
Zonal	Mm, Mf waves	Monthly, Fortnightly, Periodic variation of rotation	
Tesseral	O_1, P_1, K_1, ψ_1, ... waves	Fortnightly nutation; Semi annual nutation; Precession; Annual nutation; ...	↓ Hydrodynamical core resonance (frequency dependent)
Sectorial	M_2 wave ...	Phase lag ? →	Contribution to secular retardation

Keywords : Earth Tides, Polar Motion, Core resonance.

A. Cazenave (ed.), Earth Rotation: Solved and Unsolved Problems, 79–92.

1. INTERPRETATION OF EARTH TIDES MEASUREMENTS

To contribute to earth's rotation analysis, we have now, at our disposal, a Data Bank of Earth Tides measurements (Ducarme, 1983) which contains results from about 250 stations distributed all over the world, resulting in particular from the Trans World Tidal Gravity Profiles performed by the FAGS International Center for Earth Tides (ICET) and the Royal Observatory of Belgium (Melchior et al. 1985).

The stored results are, for each main tidal component (six waves : O_1 P_1 K_1 N_2 M_2 S_2, see Melchior, 1983), the amplitude factor (δ for gravimeters, γ for tiltmeters) and the phase, i.e. a "tidal vector" $\vec{A}(\delta A,\alpha)$.

To give a correct interpretation of such data, it is essential to start with a sufficiently good model of the earth's interior and sufficiently good models of the main components of the oceanic tides.

We have used, in this respect :

for the elastic earth tides : the Molodensky I model for an oceanless SNREI (spherical, non rotating, elastic, isotropic) Earth with liquid core. This introduces a resonance effect in the response of diurnal earth tides. The calculated vector corresponding to each tidal wave is called $\vec{R}$ (R, O), its phase being zero for a perfectly elastic earth.

for the oceans : The cotidal charts calculated by E. Schwiderski (1979) which we use to calculate, for each principal wave, the attraction and loading vector $\vec{L}$ (L, λ) by Green functions, according to Farrell procedure.

Then, the data can be easily interpreted by using the simple graphical representation as given by the figure 1.

The residue, after subtraction of the "solid" earth response and of the oceanic tides, is called the vector $\vec{X}(X, \chi) = \vec{A} - \vec{R} - \vec{L}$.

Histograms of the $X \sin \chi$ and $X \cos \chi$ (Fig. 2) components show that a large majority of the stations have residues not higher than some tens of a microgal, which corresponds to the instrumental noise. This confirms that the computations based upon an earth model with liquid core and a "Schwiderski oceanic model" are, in general, fully satisfactory.

However no error of measurement or of computation could be invoked to explain some large $X \cos \chi$ residues (up to 4 microgal for the M_2 wave) at places where the installation and maintenance proved to have been excellent.

I began to suspect a reason for this when the station Arta-Djibouti (Afar) gave us a strongly positive $X \cos \chi$ residue which meaned a body tide deformation greater than expected. The Afar region is a *hot spot* and high temperature can indeed decrease the local rigidity modulus μ which should result in a greater tidal deformation. I decided therefore to install our instruments close to a *cold spot* in Africa : three stations in this area (Arlit and Niamey in Niger, Ouagadougou in Burkina Faso) gave a non surprising negative $X \cos \chi$ residue - which means a tidal deformation less than expected. Having collected all available heat flow data, related to the places where tidal gravity measurements had been performed we have computed coefficients of correlation r and found

for Europe $r = 0.603$
for the rest of the World $r = 0.769$

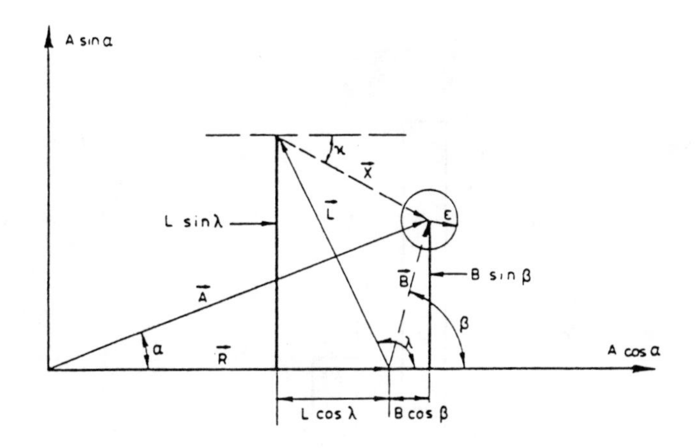

Figure 1 - Definitions and notations

$\vec{A}$ observed vector (A : amplitude, α : phase)
$\vec{R}$ tidal vector for an oceanless elastic non-viscous earth with liquid core (R : amplitude, zero phase)
$\vec{L}$ ocean attraction and loading vector (L : amplitude, λ : phase)
$\vec{B} = \vec{A} - \vec{R}$ (B : amplitude, β : phase)
$\vec{X} = \vec{B} - \vec{L}$ (X : amplitude, χ : phase)
ε instrumental noise

For the semi-diurnal wave M_2, the correct scale of this figure should be of the order of

$R \sim A \sim 400$ nms^{-1} (Europe) to 900 nms^{-1} (Equatorial zone)
$\alpha \sim 0°$ to $\pm 5°$
$L \sim B \sim 20$ to 100 nms^{-1}
$\varepsilon \sim 5$ nms^{-1} (Europe) to 10 nms^{-1} (Equatorial zone)

It is essential to check definitively this explanation because if it is really proved, we could consider that, at the level of 0.5 microgal (5.10^{-10} of g), the modelization of tidal deformations would be perfect.

The conclusions reached so far are :

1 - The Schwiderski cotidal maps are an excellent working model for the oceanic tides. Using these maps to correct the measurements for oceanic attraction and loading allows to address other geophysical effects.
2 - The noise level of spring gravimeters, correctly calibrated in phase as well as in amplitude, is not more than 0.5 or evèn 0.3 microgal (5.10^{-10} of g) on tidal amplitudes determined with six months continuous measurements.
3 - The still "unexplained" residues X cos χ seem to be correlated with heat flow. Such an important result must be confirmed by additional measurements.

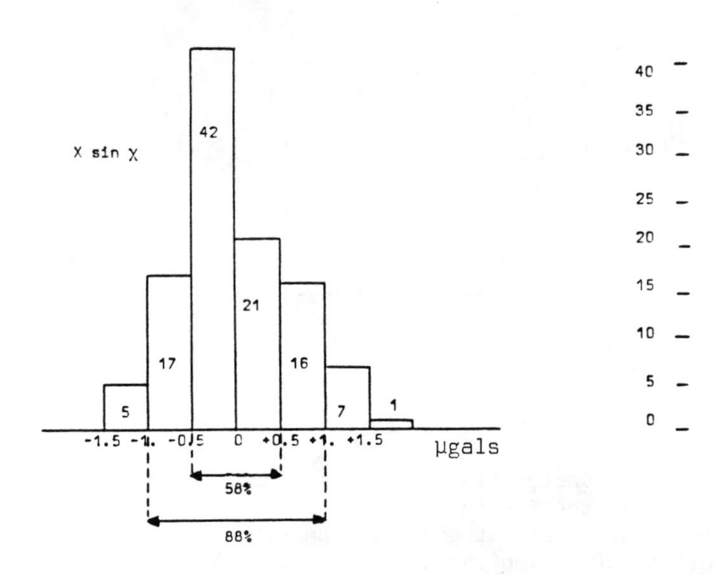

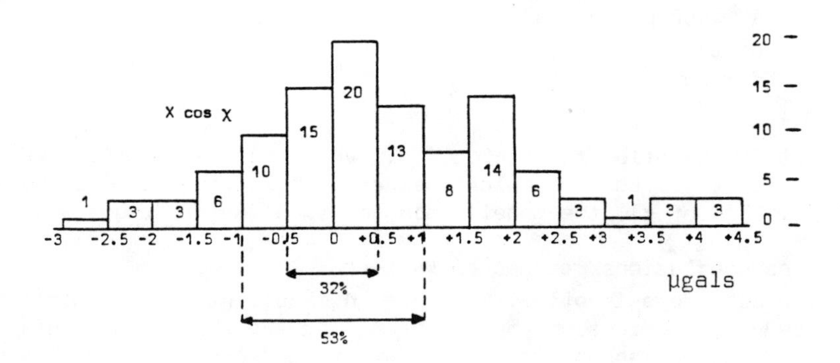

Figure 2 - Histogram of final residues for the semi diurnal tidal wave M_2 for 109 stations in Africa, Asia, Australia, South Pacific.
The broad X cos χ distribution is due to effects not taken into account in the earth model vector $\vec{R}$. The X sin χ distribution corresponds to internal noise.

We can now calculate, for all stations, excluding those with $X \cos \chi$ residues higher than 1 microgal, the vector $\vec{A} - \vec{L}$ which corresponds to the deformation of an oceanless earth.

These corrected data allow to investigate the effects related to the earth's rotation.

2. LATITUDE DEPENDENCE OF LOVE NUMBERS.

As early as 1911, in his book *Some Problems of Geodynamics*, Love discussed the effects of inertia (Chap. V) and ellipticity of the Earth (Chap. VI) on tidal deformations. He considered the perturbation of the M_2 wave and indicated the latitude dependence of the correction from the fourth-order function $(7 z^2 - r^2) / a^2$, which is equal to $(7 \sin^2 \phi - 1)$ at the surface of the Earth. His conclusion was that "without calculation we should expect the correction for inertia to be small, of the order $\omega^2 a/g$ {which is 1/289 (0.0035)} and the correction for ellipticity to be small, of the order ε {which is 1/297 (0.0033)}".

Moreover, in the case of a homogeneous incompressible Earth, Love found that the inertia effect is multiplied by a small coefficient while the coefficient by which ε is multiplied differs little from unity. The correction coefficient could thus be expected to be around 0.004.

The theoretical approach of Wahr (1981) gives an evaluation of these effects calculated with the most recent Earth models. The equations differ from the corresponding equations for an SNREI Earth by being coupled sets of *infinite* order. Of course, to perform computations it is necessary to truncate the series and, as stated by Smith (1974) "the validity of such a model is, at the present time, conjectural. A useful test of such a conjecture is to proceed on the assumption that it is correct and compare the results so obtained with observation". Our objective here is to make such a comparison.

The symmetry properties of the rotating Earth constrain the representation of normal-mode eigenfunctions by linear combinations of poloidal (σ) and toroidal (τ) vector fields :

$$\sum_{\ell=m}^{\infty} \{\sigma_\ell^m (r, \psi, \omega) + \tau_\ell^m (r, \psi, \omega)\} \tag{1}$$

the terms sharing a common functional dependence m on the longitude λ because there is invariance of a revolution-ellipsoidal Earth under any rigid rotation about OZ. Smith has shown that because of the invariance with respect to the centre of mass, σ_ℓ^m is even when ℓ is odd while τ_ℓ^m is odd when ℓ is odd. Thus, the displacement field S must have the form

$$S(r, \psi, \omega) = \sum_{\ell=m,m+2,\ldots}^{\infty} \{\sigma_\ell^m(r, \psi, \omega) + \tau_{\ell-1}^m(r, \psi, \omega)\} \tag{2}$$

or

$$S(r, \psi, \omega) = \sum_{\ell=m,m+2,\ldots}^{\infty} \{\tau_\ell^m(r, \psi, \omega) + \sigma_{\ell-1}^m(r, \psi, \omega)\} \tag{3}$$

As a result of this representation, the spheroidal term σ_2^1 (tesseral diurnal tides) is associated with a toroidal term τ_1^1 which represents the precession nutation (corresponding to the Melchior and Georis development (1968)).

Similarly, solutions for the semi-diurnal sectorial displacements are sought in the form

$$u = \sigma_2^2 + \tau_3^2 + \sigma_4^2 + \ldots \tag{4}$$

as done by Wahr, who truncated this series at σ_4^2.

Then, from Wahr (1981),

$$\delta f = -(2/r_o)\, H_\ell^m \cdot 10^2\, g_a \{\delta_o Y_\ell^m + \delta_+ Y_{\ell+2}^m + \delta_- Y_{\ell-2}^m\} \tag{5}$$

which, for $\ell = 2$, $m = 2$, gives

$$\delta f = -(2/r_o)\, H_2^2 \cdot 10^2\, g_a\, Y_2^2\, \delta \tag{6}$$

where

$$\delta = \delta_1 + \delta_2\, (Y_4^2 / Y_2^2)$$
$$\delta_o = 1 + h - (3/2)\, k \tag{7}$$

For $\ell = 2$, $m = 1$, similarly

$$\delta = \delta_1 + \delta_2\, (Y_4^1 / Y_2^1) \tag{8}$$

A comparison of computed results (Wahr, Dehant) and observed results is given in the Table I. Considering the experimental error bars, the latitude dependent coefficient δ_2 shows an excellent agreement with theory. However the first term δ_1, independent of latitude, shows a discrepancy of about 1 %. This problem is presently addressed by earth tide specialists and could probably be resolved by a correction to the calibration of gravimeters of about 0.5 % and a correction to the theoretical results of about 0.3 % by introducing attenuation effects in the earth's mantle (Dehant 1985).

TABLE I. Latitude dependence of Love numbers.

SEMI DIURNAL WAVE M_2 VERTICAL COMPONENT

$$\delta = \delta_1 + \delta_2 \frac{Y_4^2}{Y_2^2} = \delta_1 + \delta_2 \left[\frac{\sqrt{3}}{2}\,(7 \sin^2 \phi - 1)\right] \quad \text{(LOVE 1911).}$$

	δ_1		δ_2	
	WAHR MODEL	ICET DATA BANK	WAHR MODEL	ICET DATA BANK
M_2	1.160	1.1725 ± 18	-0.0045(*)	-0.0041 ± 8

DIURNAL WAVES VERTICAL COMPONENT

$$\delta = \delta_1(\omega_i) + \delta_2 \frac{Y_4^1}{Y_2^1} = \delta_1 + \delta_2 \left[\frac{\sqrt{6}}{4}(7\sin^2\phi - 3)\right] \quad \text{(LOVE 1911)}$$

	δ_1		δ_2	
	WAHR MODEL	ICET DATA BANK	WAHR MODEL	ICET DATA BANK
O_1	1.152	1.1618 ± 16	-0.0065(*)	-0.0028 ± 15
P_1	1.147	1.1522 ± 29	-0.0065(*)	-0.0039 ± 29
K_1	1.132	1.1458 ± 12	-0.0063(*)	-0.0059 ± 13

M_2 : 137 stations used
O_1 : 109 stations used
P_1 : 33 stations used (6 months registration being needed)
K_1 : 81 stations used
ω_1 : tidal frequency (see Table 2)
(*) : two digit values are given by V. Dehant.

3. POLAR MOTION OBSERVED WITH GRAVIMETERS

Gravity variations resulting from the perturbation of the centrifugal force due to the polar motion can be measured, in principle, with registrating gravimeters.

Classical spring gravimeters were not stable enough at long period scale to identify such effects with security. Cryoscopic superconducting gravimeters have a high stability and a precision almost one - may be two - digits higher than the spring instruments. The potential of the centrifugal force being

$$W_2 = \frac{1}{2}\omega^2 r^2 \cos^2\phi \tag{9}$$

its variation

$$\frac{dW_2}{d\phi} = -\frac{1}{2}\omega^2 r^2 \sin 2\phi \, d\phi = -\frac{1}{2}\omega^2 r^2 (x\cos\lambda + y\sin\lambda)\sin 2\phi \tag{10}$$

(x, y : coordinates of the pole of rotation with respect to CIO). produces a "polar gravity tide" with an amplitude

$$\Delta g = \frac{1}{2}\omega^2 r^2 (1 + h - \frac{3}{2}k)(x\cos\lambda + y\sin\lambda)\sin 2\phi \tag{11}$$

maximum at 45° latitude which should make particularly efficient the two european stations equiped with such an instrument : Bad Homburg, Germany (ϕ = 50°14') and Bruxelles, Belgium (ϕ = 50°48').

The gravity effect can reach a maximum of about 8 microgal which is, in principle, easy to measure and has been measured with these two instruments. It is however too early to conclude that superconducting gravimeters could be used to monitor the polar motion and, eventually also, the variations of the velocity of rotation because the series of measurements so far obtained are not more than 3 years and we have therefore not yet been able to analyse the yearly component which may contain not only barometric effects but also underground water table variations effects on gravity.

Spectral analysis, anyway, has clearly established the presence of the Chandlerian peak.

4. HYDRODYNAMICAL CORE RESONANCE

It is well known that Hough (1895), Sloudsky (1895) and Poincaré (1911) indicated the possible existence of a resonance in the diurnal frequency band, due to hydrodynamical oscillations in the fluid core. Jeffreys (1949) calculated it numerically on the basis of a reasonable earth's model. Molodensky (1961) developped the first theory with correct previsions for gravity and tilt measurements while the most developped theory is presently the theory of Wahr (1981) for an oceanless elliptical rotating elastic model earth.

The first experimental result showing the resonance acting on the three main diurnal waves (O_1, P_1, K_1) was obtained by Melchior in 1966 with quartz horizontal VM pendulums installed at the underground station Sclaigneaux in Belgium. Since twenty years very long series have been obtained with such quartz horizontal pendulums and some shorter but very good series with gravimeters. They confirm without any doubt the existence of such a resonance. However, and unfortunately, the error bars are still too great which does not allow to put constrains for core models or to measure a reduction of amplitude at the resonance frequency due to the core viscosity (which is in fact very small).

Global analysis of each series has been performed by a least squares method (the tidal frequencies being known with eight digits exact) using Venedikov filters. These experimental results are given in the Table II and compared to the theoretical ones obtained by Wahr (1981).

Considering the total luni-solar tidal torque exerted on the Earth :

$$N = - \iiint_V (r \wedge \text{grad } W) \, \rho \, dv \tag{12}$$

where W is the tidal potential, one can write

$$N = \iiint_V \text{rot} (\rho W r) \, dv + \iiint_V (r \wedge \text{grad } \rho) \, W \, dv$$

$$= \iint_S (n \wedge R) \, \rho W \, . \, dS + \iiint_V (r \wedge \text{grad } \rho) \, W \, dv, \tag{13}$$

and demonstrate (Melchior and Georis, 1968) that : *"Two waves of symmetric frequency with respect to the sidereal frequency* ω *(*$\omega_i = \omega + \Delta\omega_i$*,* $\omega_{-i} = \omega - \Delta\omega_i$*) form only one and the same wave of nutation"*; *the sum of their amplitudes* A_i, A_{-i} *appears in* $\Delta\psi$ *and their difference in* $\Delta\theta$ *(this is why nutations are ellipses) :*

$$\begin{aligned} \Delta\theta &= - E \sum_i \frac{\omega}{\Delta\omega_i} \left[A_i - A_{-i}\right] \cos (\Delta\omega_i t), \\ \Delta\psi &= - \frac{E}{\sin\theta} \sum_i \frac{\omega}{\Delta\omega_i} \left[A_i + A_{-i}\right] \sin (\Delta\omega_i t), \end{aligned} \tag{14}$$

with E = 0.0164".

TABLE II. Hydrodynamical Resonance effect due to the liquid core.

Diurnal waves			Vertical Component		Horizontal East West Component		
Wave	Frequency °/hour ω_i	Amplitude Unit : 1μgal	$\delta = 1 + h - \frac{3}{2}k$ Wahr model	Observed (*)	Amplitude Unit : 0"001	$\gamma = 1 + k - h$ Wahr model	Observed (**)
σQ_1 $2Q_1$ σ_1	12°9	2.0		1.1569			
Q_1	13°999	6.8	1.152	1.1570	0.57	0.689	0.668 ± 0.075
ρ_1	13°471	1.2		1.1512	0.12	0.689	0.635 ± 0.133
O_1	13°943	35.4	1.152	1.1572	3.31	0.689	0.6735 ± 0.0186
τ_1 NO_1	14°196	3.0	1.152	1.1596	0.29	0.696	0.700 ± 0.071
π_1	14°918	1.0	1.149	1.1540	0.11	0.699	0.758 ± 0.110
P_1	14°958	16.5	1.147	1.1568	1.67	0.700	0.7098 ± 0.0238
K_1	15°041	49.3	1.132	1.1449	5.29	0.730	0.7497 ± 0.0123
—	15°073	Resonance					
ψ_1	15°082	0.4	1.235	1.2392	0.02	0.523	0.430 ± 0.143
ϕ_1	15°123	0.7	1.167	1.2015	0.08	0.687	0.772 ± 0.117
θ_1	15°513	0.5		1.1759			
J_1	15°585	2.8	1.155	1.1678	0.28	0.692	0.725 ± 0.069
OO_1	16°139	1.5	1.154	1.1631	0.14	0.689	0.662 ± 0.090

(*) Ten observing stations
Twelve gravimeters (including 3 Superconducting instruments)
Total of measurements : 398.856 hourly readings.

(**) Three underground stations in Belgium and Luxemburg : Dourbes (depth 45 m), Sclaigneaux (depth 80 m), Walferdange (depth 100 m).
Eight VM Quartz Horizontal Pendulums interferometrically calibrated
Total of measurements : 806.808 hourly readings.

One can calculate in this way nutation series on the basis of earth tide experimental results including the liquid core effects as did

Melchior in 1971 (Table III). This gave an important correction for the annual nutation in longitude (which corresponds to the ψ_1 tidal wave) : from -0"0503 (rigid earth) to -0"0580 and introduces an annual nutation in obliquity (0"0056) which does not exist for a rigid earth. Wahr however obtained -0"0567 in longitude, that is a difference of +0"0013. Herring et al. (november 1983) and later Gwinn, Herring and Shapiro (1984) found, from VLBI experiments, that the Wahr amplitude had to be corrected by -0"0019 ± 0"0002 (some 2 %) which would restore practically my 1971 proposed value. They also find a phase lag of 0.8°.

This small discrepancy may be due, also, to the lack of consideration of the attenuation in the mantle (see Dehant, this workshop).

TABLE III. Nutation coefficients (units are arc s)

Term	Woolard	Optical Astronomical observations	Rigid Earth with N = 9.2050	Melchior (1971) (6)	Wahr (1979)
Principal					
Obliquity	9.2100	9.2050 ± 0.0017 (1) 9.1990 ± 0.0035 (2)	9.2050	9.2014	9.2035
Longitude	6.8584	6.8409 ± 0.0025 (1) 6.8360 ± 0.0035 (2)	6.8547	6.8386	6.8430
Annual					
Obliquity	0.0000	—	0.0000	0.0056	0.0055
Longitude	-0.0502	—	-0.0503	-0.0580	-0.0567
Semi-annual					
Obliquity	0.5522	0.578 ± 0.002 (1)	0.5528	0.5724	0.5708
Longitude	0.5066	0.533 ± 0.002 (1)	0.5072	0.5237	0.5215
Semi-monthly					
Obliquity	0.0884	0.0925 ± 0.0014 (1) 0.0897 ± 0.0007 (3) 0.0893 ± 0.0022 (4) 0.0898 ± 0.0016 (5)	0.0844	0.0910	0.0910
Longitude	0.0811	0.0853 ± 0.0010 (1) 0.0818 ± 0.0022 (4) 0.0824 ± 0.0016 (5)	0.0811	0.0831	0.0834

(1) Mc Carthy et al. (1977)
(2) Yumi et al. (1978)
(3) Gubanov and Yagudin (1978)
(4) Iijima et al. (1978)
(5) Mc Carthy (1976) : results of Washington and Hertmonceux PZTs
(6) Four models were given in my 1971 paper. I should have selected this one in 1979.

5. FORTNIGHTLY VARIATIONS OF THE EARTH'S ROTATION VELOCITY

Zonal tides deform the earth, changing its polar moment of inertia C. This in turn changes the rotation speed ω ($d\omega/\omega = -dC/C$).

Optical astrometric determinations of Universal Time have been analysed by different authors, leading to direct determinations of the Love number k which characterizes the potential variation due to the elastic body deformation (Table IV).

TABLE IV. The Love number k as derived from the Earth's speed variations.

Authors		waves M_f	M_m
Guinot	(1970)	0.331	0.265
Pilnik	(1970)	0.300	0.282
Guinot	(1974)	0.334	0.295
Djurovic	(1975)	0.343	0.301

Clearly one obtains a systematic difference between fortnightly Mf wave and monthly Mm wave, a difference which could be due to an effect of the corresponding oceanic and atmospheric tides.

It is therefore interesting to consider the results obtained for the corresponding earth tides. They are given in the Table V for the wave Mf.
(Note that the zonal tides give no contribution in the East-West component).

It has not yet been possible to obtain a correct determination of the Mm waves from earth tide measurements.

TABLE V. The zonal fortnightly tide Mf.

Argument (2☾, Period 13.661 days (half tropic month).

North South Component : Amplitude at 50° Latitude : 0"00156

Station Dourbes 1	169560 hours,	Pend. VM7	γ = 0.79 ± 0.07	α = -2°8 ± 5°4
Station Dourbes 2	54000 hours,	Pend. VM29	γ = 0.79 ± 0.06	α = +1°7 ± 4°8

Vertical Component : Amplitude at 50° Latitude : 5.8 μgal
at Alice Springs : 3.4 μgal

Station Bruxelles	11160	Superconduct.	δ = 1.144 ± 0.016	α = -0°4± 0°8
Station Bad Homburg	10080	Superconduct.	δ = 1.150 ± 0.009	α = +0°4± 0°4
Station Alice Springs	8280	Geodynamics 84	δ = 1.03 ± 0.18	α = 2°5± 10°7

Oceanic attraction and loading effect in the vertical component :

$\vec{L}$ (Mf) ≡ 0.164 μgal, 5°1 at Bruxelles
0.122 μgal, 3°3 at Bad Homburg

6. SECULAR RETARDATION OF THE EARTH'S ROTATION

It is generally considered that the body tide do not contribute very much to the secular retardation which is essentially due to bottom pressure effects by the oceanic tides (Schwiderski 1983).

According to Schwiderski however the body tide contribution is not absolutely negligeable and would involve a tidal bulge phase lag of about 0.44°. It is therefore worthwile to consider which experimental value one obtains for the sectorial M_2 wave phase lag, after correction for the oceanic effects.

The analysis of our data gives, for the main waves M_2 and O_1 a small negative phase of about 0.1° (Table VI) which could indeed correspond to a viscous bulge phase lag of about 0.5°.

TABLE VI. General mean δ and phase after oceanic attraction and loading correction (Schwiderski Maps).

	Number of stations	Criterion of selection	$\bar{\delta}$	$\bar{\alpha}$
WAVE O_1				
	77	1 + 2	1.1568 ± 0.0045	-0°.11 ± 0°.05
	78	1 + 2	1.1563 ± 0.0044	-0°.09 ± 0°.05
	86	1 + 2	1.1576 ± 0.0041	-0°.20 ± 0°.06
	187	1	1.1603 ± 0.0024	-0°.13 ± 0°.03
WAVE M_2	Number of stations	Criterion of selection	$\bar{\delta}$	$\bar{\alpha}$
	93	1 + 2	1.1635 ± 0.0015	-0°.12 ± 0°.04
	204	1	1.1690 ± 0.0028	-0°.11 ± 0°.03

Criterion (1) : corrected phase in absolute value < 1°
Criterion (2) : distance to the sea > 200 km
Total number of tidal gravity stations in the Data Bank : 245.

However, the integration of the differential equations for the deformations of an ellipsoidal rotating earth with *inelastic* mantle and liquid core shows that the observed phase lag cannot exceed few hundredths of a degree (Zschau 1985, Dehant 1985).

REFERENCES

Dehant V., 1986. Body Tides for an Elliptical Rotating Earth with an Inelastic Mantle. This Symposium.

Ducarme B., 1983. A Data Bank for Earth Tides, IUGG General Assembly, Hamburg - Symposium 6 : Data Management, Bull. Inf. Marées Terrestres 91 : 5963-5980.

Gwin C.R., Herring T.A. and Shapiro L.I., 1984. Geodesy by Radio Interferometry : Corrections to the IAU 1980 nutation series, EOS, 65, n° 45, 859.

Herring T.A. et al., 1983. Geodesy by Radio Interferometry : Studies of the nutations of the Earth's rotation axis, EOS, 64, n° 45, 674.

Hough S.S., 1895. The oscillations of a rotating ellipsoidal shell containing fluid, Phil. Trans. Royal Soc. London, 186, 1, 469-506.

Jeffreys H., 1949. Dynamic effects of a liquid core, Monthly Not. R. Astr. Soc. 109, n° 6, 670-687 and 110, n° 5, 460-466.

Love A.E.H., 1911. Some problems of Geodynamics, Dover Publications, Inc. New York.

Melchior P., 1966. Détermination expérimentale des effets dynamiques du noyau liquide de la Terre dans les marées terrestres diurnes, Bull. Acad. Roy. Belg. LII, 93-100.

Melchior P., 1980. Luni Solar Nutation Tables and the Liquid Core of the Earth, Astron. and Astrophys. 87, 365-368.

Melchior P., 1981. An effect of the earth ellipticity and inertial forces is visible from M_2 and O_1 tidal gravity measurements in the Trans World Profiles, 9th Intern. Symposium on Earth Tides, New York. Comm. Obs. R. Belg., A 63, S. Geoph. 141 : 1-9.

Melchior P., 1983. The Tides of the Planet Earth, Pergamon Press, 2nd edition, 641 pages.

Melchior P. and Georis R., 1968. Earth Tides, precession-nutation and the secular retardation of earth's rotation, Phys. of the Earth Planet. Int. 1, 4 : 267-287.

Melchior P. and De Becker M., 1963. A discussion of world-wide measurements of tidal gravity with respect to oceanic interactions, lithosphere heterogeneities, Earth's flattening and inertial forces, Phys. Earth Planet. Inter., 31 : 27-53.

Melchior P., Ducarme B., Van Ruymbeke M., Poitevin C. and De Becker M., 1984. Interactions between oceanic and gravity tides, as analysed from world-wide earth tide observations and Ocean Models, Marine Geoph. Res. 7 : 77-91.

Molodensky M.S., 1961. The theory of nutations and diurnal earth tides, Obs. Roy. Belg. Comm. n° 188, S. Geoph. 58 : 25-26.

Poincaré H., 1910. Sur la précession des corps déformables, Bull. Astron., 27 : 321-356.

Schwiderski E.W., 1979. Global Ocean Tides. Part II. The Semidiurnal Principal Lunar Tide (M_2), Atlas of Tidal Charts and Maps, Naval Surface Weapons Center, Dahlgren Laboratory TR 79-414, Dahlgren, VA.

Schwiderski E.W., 1980a. On charting global ocean tides, Rev. Geophys. Space Phys., 18 : 243-268.

Schwiderski E.W., 1980b. Ocean tides. Part I. Global ocean tidal equations. Part II. A hydrodynamical interpolation model, Mar. Geodesy, 3 : 161-255.

Schwiderski E.W., 1983. The Braking of the Earth's rotation, Naval Surface Weapons Center, Dalhgreen, Virginia (703) 663-8454, 31 pages.

Smith M.L., 1974. The scalar equations of infinitesimal elastic-gravitational motion for a rotating, slightly elliptical Earth, Geophys. J.R. astr. Soc. 37, 491-526.

Wahr J., 1981. Body tides on a elliptical, rotating, elastic and oceanless earth, Geophys. J. R. Astron. Soc. 64 : 677-704.

Zschau J., 1978. Tidal Friction in the Solid Earth : Loading Tides Versus Body Tides, Tidal Friction and the Earth's Rotation, Springer-Verlag Berlin : 62-94.

Zschau J., 1986. Tidal Friction in the Solid Earth. This Symposium.

TIDAL AND NON TIDAL ACCELERATION OF THE EARTH'S ROTATION

F. MIGNARD
C.E.R.G.A.
Avenue Copernic
06130, Grasse
France

ABSTRACT. Our present knowledge of the long time evolution of the Earth's rotation rate is reviewed. First I present various determinations of $\dot{\omega}/\omega$ with respect to UT and several solutions where the rate of change of the length of the day was determined simultaneously with the secular acceleration of the Moon mean longitude. Quite recently it has been possible to rely on independent measurements of $\dot{n}$, the value of which has become central in studies of the evolution of the Earth's rotation rate. Thus, this paper includes also a section devoted to the determinations of the secular acceleration of the Moon. Then, I show how the use of observations of timed and untimed lunar and solar eclipses carried out by the Babylonians and Arabs has allowed to derive the variation of the length of the day during the last two millenia. The existence of a non tidal acceleration of the Earth's rotation is confronted to the recent discovery of a secular variation in the Earth's J_2.

I. INTRODUCTION

The long term evolution of the Earth rotation is intimately connected to the evolution of the Earth-Moon system through the tidal dissipation in the oceans and to a lesser degree in the solid Earth as well. Over the period encompassed by the historical records, a few millenia, this effect is likely to have caused a steady lengthening of the day, at a rate of the order of 2 ms per century, and should be visible in any long term analysis of the difference between a uniform timescale and the solar time. Until very recently, it was not possible to investigate the history of the Earth's rotation independently of that of the Moon's orbit, which resulted in solutions where the rate of the slowing down of the Earth and the secular acceleration of the Moon were intricately entangled. We will show below that improvements in satellite tracking, as well as the advent of the lunar laser ranging has brought us a satisfactory solution to separate the two parameters.

A. Cazenave (ed.), Earth Rotation: Solved and Unsolved Problems, 93–110.

2. THE ACCELERATION OF THE MOON'S MOTION WITH RESPECT TO U.T.

The discovery of the acceleration in the Moon's mean longitude was an observational fact well before any theory could account for it. It was first suspected by E. Halley who achieved this conclusion from a comparison of ancient eclipses observations with those made at his own epoch, although he was not able to draw a quantitative conclusion from his investigation. This was left to the British astronomer R. Dunthorne who demonstrated in 1749 from a comparison between the recorded and the computed times of eclipses distributed of 2000 years that such an acceleration existed and assigned to it a value of + 20" of arc per (century)². (the sign of this acceleration is actually positive because it is refered to the solar time and not to a uniform time). The French mathematician and astronomer P.S. Laplace showed in 1787 that such an acceleration is very precisely accounted for by the gravitational effect of the secular diminution of the eccentricity of the Earth's orbit ; but according to J.C. Adams and C. Delaunay, the acceleration due to this cause amounted to only + 12" of arc and the residue must be explained by other causes, primarily tidal friction.

Until the beginning of the 20th century, the reliability of the Earth clock was not questioned, till Fotheringham (1920) solved simultaneously in a series of papers (1908-1921) the equations of observation for both the acceleration of the Moon (ν) and the Sun (ν') with respect to the solar time. In spite of the fact he did not compute (he had no reason to do so), the corresponding secular acceleration in the Moon's motion and the secular decrease of the Earth's rate of rotation, his results contained such an information . Indeed we have the following relationships between the acceleration in the mean motion of the Moon and Sun with respect to UT, and the secular acceleration of the Moon ($\dot{n}$) and the deceleration of the Earth's rate of rotation ($\dot{\omega}$) with respect to ET,

$$\nu' = - n_{\odot} \ (\dot{\omega}/\omega)$$

$$\nu = \dot{n} - n(\dot{\omega}/\omega)$$

where $n_{\odot}$ is the mean motion of the Sun $= 0.130\ 10^{9}{}''/\mathrm{cy}$
n is the mean motion of the Moon $= 1.733\ 10^{9}{}''/\mathrm{cy}$
ω is the rate of rotation of the Earth $= 47.47\ 10^{9}{}''/\mathrm{cy}$
$= 7.3 10^{-5}\ \mathrm{rad/s}$

(throughout this paper an acceleration always means twice the coefficient of t^2).

3. THE VARIABLE EARTH RATE OF ROTATION

The recognition that the Earth's rate of rotation was in fact variable was first reported in a convincing manner by Glauert (1915) ; however, the general acceptance that our Earth was not the perfect clock it was assumed to be had to wait the decisive contribution of Spencer John (1939) who established fully the occurence of a large fluctuation in the length of the day during the mid XIXth century. It is curious to note that it is about at the same time that N. Stoiko (1937) came up with an unexpected result : the Earth was not rotating at the same rate in winter and summer. This result combined with Spencer John's helped to convince astronomers that the dogma of the invariability of the rate of rotation of the Earth must be definitely rejected, along with the concept of UT as a uniform timescale.

Since then the accelerations $\dot{n}$ and $\dot{\omega}$ were searched for with respect to the ephemeris time and quite recently to the atomic time. For example, Clemence (1947) derived from the works of de Sitter (1927) and those of Spencer John, who respectively determined ν and ν', a secular acceleration of the Moon $\dot{n} = -22''.44/cy^2$ which was incorporated in the ephemeris of the Moon based on the theory of Brown. Although the idea of an ephemeris timescale is based on the orbital motion, its practical realization relies heavily on the Moon's motion, which renders the knowledge of an accurate value of $\dot{n}$ an absolute necessity. To my knowledge, there has been no attempt to check the adequacy of Spencer John's value of $\dot{n}$ to construct a uniform timescale, until the early 70's, fifteen years after the availability of the atomic time. In their studies, van der Waerden (1961) and Curott (1966) have adopted the standard value of $\dot{n}$ in analyzing the ancient observations in order to find a value of $\dot{\omega}$. As we shall see later in this paper, this procedure is being to become the common use now.

In principle, the measurement of the rate of rotation of the Earth amounts to the determination of the difference between ET and UT over the past centuries. Provided ET is effectively a uniform timescale, the presence of a quadratic term in $\Delta T = ET - UT$ must originate from the secular change in the length of the day. Unfortunately the most useful observations to trace back changes in UT involve the motion of the Moon and its poorly known secular acceleration. Examples of such observations are lunar conjonctions and occultations, magnitude and/or timing of lunar and solar eclipses, location of points where total solar eclipses were observed. As a consequence, the equations of conditions must be solved simultaneously for the variation in the Earth's rotation and the mean lunar motion. This is a major difficulty which cannot be avoided, unless one of the two accelerations is assumed to be known. A possible solution could be to resort only to planetary observations dated in some variety of solar time, since the theory of planetary motion has reached a satisfactory level. Values of ΔT might be estimated from the difference between the ephemeris time of a specific observation (e.g. transit of Mercury) computed from the available theories and that of the recorded UT time. However, valuable timed ancient planetary observations are only few centuries old, while the decrease of the tidal slowing down of the Earth's rate of rotation over the past three centuries is obscured by the much larger fluctuations in rate occuring over periods of several decades. These fluctuations are thought to originate from an exchange of angular momentum between the Earth's core and mantle, the understanding of which is far from being complete (Lambeck, 1980).

Until quite recently workers in this field determined from the ancient records both the solar and lunar accelerations, with respect to

UT, which is equivalent to the knowledge of $\dot{n}$ and $\dot{\omega}$. Newton (1970-1979) undertook an extensive analysis of ancient and medieval observations, including times and magnitudes of large solar eclipses, times and magnitudes of lunar eclipses, conjunctions and occultations preserved by Ptolemy and the arabian astronomer Ibn Junis. He processed the data in such a way that he was able to derive accelerations at different epochs. His results were :

$$\dot{n} = -41''.6 + 4''.3 \ / \ cy^2$$

$$10^9 \ (\dot{\omega}/\omega) = -22.7 \pm 3.4 \ / \ cy \qquad \text{for the mean epoch 200 B.C.}$$

$$\frac{d(lod)}{dt} = +2.4 \pm 0.3 \ ms/cy$$

and

$$\dot{n} = -42''.3 \pm 6.1 \ / \ cy^2$$

$$10^9 \ (\dot{\omega}/\omega) = -22.5 \pm 3.6 \ / \ cy \qquad \text{for the mean epoch A.D. 1000}$$

$$\frac{d(lod)}{dt} = 1.94 \pm 0.3 \ ms/cy$$

where $\frac{d(lod)}{dt}$ is the rate of change of the length of the day, deduced from $\dot{\omega}/\omega$,

$$\frac{d(lod)}{dt} = -0.864 \ 10^8 \ (\dot{\omega}/\omega) \text{ , with the same unit as above}$$

A detailed investigation of ancient and medieval solar eclipses made by Müller and Stephenson (1975) has yielded :

$$\dot{n} = -37''.5 \pm 5'' \ / \ cy^2$$

$$10^9 \ (\dot{\omega}/\omega) = -29.0 \pm 0.3 \ ms/cy$$

$$\frac{d(lod)}{dt} = 2.5 \pm 0.3 \ ms/cy$$

They concluded that there was no reason to suspect a change in the tidal dissipation over the past three millenia, as the curve ET - UT can be satisfactorily fitted by a single parabola. The validity of this statement is questioned in a recent paper of Stephenson and Morrison (1984), which we shall discuss in detail below. In a more recent investigation, Stephenson (1978) has beforehand sifted the available reports of solar eclipses to retain only a comparatively small number of reliable sightings of central and near central eclipses. He argues soundly about the difficulty of utilizing the numerous eclipses observations of specified magnitude. In addition, in dealing with a rather small set of data, it was possible to enquire in detail into the historical circumstances relating to each individual observation. In contrast with the work of Newton, all but one the ancient eclipses (prior to 800 AD) have been obtained from chinese records, starting with an untimed eclipse in July 17, 709 BC. The only eclipse from the middle East is the most accurately reported total solar eclipse until modern times, observed at Babylon on April 15, 136 BC, the description of which is attested in two distinct babylonian tablets. Both stars and planets were visible, which bears witness without any doubt of the totality. Not only is the date given, but likewise the instant of the beginning of the totality refers to the sunrise.

For the main part, the medieval eclipses are European, as a result of the vast number of monastery chronicles. The result of this investigation were :

$$\dot{n} = -30.0 \pm 3''.0 \ / \ \mathrm{cy}^2$$

$$10^9 \ (\dot{\omega}/\omega) = -24 \pm 2 \ / \ \mathrm{cy}$$

$$\frac{d(\mathrm{lod})}{dt} = 2.07 \pm 0.17 \ \mathrm{ms/cy}$$

It appears clearly from the above results that the estimates of standard deviation stem mostly from the scattering of individual results rather than an external accuracy. As pointed out by Newton (1970), the error in a measurement has a strong tendency to be larger than the measurer's estimate. On the other hand, the derivation of a reliable external error requires a right assessment of systematic errors, which proves generally to be unfeasible.

Moreover all these results based on solar and lunar eclipses are plagued by an intrinsic shortcoming of the method. The parameter which is actually well determined by eclipses is the secular evolution of the mean elongation of the Moon from the Sun $D = L - L_\odot$, and for the corresponding secular rate of change :

$$\ddot{D} = \dot{n} - (n + n_\odot) \ (\dot{\omega}/\omega) = \dot{n} - 1.6 \ 10^9 (\dot{\omega}/\omega)$$

(this fact was discovered empirically by R. Newton (1970) from his least square system before he realized fully its physical meaning). Then, the coefficient in the partial derivatives with respect to n and $(\dot{\omega}/\omega)$ are nearly always proportional, which gives in the solution two highly correlated variables.

There are two ways to improve the situation. We could give up the idea of using records of solar eclipses for the purpose of deriving the two accelerations. But as we have already mentioned, no reliable ancient planetary observation is available to replace the eclipse data. A second solution consists in relying on an independent determination of the Moon's secular acceleration and to solve the observation equations only for the lengthening of the day. Such a conclusion was reached by Stephenson and Morrison (1982, 1984). Within this framework, the precise value of the secular acceleration of the Moon to be used is of paramount importance, first because the data will be analysed for $\dot{\omega}/\omega$ alone and consequently the determined rate of change in the Earth's rotation rate will be $\dot{n}$-dependent. Second, the result of a discussion related to the conservation or non conservation of the total angular momentum of the Earth-Moon system, that is to say the existence or absence of a non-tidal acceleration of the Earth's rotation, will be likewise affected by the specific choice of $\dot{n}$.

4. THE SECULAR ACCELERATION OF THE MOON

Before proceeding further in this section, we must first wonder whether this quantity is meaningful and over which timespan. The lunar acceleration is produced by the action on the orbit of the Moon of the tidal bulge on the Earth raised by the Moon. As a result of this action, there is an exchange of angular momentum between the Earth's rotational angular momentum and the Moon's orbital angular momentum, which causes the Moon to recede from the Earth at a rate of few centimeters a year, and the Earth's rotation to slow down. The dissipation within the ocean of the Earth is primarily responsible for this situation. The dissipation within the solid Earth must account for less than 10 % of the whole effect. The key to the explanation of the momentum transfer therefore lies in the oceans. Over periods of several hundred million years, there are several pieces of evidence that the rate of transfer of angular momentum from the Earth to the Moon has not been constant. First comes the timescale problem (Mignard, 1982). If the present secular acceleration is assumed not to have changed in the past, the Moon was at only 5 $R_{\oplus}$ 1.3 billion years ago, which is untenable on the basis of biological and geophysical arguments. As for the rate of dissipation, data reviewed by Scrutton (1978) indicate a relatively low dissipation rate up to the permian. This is consistent with the idea of continental drift and sea floor spreading, which involves large modifications in the shape of the ocean and in the level of the coastlines above the continental shelves. The paleodissipation in the permian ocean computed by Sunderman and Brosche shows a number half as large as the present one. This implies a significantly smaller transfer of angular momentum from the Earth to the Moon. For all these reasons, it would not be reasonable to extrapolate the present value of the secular acceleration of the Moon in the distant past.

However if we restrict our investigation to historical time, there is no reason to suspect $\dot{n}$ of variability over such a short timespan. There is strong presumptions that the global sea-level was never more than about 1 or 2 m below or above the present level during the historical period. Translated in terms of change of the area of the continental shelves, such a small variation for sea-level would cause variations of only one percent.

As a conclusion, it is justified to analyze ancient astronomical observations by assuming $\dot{n}$ to be constant. A question arises : which of the published values should be used ?

Since the early seventies, numerous investigations related to the tidal evolution of the Moon's orbit have been published. The set of available values of $\dot{n}$ is summarized in Fig.(1), where all the determinations are listed in chronological order, so as to show the convergence as a function of time. It must be noted that more than half of these determinations are posterior to 1978, when Calame and Mulholland pointed out that the search for a definite value of $\dot{n}$ was not converging and depended heavily on the data analysed and the determination methods used. At the time Calame and Mulholland were reviewing, lunar laser determinations were only provisional. It is visible in Fig.(1) that this method provides us now with the most consistent determinations. Dickey et al (1983) show that the accuracy of $\dot{n}$ from LLR must improve as a function of time as $t^{-5/2}$, where t^{-2} results from the t^2 change in orbital longitude induced by $\dot{n}$ and $t^{-1/2}$ comes from the statistics of evenly distributed measurements of equal accuracy. It is expected that the standard deviation in $\dot{n}$ will fall below 0.5"/cy² by the end of 1988.

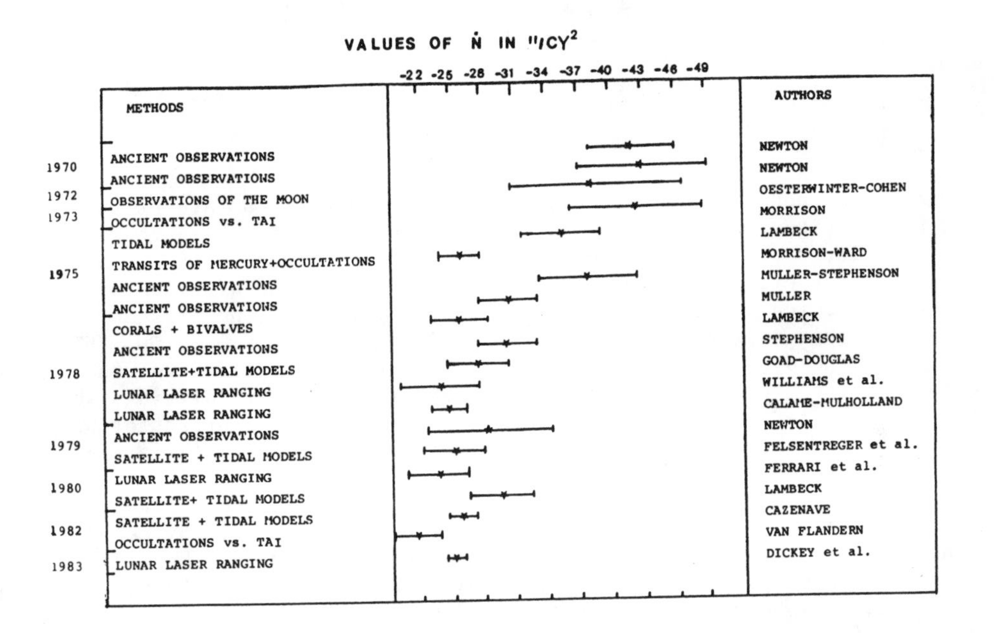

Figure 1. Value of the secular acceleration of the Moon's $\dot{n}$ obtained during the last fifteen years.

Also Fig.(1) indicates that a variety of techniques have been recently used to estimate the value of $\dot{n}$, like the already mentioned LLR, analysis of ancient and telescopic observations, observations of occultations since 1955 dated in atomic time, tidal models and satellite determinations of the amplitude and phases of oceanic tides from the motion of geodynamic satellites. In view of these modern and accurate determinations, the high quality of the tidal acceleration of the Moon obtained by Morrison and Ward (1975) must be aknowledged. Van Flandern

(1975) has advocated the possibility for $\dot{n}$ to be different according as the determination is performed with respect to ET or TA. He interpreted the discrepancy existing between the two sets of determination as an evidence of a secular change in the gravitational constant G and gave $10^{11}\ \dot{G}/G = -\ 3.6 \pm 1.8$/yr. Independent and more accurate observations of $\dot{G}/G$ have been obtained with the Viking orbiters and landers and lead to the conclusion that $\dot{G}/G < 0.2\ 10^{-11}$/yr (Hellings et al, 1983), which implies that at the present level of accuracy, there is no reason to make a distinction between $\dot{n}$ refered to either ET or TA, since the expected difference reaches at most 0.6"/cy² for the upper limit of $\dot{G}/G$ quoted above. To conclude this section, I would say that the agreement in Fig.(1) between recent results obtained with different methods allows to assign confidently the secular acceleration of the Moon a value $\dot{n}$ = - 25" ± 1"/cy². In their investigations of the history of the Earth's rotation, Morrison and Stephenson have used $\dot{n}$ = - 26"/cy as being the central value of the analysis of Morrison and Ward, which covered the longest timescale and happens to be identical to the mean of the other recent determinations.

4. MONITORING THE EARTH'S ROTATION RATE.

Astronomers do not observe directly changes in the Earth's rate of rotation, but only the cumulative effect of the departure of the length of the mean solar day with respect to the ephemeris SI day of 86 400 s SI. This cumulative effect becomes visible in the past on the difference between a dynamical timescale and the Universal Time. In the present discussion, this difference will always be ΔT = ET - UT or ΔT = TD - UT, where TD is the dynamical time used in the modern ephemeris. Given the accuracy of the present investigations, there is no need to distinguish between the different variety of dynamical time, either referred to the Earth or to the barycenter of the solar system. The quantity ΔT is expressed in seconds. Astronomers actually compare the time read at the same physical instant on two different clocks, the difference of which being precisely ΔT. The ephemeris clock is conveniently built by means of theory of the motion of the Sun, the planets or the Moon. The stars are the digits on the dial, while the radius vector from the observer to the celestial body selected is the hand of the clock. The UT clock does not differ fundamentally from the ephemeris clock in that sens that instead of modelling the translation of a body, the Earth's rotation is used. To change from an angle (the orientation of the Earth with respect to the stars) to a time, it suffices to define 86 400 s(TU) = 1 revolution of the Earth. This amount to making a wrong theory of the Earth rotation by identifying UT to the dynamical time. The first derivative of ΔT with respect to the uniform time t, multiplied by the length of the day (lod), will give the difference between the length of the day at the particular epoch t with the nominal length of day of 86 400 s SI. This information is the same as the rate of the Earth's rotation at the specific epoch. Most investigations have in view the geophysical origin of ΔT and are concerned with the average acceleration in rotation expressed either by the rate of change in the lod per century or by the rate of change of the rotational speed of the Earth. Between the various quantities mentioned above, we have the following useful numerical relationships :

$\Delta T = \Delta T(t)$, a function of time expressed in seconds with t usually in centuries.

$\frac{d(\Delta T)}{dt}$, in s/cy

$$\Delta(\mathrm{lod}) = \frac{d(\Delta T)}{dt} \times \frac{\mathrm{lod}}{1s} \times \frac{1s}{1cy} \times 1000 = 2.738\ 10^{-2}\ \frac{d\Delta T}{dt} ,$$

excess in the length of the day in ms. The same number is also the rate of change of ΔT in s/d

$\frac{d^2 \Delta T}{dt^2}$, in s/cy²

$$\frac{d(\mathrm{lod})}{dt} = \frac{d^2 \Delta T}{dt^2} \times \frac{\mathrm{lod}}{1s} \times \frac{1s}{1cy} \times 1000 = 2.738\ 10^{-2}\ \frac{d^2 \Delta T}{dt^2} ,$$

rate of change in the length of the day in ms/cy

$$\frac{\dot{\omega}}{\omega} = \frac{1}{\mathrm{lod}} \times \frac{d(\mathrm{lod})}{dt} = 1.157\ 10^{-8}\ \frac{d(\mathrm{lod})}{dt} = \frac{d^2 \Delta T}{dt^2} \times 1/(3.16\ 10^9 s/cy),$$

fractional rate of change in rate of rotation in cy^{-1}.

To obtain ΔT from astronomical observations, we must collect a well recorded astronomical phenomenon, possibly dated in solar time, as well as a theory to compute the time of occurence of this phenomenon in dynamical time. The simplest cases deal with timed lunar eclipses, the times and circumstances of which are the same for all points of the Earth from which the Moon is visible. Stephenson and Morrison(84) have reported of about 40 accurately timed contacts of lunar eclipses recorded by Babylonians. These observations lead to a direct evaluation of ΔT from a comparison of the computed time in TD with the recorded time. In the case of a solar eclipse or a stellar occultation, there is a small difficulty because the circumstances depend on the precise location of the observer, whose evaluation of coordinates with respect to the stellar reference frame requires the knowledge of the orientation of the Earth in space, which is precisely what is searched. There are several techniques to overcome this difficulty. An untimed large solar eclipse gives only a range of possible ΔT due to the fact that the umbra on the Earth has a small but finite size of few hundreds kilometers. In practice, a path of totality is calculated with the assumption of an unvariable rate for the Earth's rotation. After this is done, it turns out that the computed path crosses the latitude where the eclipse was observed with a certain departure in longitude, which is precisely ΔT, the angle the Earth must be rotated to reconcile the computed path

with the observations.
As the data analyzed by Morrison and Stephenson are reported elsewhere in this book, I only summarize the main results. The overall result is shown in Fig.(2) which gives ΔT for the period 700 BC to AD 1980.

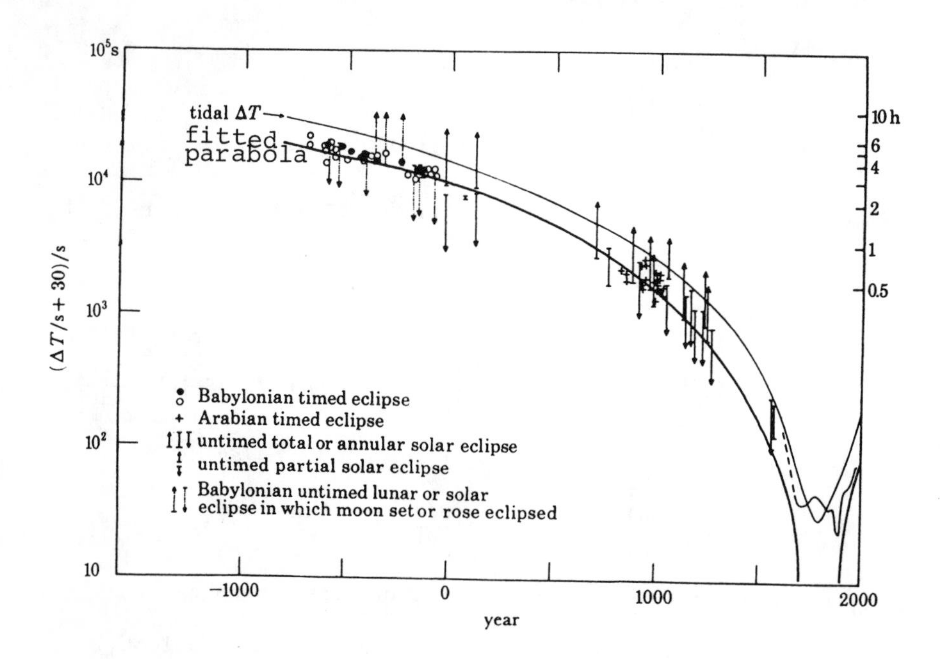

Figure 2. Plot of ΔT + 30 s for the period 700 BC to AD 1980. The tidal ΔT curve refers to the computed tidal deceleration of the Earth with $\dot{n} = -26''/cy^2$. The fitted parabola is plotted according to Eq.(9) (adapted from Stephenson and Morrison, 1984).

For the arabian and modern data, Morrison and Stephenson have fitted the following parabola to individual ΔT

$$\Delta T = (-\ 25.5 \pm 1.1)\ t^2 \tag{1}$$

with t in centuries from AD 1800.
Earlier Babylonian observations yields :

$$\Delta T = 43.8\ t^2 \tag{2}$$

As for the increase in the lod, it can be computed from Eqs (1)(2) as :

$$\frac{d(lod)}{dt} = 1.4 \text{ ms/cy} \quad \text{(from AD 950 onwards)} \tag{3}$$

$$10^9\ (\dot{\omega}/\omega) = -\ 16.2\ \text{cy}^{-1} \qquad \text{(from AD 950 onwards)} \qquad (4)$$

$$\frac{d(\text{lod})}{dt} = 2.4\ \text{ms/cy} \qquad (5)$$

$$10^9\ (\dot{\omega}/\omega) = -\ 28\ \text{cy}^{-1} \qquad \text{(up to AD 950)} \qquad (6)$$

Unfortunately, no data is available between AD 200 and 800. Stephenson and Morrison state that although no single parabola can be fitted to the whole set of observations, this results does not imply the occurence of an abrupt change in the history of the Earth's rotation rate around AD 950. This change, if real, could have been quite gradual between AD 200 and AD 800, without any possibility to notice it with the available data.

Unless there are large fluctuations of the Earth's rotation with a characteristic time of the order of a thousand years, it seems very unlikely that the secular variation in the rotation of the Earth has suffered such considerable change over the historical periods. The statement of Morrison and Stephenson might result from an artifact of the data processing. It must be noted that with a data set less rich in medieval observations, Stephenson and Morrison (1982) came up with a ΔT trend well represented by a single quadratic term $\sim 31\ t^2$ over the medieval and ancient period. This figure corresponds to :

$$\frac{d(\text{lod})}{dt} = 1.7\ \text{ms/cy} \qquad (7)$$

$$10^9\ (\dot{\omega}/\omega) = -\ 19.6\ \text{cy}^{-1} \qquad (8)$$

Using the data published by Stephenson and Morrison (1984), I also attempted to fit a parabola over the whole set of data. I obtained :

$$\Delta T = (-\ 113 \pm 273) + (44 \pm 55)t + (33.3 \pm 2)t^2 \qquad (9)$$

with $t = 0$ in 1800 and ΔT in seconds.

While the first two coefficients are not statistically different from zero, the quadratic term is fairly well determined. We must finally correct this result using an improved estimate for $\dot{n}$ ($\dot{n} = -\ 25.2''/\text{cy}^2$ instead of $-\ 26''/\text{cy}^2$). If $\delta\dot{n}$ is the error on $\dot{n}$, the quadratic term in (9) has the following meaning :

$$ct^2 = (-\ \frac{\dot{\omega}}{2\omega} - \frac{\delta n}{2n})\ t^2$$

whence the correct value for the coefficient of t^2 is $32.5 \pm 2\ \text{s/cy}^2$, which yields :

$$\frac{d(\text{lod})}{dt} = 1.8 \pm 0.1\ \text{ms/cy} \qquad (10)$$

$$10^9\ (\dot{\omega}/\omega) = -\ 20.4 \pm 1.2\ \text{cy}^{-1} \qquad (11)$$

I will use this latter value of $\dot{\omega}/\omega$ in the following section to discuss the geophysical implications of this result.

The fitted parabola goes well through the scattered Babylonian and Arabian data but, for the modern data, the computed values are about 300 s smaller than the observed values. This figure does not represent a considerable discrepancy. It must be pointed out that the scatter of the data in the Babylonian period is of the order of 4000 s, and 1000 s in the medieval period. The adjustment of a parabola by means of a least square fitting tends to minimize the residuals where the points are the most scattered.

The fact that a single parabola cannot fit perfectly all the distribution might correspond to an underlying physical phenomenon. If the Earth's deceleration had not been as pronounced during the last three centuries as it was in the historical period, it would result in ΔT larger in the modern time than expected. A century or a two-century long fluctuation in the Earth's rate of rotation of magnitude few 10^{-8} would give rise to variations in ΔT of hundreds seconds. This might have been the case between 1650-1900, when the ΔT curve is rather flat, as if the tidal deceleration of the Earth's rotation had been almost exactly compensated by an acceleration comparable in magnitude.

Any attempt to fit a parabola over only medieval and modern time would lead to a parabola different from Eq.(11). But Figure 2 shows clearly that the observations are packed with concentrations around AD 1000 and AD 1800. In the fitting these concentrations act as two weighted observations through which a parabola must go. This quadratic curve will be badly determined with such a distribution of observational points. Moreover, the possible fluctuations of the Earth's rotation would become comparatively important over this short timespan and they could lead to a spurious parabola with the wrong quadratic trend.

To conclude this section, it is of interest to compare the evolution of the Earth's rotation rate based upon historical records with similar determination based on bivalves and corals changes in growth rythms of these organic structures. The combined data reviewed by lambeck (1978) have yielded :

$$10^9 \; (\dot{\omega}/\omega) = - \; 22.2 \pm 1 \; cy^{-1}$$

a value fairly close to the astronomical determination. This strengthens our conviction that the data collected and analyzed by Morrison and Stephenson provide us with a good evaluation of the lengthening of the day.

5. GEOPHYSICAL IMPLICATIONS

To draw useful geophysical conclusions from the above result, it is necessary to compare the observed $\dot{\omega}/\omega$ to the value expected from the tidal friction alone. The deformation of the plastic Earth under tidal stresses results in frictional dissipation. In this process, there is a transfer of angular momentum from the spinning Earth to the orbital angular momentum of the Moon, in such a way that the total angular momentum of the Earth-Moon system is preserved. The whole process will manifest directly by a loss of spin angular momentum, that is a slowing down of the Earth's rotation rate in space or a lengthening of the day. Since we may be rather confident in the current value of the secular acceleration of the Moon, we can infer a tidal $\dot{\omega}_t/\omega$ for the Earth by expressing the conservation of angular momentum. Taking into account the various tides, Lambeck (1980) demonstrated that :

$$10^9\ (\dot{\omega}_t/\omega) = 1.07\ \dot{n}$$

where the units are cy^{-1} and "/cy² respectively for $\dot{\omega}/\omega$ and $\dot{n}$. With $\dot{n} = -\ 25.2''/cy^2$, we obtain :

$$10^9\ (\dot{\omega}_t/\omega) = -\ 27/cy^{-1} \qquad (12)$$

$$\frac{d(lod)_t}{dt} = 2.3\ ms/cy \qquad (13)$$

or for the quadratic trend in the Earth's rotation rate :

$$\Delta T = -\ 42.7\ t^2$$

The corresponding parabola is plotted in Fig.(2).

A comparison of Eqs (12)(13) with Eqs (10)(11) shows it must exist a secular (at least for the last 3000 years) positive acceleration in the Earth's rotation rate superimposed to the tidal secular decrease. More precisely, we have for the non tidal component :

$$10^9\ (\dot{\omega}_{nt}/\omega) = +\ 6.6\ cy^{-1} \qquad (14)$$

As pointed out by Stephenson and Morrison (1982), the errors in $(\dot{\omega}/\omega)$ (Eq. 11) and in $\dot{\omega}_t/\omega$ (Eq. 13) depend primarily on the error in $\dot{n}$. These uncertainties tend to cancel out in the non tidal acceleration of the Earth's rotation (Eq. 14). The error in Eq.(13) should be of the order of the formal error quoted in Eq.(11) which does not take into account the uncertainty in $\dot{n}$.

The non tidal variation in the Earth's rotation rate arises probably from a change in the moment of inertia of the Earth resulting from a redistribution of mass over the Earth's surface.

It turns out that recently a secular decrease in the Earth's J_2 has been detected from the analysis of 5.5 years of range measurements carried out on LAGEOS (Yoder et al., 1983).

The idea with LAGEOS was to monitor the rotation of the Earth with respect to the orbital node of the satellite, provided that the orbit of the satellite is well known. But Yoder et al. (1983) have demonstrated that UT_1 and LAGEOS' node show proportional, comparable in magnitude responses to J_2 variations. So, if such variations are present the UT_1 solution derived from LAGEOS will be systematically in error.

From a steady change in J_2, an acceleration in the motion of LAGEOS' note is expected with :

$$\ddot{\Omega} = -(3/2)n\ (R_\oplus/a)^2 \cos i\ \dot{J}_2$$

Here n is the orbital mean motion of LAGEOS, $R_\oplus$ = 6378 km is the Earth's equatorial radius and i its orbital inclination. Yoder and his colleagues have found a pronounced acceleration in the residuals of the node which they interpret as the result of J_2 variation, the magnitude of which is :

$$\dot{J}_2 = -3\ 10^{-9} cy^{-1}$$

This value slightly differs from a more recent determination by Tapley et al.(1984) :

$$\dot{J}_2 = (-2.6 \pm 0.6)\ 10^{-9} cy^{-1}$$

which we shall adopt hereafter.

Variations in J_2 are directly proportional to variations in the polar moment of inertia C, provided the deformations involved preserve the volume. Such are tidal deformation, variation of mean sea level due to deglaciation and viscous post-glacial uplift of the crust. More generally, any process involving only a redistribution of mass on the Earth's surface belongs to this category. With this assumption, it can be shown that :

$$\dot{A} + \dot{B} + \dot{C} \cong 0$$

where A, B, C are the three principal moments of inertia of the Earth.

With :

$$J_2 = \frac{C-A}{MR^2} \quad \text{and} \quad A = B$$

we obtain :

$$\dot{J}_2 = \frac{3}{2}\frac{\dot{C}}{MR^2}$$

By expression the conservation of angular momentum $C\omega$, we have :

$$\frac{\dot{\omega}}{\omega} = -\frac{2\dot{J}_2}{3}\frac{MR^2}{C} = (5.2 \pm 1.2)\ 10^{-9}\ cy^{-1} \tag{15}$$

which compares quite favorably with (14).

6. A POSSIBLE EXPLANATION

While the mere existence of an acceleration of the Earth's rotation of a non tidal origin has been for several decades the subject of debates, it is now firmly established from two independent bases. First astronomical observations show conclusively that the rate of lengthening of the day is smaller than it is expected from the observed secular acceleration of the Moon. Secondly accurate range measurements on LAGEOS have allowed to detect change in the Earth's J_2, which we interpret as change in its polar moment of inertia, responsible at last for variations in the rotation rate. In their discovery paper, Yoder and his colleagues provide a first explanation for a non zero J_2, which rests on the response of the Earth to the deglaciation. It is now widely accepted that until 10 000 years ago, ice sheets covered all of Canada and much of the United States. Existing glaciers in Alaska and Rocky Mountains were greatly enlarged. In Europe, the Scandinavian countries, England, Northern Germany and the Baltic area were glaciated. This was equally true for European part of Soviet Union and much of Siberia. The average temperature of the Earth's surface was about 5° less than at present. For every 1000 meters of ice, the crust is depressed by 330 meters. But deformation of the crust because of glacial loading is not a permanent condition and after the ice has melted away, the crust tends towards a new isostatic equilibrium. Also the water originating from the solid ice tends to migrate from the northern region to the tropical region. It is estimated that before the latest deglaciation, about 5 percent of all the water was in form of ice and that its melting gave rise to an 80 meter change in sea level. Due to the viscous behavior of the Earth's mantle, the crustal uplift succeeding the melting did not occur immediatly after the melting and has lasted up to the present days. Crustal upwraping is going on in Scandinavia, North America and the British Isles as a result of the disappearance of the ice sheets from these areas about 10 000 years ago. In Scandinavia the rate of uplift is between 40 cm to 50 cm per century depending on the place. Soon after the melting, J_2 must have increased because of a net flux of material from boreal regions towards equatorial regions, making the Earth's flattening larger. As the isostatic balance of the crust-mantle proceeds, it must give rise to a decrease in J_2, the phase that we actually observed.

Recently, Yoder and Ivins (1985) came up with a more sophisticated explanation. The retreat of mid-latitude glaciers as it is observed today should cause an increase in J_2, comparable to the one caused during the deglaciation period. The predicted $\dot{J}_2$ from the melting of present day glaciers is + 3.5 10^{-9} cy^{-1}. On the other hand the expected rate of J_2 due to the post glacial rebound is near $-3\ 10^{-9}$ cy^{-1} for mantle viscosity of 2 10^{22} P or 5 10^{24} P and reaches a maximum value of $-6\ 10^{-9}$ cy^{-1} at about 1.5 10^{23} P. The combination of these two effects would yield the observed J_2 only if the contribution from mantle rebound is near its maximum possible value. It is probably too early to draw a definite conclusion, as to which process is the most relevent. In any case the search for a probable origin of J_2 turns out to be more complex than it was originally envisioned by Yoder et al. in their discovery paper.

CONCLUSION AND SUMMARY

In this paper, we have shown that the investigations of the long term evolution of the Earth's rotation have recently taken a new pace. First, it is now possible to use a fairly reliable value for the secular acceleration of the Moon's mean longitude, owing to the recent progress made with the lunar laser ranging technique. At the moment a value of $\dot{n}$ of the order of $-25''/cy^2$ is a reasonable guess, with an accuracy of about $\pm\ 1''/cy^2$. This determination should improve in the coming years as more operating stations provide range data. Then the once unavoidable problem of correlation between $\dot{n}$ and $\dot{\omega}$ is on the verge of being cleared off. Secondly, it seems to the author of the present review, although his opinion is not shared by all workers in this field, that eclipse data allow to determine fairly well $\dot{\omega}/\omega$ at $(-21 \pm 1)\ 10^{-9}\ cy^{-1}$, as an average over the past two millenia. The reality of a non tidal component in the acceleration of the Earth's spin seems then fairly secured. It arises from a change in the moment of inertia of the Earth, such that $\dot{C}/C \sim (5 \pm 1)\ 10^{-9}\ cy^{-1}$. This result is consistent with the recently discovered secular variation of the Earth's gravitational harmonic J_2, the geophysical origin of which is thought to stem from the deglacial unloading of the Earth's crust.

REFERENCES

Calame, O., Mulholland, J.D., 1978, 'Lunar Tidal Deceleration Determined from Lunar Laser Range Measurements', *Science*, **199**, 977-978.

Cazenave, A., 1982, 'Tidal Parameters from Satellite Observations, in *Tidal Friction and the Earth's Rotation II*, P. Brosche and J. Sundermann Eds, Springer Verlag.

Clemence, G.M., 1947, 'On the System of Astronomical Constants', *Astron. J.*, **53**, 169-179.

Curott, D.R., 'Earth Deceleration from Ancient Solar Eclipses', *Astron. J.*, **71**, 264-269.

De Sitter, W., 1927, 'On the Secular Acceleration and Fluctuations on the Longitude of the Moon and Sun', *Bull. Astron. Inst. Neth.*, **4** , 21-38.

Dickey, J.O., Williams, J.G., Newhall, W.W., Yoder, C.F., 1983, 'Geophysical Applications of the Lunar Laser Ranging', presented at the XVIIIth Gen. Ass. of IUGG, Hamburg.

Dunthorne, R., 1749, 'A Letter Concerning the Acceleration of the Moon', *Phil. Trans.*, **46**, 162-172.

Felsentreger, T.L., March, J.G., Williamson, R.G., 1979, 'M2 Ocean Tide Parameters and the Deceleration of the Moon's Mean Longitude from Satellite Orbit Data', *J. Geophys. Res.*, **84**, 4675-4679.

Ferrari, A.J., Sinclair, W.S., Sjogren, W.L., Williams, J.G., Yoder, C. F., 1980, 'Geophysical Parameters of the Earth Moon System', *J. Geophys. Res.*, **85**, 3939-3951.

Fotheringham, J.K., 1920, 'A Solution of Ancient Eclipses of the Sun', *Mon. Not. R. astr. Soc.*, **81**, 104-126.

Glauert, H., 1915, *Mon. Not. R. astr. Soc.*, **75**, 489-495.

Goad, C.C., Douglas, B.C., 1978, 'Lunar Tidal Acceleration Obtained from Satellite-derived Ocean Tide Parameters', *J. Geophys. Res.*, **83**, 2306-2310.

Hellings, R.W., Adams, P.J., Anderson, J.D., Keesey M.S., Lau, E.L., Standish, E.M., 1983, 'Experimental Test of the Variability of G using Lander Ranging Data', *Phys. Rev. Let.*, **51**, 1609-1612.

Lambeck, K., 1975, 'Effect of Tidal Dissipation in the Oceans on the Moon's Orbit and the Earth's Rotation', *J. Geophys. Res.*, **80**, 2917-2925.

Lambeck, K., 1978, 'The Earth's Paleorotation in Tidal Friction and the Earth's Rotation', in *Tidal Friction and the Earth's Rotation'*, P. Brosche and J. Sundermann Eds, Springer Verlag.

Lambeck, K., 1980, *The Earth Variable Rotation*, Cambridge University Press.

Mignard, F., 1982, 'Long Time Integration of the Moon's Orbit', in *Tidal Friction and the Earth's Rotation II*, P. Brosche and J. Sundermann Eds, Springer Verlag.

Morrison, L.V., 1973, 'Rotation of the Earth and the Constancy of G', *Nature*, 241, 519-520.

Morrison, L.V., Ward, C.G., 1975, 'Analysis of the Transits of Mercury', *Mon. Not. R. astr. Soc.*, 173, 183-206.

Muller, P.M., 1975, 'An Analysis of the Ancient Astronomical Observations with the Implications for Geophysics and Cosmology', Ph.D. Thesis, University of Newcastle upon Tyne.

Muller, P.M., Stephenson, F.R., 1975, 'The Acceleration of the Earth and Moon from Early Astronomical Observations', in *Growth Rythms and the History of the Earth's Rotation*, G.D. Rosenberg and S.K. Runcorn Eds, John Wiley and Sons, 459-533.

Newton, R.R., 1970, *Ancient Astronomical Observations and the Accelerations of the Earth and Moon*, John Hopkins Press.

Newton, R.R., 1979, *The Moon Acceleration and its Physical Origin*, 1 , John Hopkins University Press.

Oesterwinter, C., Cohen, C.J., 1972, 'New Orbital Elements for Moon and Planets', *Celest. Mech.*, 5, 317-395.

Scrutton, C.T., 1978, 'Periodic Growth Features in Fossil Organisms and the Length of the Day and Month', in *Tidal Friction and the Earth's Rotation II*, P. Brosche and J. Sundermann Eds, Springer Verlag.

Spencer John, H., 1939, 'The Rotation of the Earth and the Secular Accelerations of the Sun, Moon and Planets', *Mon. Not. R. astron. Soc.*, 99, 541-558.

Stephenson, F.R., 1978, 'Pretelescopic Astronomical Observations', in *Tidal Friction and the Earth's Rotation*, P. Brosche and J. Sundermann Eds, Springer Verlag.

Stephenson, F.R., Morrison, L.V., 1982, 'History of the Earth Rotation since 700 BC', in *Tidal Friction and the Earth's Rotation*, P. Brosche and J. Sundermann Eds, Springer Verlag.

Stephenson, F.R., Morrison, L.V., 1984, 'Long Term Changes in the Rotation of the Earth : 700 BC to AD 1900', *Phil. Trans. R. Soc. London*, A 313, 47-70.

Stoiko, N., 1937, 'Variations saisonnières de la rotation de la Terre', *C.R. Acad. Sc.*, 205, 79.

Sundermann J., Brosche, P., 1978, 'The Numerical Computations of Tidal Frictions for Present and Ancient Oceans', in *Tidal Friction and the Earth's Rotation*, P. Brosche and J. Sundermann Eds, Springer Verlag.

Tapley, B.D., Schutz, B.E., Eanes, R.J., 1984, 'Geodetic and Geophysical Parameter Determination from Satellite Laser Ranging', Presented at the XXV COSPAR Gen. Ass., Graz, 1984.

Van der Waerden, B.L., 1961, 'Secular Changes and Fluctuations in the Motions of the Sun and the Moon', *Astron. J.*, 66, 138-147.

Van Flandern, T.C., 1975, 'Determination of the Rate of Change of G', *Mon. Not. R. astr. Soc.*, **170**, 333-342.
Van Flandern, T.C., 1982, 'Is the Gravitational Constant Changing ?', *Astron. J.*, 248.
Williams, J.G., Sinclair, W.S., Yoder, C.F., 1978, 'Tidal Acceleration of the Moon', *Geophys. Res. Lett.*, **5**, 943-946.
Yoder, C.F., Williams, J.G., Dickey, J.O., Schutz, B.E., Eanes, R.J., Tapley, B.D., 1983, 'Secular Variations of the Earth Gravitational Harmonic J2 Coefficient from Lageos and non Tidal Acceleration of the Earth's Rotation', *Nature*, **303**, 757-762.
Yoder, C.F., Ivins, E.R., 'Changes in the Earth's Gravity Field from Pleistocene Deglaciation and Present Day Glacial Melting', presented at the AGU SPRING Meeting, Baltimore, 1985.

GLOBAL CLIMATIC CHANGES AND ASTRONOMICAL THEORY OF PALEOCLIMATES

A. Berger and Chr. Tricot
Catholic University of Louvain
Institute of Astronomy and Geophysics G. Lemaître
2 Chemin du Cyclotron
B-1348 Louvain-la-Neuve
Belgium

ABSTRACT. According to the astronomical theory of paleoclimates, the long-term variations in the geometry of the Earth's orbit are the fundamental cause of the succession of Pleistocene Ice Ages. Spectral analysis of geo-climatic records of the past 800,000 years has provided substantial evidence that a considerable fraction of the climatic variance is driven, in some way, by insolation changes accompagnying changes in the Earth's orbit, at least near the frequencies of variation in obliquity (main period of 41 kyr) and climatic precession (main periods of 23 and 19 kyr). Moreover, the variance components centered near a 100 kyr cycle which dominates all climatic records, seems in phase with the eccentricity cycle, although the exceptional strength of this cycle needs a nonlinear amplification by the glacial ice sheets themselves.

Recent climatic models have shown that the astronomical model has not only past several statistical tests but also the test of physical plausibility. This evidence, both in the frequency and in the time domains, that orbital influences are felt by the climate system, implies that the astronomical theory might provide a clock with which to date old sediments with a precision several times greater than now possible.

1. INTRODUCTION

When climatic variations and variability have to be explained, it is fundamental to first clearly state the scale one would like to consider, both in time and space. Indeed, climate does fluctuate significantly from one year to another, but also it gradually changes in such a way as to make one decade, one century, one millenium or larger time period different from the one before. Even if the longer-term records are still being laboriously reconstructed, there is much geo-ecological evidence that climate has fluctuated in the more distant past (Berger, 1979, 1981). Ice Ages are defined as time intervals of the Earth's history during which continental glaciations have occurred. Such Ice Ages

A. Cazenave (ed.), Earth Rotation: Solved and Unsolved Problems, 111–129.

have occurred episodically on Earth during at least the latter fifty percent of its history (Crowley, 1983). Seven major glacial events have been recorded during the Earth's history : the first one during the Early Proterozoic (around 2.3 Byr), three during the Middle and Late Proterozoic (between 0.6 and 1.0 Byr), one during the Ordovician-Silurian (420 Myr ago), one during the Permo-Carboniferous (300 Myr ago) and the last one during the Late Cenozoic (from about 30 Myr ago to present). A generally warm climate seems to have prevailed during the Archean, much of the Early to Middle Proterozoic, and the Mesozoic to Early Cenezoic. The last important global cooling began around 40 Myr ago. The date of the formation of Antarctic ice sheet is controversial, situated between 40 Myr and 12 Myr ago whereas the Northern Hemisphere glaciation was apparently initiated only about 3.2 Myr ago. As far as global average is concerned for the last 100 Myr, gross estimates by climatic models of the global temperature changes between warm periods and Ice Ages yield a total range of 10°C (Crowley, 1983). The present geological period, called Quaternary, began around 2.5 Myr and displays quite a lot of natural climatic variations which are essentially characterized by their respective amplitudes and frequencies : the mean rate of temperature variation over long extended periods is of the order of 5°C/10.000 yrs, while over the historical period, it reaches around 1.5°C/100 yrs.

From geological data, curves of global ice volume can be drawn, especially for the last 850 kyr, and air temperature can be inferred from these curves. Periods of global mean temperature lower than today mean temperature (15°C) are called Glacials and periods with mean temperature equal or higher than the today value are called Interglacials. Presently, we are in an Interglacial (called Holocene, which has begun around 10 kyr BP). The previous Interglacial (called the Last Interglacial or Eemian) was situated around 120 kyr BP and the last glacial maximum was reached around 18 kyr BP. In this paper, we will focus on climatic variations with characteristic periods of 10,000 to 100,000 years which represent glacial and interglacial oscillations during the Quaternary Ice Age characterized, respectively, by waxing and waning of continental ice sheets : roughly speaking, a glacial time comes back once every 100 kyr (at least over the last one million years), but the peaks have an asymmetric, sawtooth shape, indicating that the ice takes much longer to build up than it does to disappear; in addition, many shorter fluctuations are superimposed on this dominant 100 kyr cycle.

These glacial and interglacial periods were not only characterized by changes in the ice loading at the Earth's surface but also by subsequent changes in sea-level (Table 1) and general circulation of the atmosphere and oceans, all acting directly on the Earth's rotation.

Table 1. Change in ice loading (area and volume of ice sheets) and in sea-level between today and maximum glacial extent.
Earth surface : 510 10^6 km^2; Surface of the continents : 150 x 10^6 km^2

	Area ($10^6 km^2$)	Volume ($10^6 km^3$)	Sea level equivalent (m)
Today	14.9	26	65
Maximum glacial	43.7	77	197
Difference	-28.8	-51	-132

This difference represents 10^{-5} of the Earth's mass (6 x 10^{21} tons) or 0.04 of the oceanic mass (1.4 10^{18} tons) and one fifth of the total present surface of the continents.

From an idealized spectrum of climatic changes extending from periods comparable to the age of the Earth to about 1 hr, these inter-Quaternary ice-volume cycles seem to be related to only one external forcing, as explained by the Milankovitch astronomical theory of paleoclimates.

2. MILANKOVITCH ASTRONOMICAL THEORY OF PALEOCLIMATES

The astronomical theory of paleoclimates relates the climatic variations to those in the solar energy which would be available at the earth's surface for a completely transparent atmosphere (for a general introduction confer Berger, 1980 or Covey, 1984 or Berger, 1985; for an historical survey : Imbrie and Imbrie, 1979; for a complete survey : Berger et al., 1984). For any latitude on the Earth, this insolation is a single-valued function of the total radiative energy emitted by the sun S_0, of the semi-major axis a of the ecliptic (the elliptical orbit of the Earth around the sun), of its eccentricity e (which gives its flattening), of its obliquity ε (the angle between the rotational axis of the Earth and the normal axis at the ecliptic) and of the longitude $\tilde{\omega}$ of the perihelion measured from the moving vernal equinox (Berger, 1975; Fig. 1).

Although the idea of a cause-to-effect relationship between the long term variations of the Earth's orbital elements and those of the climate originates from the early part of the 19th century, it is only during the first decades of the 20th century that Köppen and Wegener (1924), Spitaler (1943) and mainly the Yugoslav astronomer Milankovitch (1920, 1930, 1941) indicated that under the assumption of a perfectly transparent atmosphere, a minimum in the northern hemisphere summer insolation at high latitudes would generate a glacial period. In fact, this theory requires that the northern high-latitude summers must be cold to prevent the winter snow from melting in order to allow a positive value in the annual budget of snow and ice which will initiate a positive feedback : the subsequent increase of the surface albedo will implify the initial radiative perturbation, resulting in a further extension of the snow cover.

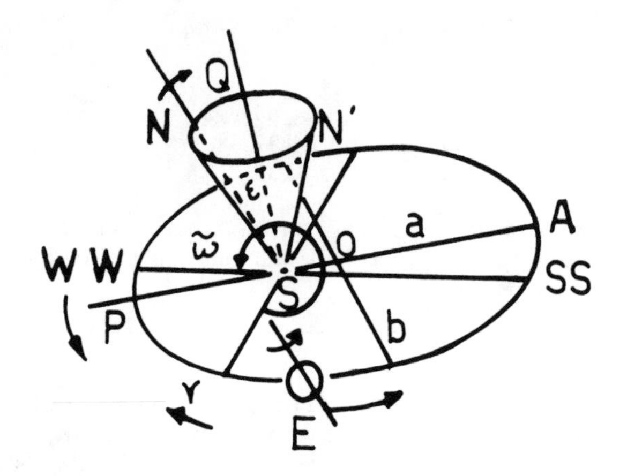

Figure 1. Elements of the Earth's orbit. The orbit of the Earth, E, around the Sun, S, is represented by the ellipse PγEA, P being the perihelion and A the aphelion. Its eccentricity, e, is given by $(a^2-b^2)^{1/2}/a$, a being the semi-major axis and b the semi-minor axis. WW and SS are respectively the winter and the summer solstice, γ is the vernal equinox; WW, SS and γ are located where they are today. SQ is perpendicular to the ecliptic and the obliquity, ε, is the inclination of the equator upon the ecliptic - i.e. ε is equal to the angle between the Earth's axis of rotation SN and SQ. $\tilde{\omega}$ is the longitude of the perihelion relative to the moving vernal equinox, and is equal to $\Pi+\psi$. The annual general precession in longitude, ψ, describes the absolute motion of γ along the Earth's orbit relative to the fixed stars. Π, the longitude of the perihelion, is measured from the reference vernal equinox of 1950 A.D. and describes the absolute motion of the perihelion relative to the fixed stars. For any numerical value of $\tilde{\omega}$, 180° is subtracted for a practical purpose: observations are made from the Earth, and the Sun is considered as revolving around the Earth (Berger, 1976).

The underlying assumptions of this Milankovitch theory are that the surface air temperature is directly related to the insolation available at the Earth's surface for a completely transparent atmosphere and that the northern high latitudes are the most sensitive to climate because of maximal continentality of these regions.

Consequently, the Milankovitch theory of paleoclimates involves three consecutive steps (Berger, 1980) :

1. to build an astronomical solution which provides us with the variations of the orbital parameters during Quaternary.
2. to compute insolations from this astronomical solution.
3. to relate these insolations to climate by using a climate model.

About the long-term variations of the Earth's orbital elements over the Quaternary, provided the semi-major axis and the solar constant are considered as constant, three astronomical parameters are of primary importance in paleoclimatology and are still at the basis of modern versions of the Milankovitch theory (it means versions where numerical values have been greatly improved and where details of the explanation of why the astronomical cycles should affect climate have been deeply modified). First, the eccentricity, e, affects the total amount of energy received by the Earth over one year, by changing the mean distance from the Sun to the Earth : $r_m = a(1-e^2)^{1/4}$. This effect is very small : for the range of changes of e between 0 and 0.06, it amounts only up to 0.2%. However, it affects also the duration of the seasons in the different hemispheres and the difference between maximum and minimum insolation received in the course of one year, a difference which can amount as much as 30% for the most extreme elliptical orbit. The other two parameters, e sin $\tilde{\omega}$ and ε, mainly influence the geographical and the seasonal patterns of irradiation. The variation of the longitude of the perihelion relatively to the moving equinox shows that the season of closest approach to the Sun varies. In fact, $\tilde{\omega}$ is mainly used through the climatic precession parameter e sin $\tilde{\omega}$. This parameter plays an opposite role in both hemispheres and is namely a measure of the difference in length between half-year astronomical seasons and of the difference between the Earth-Sun distances at both solstices. Finally, the effect of the varying obliquity ε, a super nutation, must be considered. A greater obliquity reduces the geographical contrasts of the annual irradiations and intensifies the difference between summer and winter in both hemispheres.

A caloric half-year was introduced in the original theory of Milankovitch because the length of the astronomical seasons shows secular variation (a caloric summer is exactly half-a-year which comprises all the days receiving a geometrical insolation larger than on any day belonging to the caloric winter). Consequently the mathematical definition of such seasons implies that glaciation will occur when :

1) the longitude of the perihelion is such that the northern hemisphere summer begins at the aphelion ;

2) the eccentricity is maximum, which means that the Earth-Sun distance at the aphelion will be the largest.

3) obliquity is low, which means that the difference between summer and winter is weak and the latitudinal contrast is large.

Moreover, as the high-latitude caloric insolations are mainly dependent on ε , whereas those of low latitudes are essentially dependent on

$e \sin \tilde{\omega}$, and as the ε-effect is the same in both hemispheres, whereas the $e \sin \tilde{\omega}$ effect is opposite, the nature itself of this model implies compensation of negative summer insolation deviations by positive winter deviations and an antisymmetry between hemispheres which becomes minimal for all latitudes higher than 70°.

As a consequence, following all these requirements, not only would the summer temperature in northern high latitudes be "fresh" enough to prevent snow and ice from melting, but also mild winters would allow a substantial evaporation in the intertropical zone and abundant snowfalls in temperate and polar latitudes, the humidity being supplied there by an intensified general circulation due to a maximum latitudinal "thermal" gradient. For the present time, a relatively high obliquity (23°27') and a weak eccentricity (0.016) are not favourable to Glacial conditions, whereas the aphelion reached almost at the Northern Hemisphere summer solstice would be favourable.

3. MODERN VERSION OF THE MILANKOVITCH THEORY

Till around 1973, this theory has been largely disputed because the discussions were based on fragmentary geological sedimentary records and on Milankovitch summer radiation curves at 65° north, the absolute accuracy of which was not demonstrated. Moreover, this theory was in conflict with some well-admitted observations, namely the quasi-simultaneity of glacial ages in both hemispheres.

Despites improvements in dating and in interpretation of the geological data in terms of paleoclimates (Imbrie and Kipp, 1971), the following four fundamental questions related to the Milankovitch theory were not answered yet :

1) Are the long-term variations of the Earth's orbital elements and of the insolation reliable ?

2) Are the quasi-periodicities of the Earth's orbital elements significantly present in the geological records ?

3) Is there any significant correlation between insolation curves and geological data ?

4) Can these insolation changes have induced climatic changes of a magnitude similar to those which have been recorded in the geo-ecological data ?

4. ACCURACY AND SPECTRUM OF ASTRONOMICAL AND GEOLOGICAL DATA

The procedure used for obtaining long-term changes in the elements of planetary orbits involves a drastic simplification of the astronomical problem and the results obtained are admittedly of a limited accuracy (Berger, 1976, 1977a).

A study of the accuracy and the frequency stability of the Earth's orbital elements using the most accurate solutions of the astronomical problem shows that the long term variations of the precession, eccentricity and obliquity can be considered reliable for, respectively, $1.5x10^6$, $3x10^6$ and $4x10^6$ yrs from now. Moreover, all their frequencies are stable enough for paleoclimate studies over $5x10^6$ years at least, but their relative importance is a function of the time-period considered (Berger, 1984). Numerical experiments conducted over the previous 5 million years show that the eccentricity varies between 0 and 0.0607 (present value 0.0167) with an average quasi-period of 95 kyr (main spectral components at 410, 95, 120 and 100 kyr). The obliquity varies between 22° and 24°30' (today's value 23°27'), with a very prominent and stable quasi-period of 41 kyr, although periods of 54 and 29 kyr are not negligible. The revolution of the vernal point relative to the moving perihelion (climatic precession) has an average quasi-period of 21.7 kyr (main periods are about 23 and 19 kyr, although a higher spectral resolution does exist) whereas relative to the fixed perihelion of reference, this quasi-period is 25,700 years (astronomical precession of equinoxes). Climatic precession is presently equal to 0.01635 and oscillated roughly between -0.05 and 0.05 (Berger, 1978).

As far as the insolations are concerned, numerical experiments using these most reliable variations of the Earth's orbital elements have shown (Berger and Pestiaux, 1984) that improvements are necessary for times older than 1.5 Myr BP if insolation time-series are going to be compared with geological records in the time domain. On the other hand, any calibration of geological time scale in the frequency domain seems perfectly allowed for the last 5 Myr, because the fundamental quasi-periodicities around 40, 23 and 19 kyr do not deteriorate with time over this time interval (or at least their accuracy remains well beyond the accuracy of the geological time scale).

It is in 1976 that Hays et al. demonstrated, for the first time, that the astronomical frequencies independently found in celestial mechanics by Berger (1977b) were significantly present in paleoclimatic data. Indeed, from a spectral analysis of records taken from various deep-sea cores, these workers have shown that the following quasi-periods were statistically significant : 105,000, 41,000, 23,000 and 19,000 years. All these quasi-periods have been confirmed since then (Imbrie et al., 1984).

Later, Kominz et al. (1979) have shown that records from one of the best dated deep-sea core V28-239 (Shackleton and Opdyke, 1976) display spectral peak periodicities centered on 104,000, 92,300, 58,500, 52,200, 41,000, 30,000, 23,000 and 19,000 years. In order to be sure that this observation was not simply an artifact of visual curve matching, the coherency between the $\delta^{18}O$ records and the orbital variations has also been determined by cross-spectral analysis. Significant peaks with a coherency greater than 0.40 are only related to precession and obliquity, with obliquity consistently leading the $\delta^{18}O$ record by about 10,000 years.

More recently, Ruddiman et al. (1985) have analysed deep sea core data ranging from 600 kyr to 1.1 Myr and found a periodicity centered around 54 kyr which can be related to one of the spectral peak periodicities of the obliquity which has been forecasted by Berger as early as in 1977.

For the eccentricity, the situation is far more complicated. The most important period of 412,000-year (found by Berger in 1976) needs a time series record long enough to allow its discovery and has been found in geological records only quite recently (e.g., Moore et al., 1982). Moreover, its interpretation and that of peaks in the range of 100,000-year containing most of the climatic variance, is difficult (Imbrie and Imbrie, 1980). The distribution of insolation is indeed affected by tilt and precession much more than it is by eccentricity. Therefore, the relative importance of the 100 kyr peak is very weakly predicted by the astronomical theory if we suppose a linear response of the climatic system to the external solar forcing. As a consequence, these quasi-periods in geological data may be related either to the eccentricity periods themselves, or to a beat effect due to non-linear interactions between the two precession peaks (Wigley, 1976).

Assuming either a linear or a non-linear response of the climate system to the external solar forcing, the frequencies found in the various insolation parameters are expected to help understanding the time and geographical distributions of the frequencies found in geological data. Indeed, since a few years, there are more and more facts which prove that the variance and spectrum of geo-climatic time series vary in space and time. For example, some deep sea cores show clearly two distinct climatic regimes within the Quaternary (Start and Prell, 1984) : the late Quaternary spectra display periods at around 100 kyr and 23 kyr, whereas the Early Quaternary spectra have more power at 41 kyr and 19 kyr. To study in a more systematic way this time evolution of the spectral characteristics of paleoclimatic records, an evolutive maximum Entropy Spectral Analysis has been applied by Pestiaux on the whole V28-239 oxygen isotope record, over 2 Myr (Berger and Pestiaux, 1985). This Evolutive Spectral Analysis is performed by a successive application of the Maximum Entropy Spectral Analysis over a moving restricted part of the record. A shift of 12 kyr was used at each step. A three-dimensional diagram representing the power spectra corresponding to each of these time shifts shows a progressive decay of the 100 kyr quasiperiodicity, mainly if the time interval 0-700 kyr BP is compared to the 900-1300 kyr BP and 1300-1900 kyr BP ones (Fig. 2).

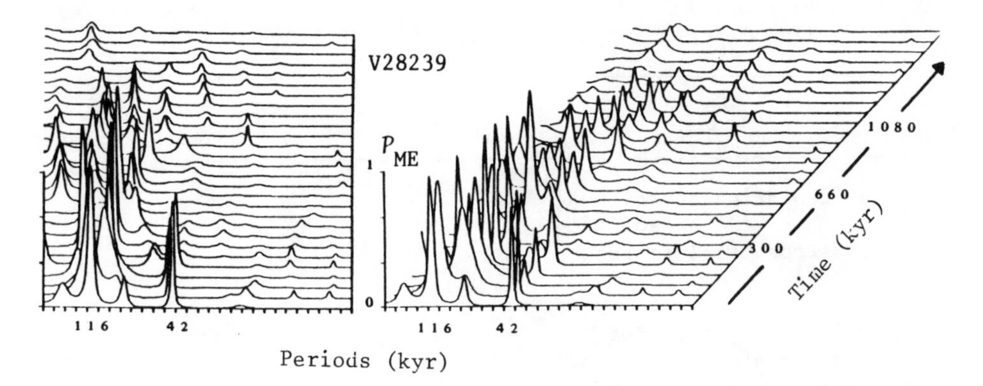

Figure 2. Evolutive Maximum Entropy Spectral Analysis of deep sea core V28-239, geological data from Shackleton and Opdyke (1976) (Berger and Pestiaux, 1985).

5. MONTHLY INSOLATION AND PALEOCLIMATES

In the modern version of the astronomical theory of paleoclimates, not only the sensitive latitudes play an important role in climate modelling, but also the critical season.

The initial use by Milankovitch of the caloric half-years mask the intra-annual variability and its variations; moreover, the whole summer season is not necessarily the most sensitive period of the year to explain the advance and retreat of ice sheets and glaciers. As a consequence, the simulation of the past climate will thus need the knowledge of the past daily or monthly insolation instead of, or in addition to, the Milankovitch caloric season insolation.

This is why the differents kinds of insolation which are used for modelling the climate or for simulating the climatic variations were calculated and their astronomical dependance underlined in Table 2.

As an illustration of all the computations, it could be very interesting to show first how the numerical values of the monthly insolation pattern compare with and complement the classical Milankovitch caloric insolation. For example, 18,000 years ago when a glacial maximum occurred with an Arctic ice-cap extending to the Netherlands, central England and New York, the obliquity, the precessional parameter and the eccentricity, whose values were respectively 23°27', 0.00544 and 0.01945, lead to an annual insolation and a caloric summer insolation close to their present-day value. However, the monthly insolation anomalies for the 60-70°N belt amounted to -35 cal/cm^2/day (-17 W m^{-2}) in August-September and +40 cal/cm^2/day (+19 W m^{-2}) in April-May, which represent respectively 8 and 5% of the actual corresponding daily inso-

Table 2. Insolations as a function of astronomical parameters (++ means stronger dependancy) (Berger and Pestiaux, 1984).

	ε	$e \sin \tilde{\omega}$
Mid-month insolation at equinoxe		+
at solstice	+	++
Half-year astronomical seasons		
- total insolation	+	
- length		+
- mean in polar latitudes	++	+
in equatorial latitudes		+
Caloric seasons polar latitudes	+	
equatorial latitudes		+
Meteorological seasons (astronomical definition)		
- total insolation	+	
- length		+
Meteorological seasons (monthly mean)	+	++

lation. Similar comparisons made over the last 1,000,000 years show that the deviations of the caloric insolations from their present-day values are always below 5%, whereas the deviations for the monthly insolations can reach up to 12% (Berger, 1978).

To figure out how the changes in the annual cycle from one 1000-year period to the following one are related to climatic variations, the dynamic behaviour of some insolation features must be analysed. Many times during the last million years, a maximum in May shifts progressively towards July-August at the same time that a minimum appears in February and moves towards April. From time to time, this shift is faster, the deepening in the spring season is steeper, and is then rapidly replaced by a maximum, the spring minimum being shifted towards summer. This feature, called insolation signature (Berger, 1979), is thought to be related to a warm phase going into a cool one, a shortening of the lag times for the oceanic and cryospheric responses being expected due to the power of such insolation changes.

Over the last 500,000 years, such insolation signatures started around 505, 486, 465, 315, 290, 243, 220, 199, 127, 105, 84 and 13 kyr BP. At these dates, a weak minimum is observed in late winter and early spring. A strong maximum culminates in June-July and is sometimes replaced by a deep minimum in some 7000 years. It is remarkable how well these dates fit with the maxima of the geological curves, namely that deduced from deep-sea core RC13-229 whose time scale has recently been revised, the boundaries from cold to warm main stages being now located 504, 422, 335, 245, 128 and 10 thousand YBP (Morley and Hays, 1981).

This analysis of the deviations of mid-month daily insolation (preferably from the mean state calculated over the last million years) clearly indicates that variations in the annual cycle of the insolation and in the amplitude of the monthly insolations are related to Quaternary climatic changes. It is not a linear real-time relationship, but it is their degree of steadiness in time which is thought to be responsible for the appearance, or not, of a full glacial or interglacial. When the speed of change of the annual cycle pattern from one 1000-year to the next is fast, going from a large maximum summer insolation to a deep minimum and back to a maximum in a very limited time span, a short cooling or "abortive" glaciation has to be expected. It is the case from 127 kyr BP (maximum) to 118 kyr BP (minimum) and to 107 kyr BP (maximum again).

As far as the present Holocene interglacial is concerned, the 11 kyr BP July insolation maximum is related to its beginning although higher frequency deglaciation phenomena (Duplessy and Ruddiman, 1984) can not be explained by the astronomical theory (at least to the present stage of our knowledge). From orbital geometry, it is also evident that the present-day insolation during summer months has been decreasing since 3 kyr BP, will reach its minimum in July-August around 3 kyr AP, and will not be significantly larger than the mean state before 24 kyr AP.

6. MODELLING THE DYNAMICS OF CLIMATIC CHANGES

In lights of all the results presented so far and many others relating Milankovitch theory to climate variations (Berger et al., 1984), we can conclude that this theory has passed both severe statistical tests and the test of physical plausibility. But there still remain difficulties in explaining how these changes in insolation could be sufficient to initiate or end glacial ages. Since a few years numerous climate models have tackled this problem. Simple climate models (such as energy balance models) cannot take into account in a correct way all relevant processes and feedbacks but they are interesting namely for sensitivity studies, giving valuable information on the relative importance of different processes and assumptions (North et al., 1981).

Results from more complex numerical models (e.g., general circulation models) are often more complex to analyse because of the inherent complexity of such models. Also the use of GCM's requires large amounts of computer time, which seriously limit their practical use for the study of long-term climatic changes including the analysis of the transient response of the climate system. Presently, GCM's are used to give instantaneous climatic patterns for selected particularly significant dates such as 122 (Royer et al., 1983), 18 and 6 kyr BP (Kutzbach and Guetter, 1984). Such experiments allow to test the validity of simpler models (e.g., Adem et al., 1984), at least for a few cases in the past.

Such a sensitivity experiment to astronomical forcing was performed with a GCM to investigate what can be the primary response of the atmosphere to the anomalies in the distribution of incident solar radiation by latitudes and seasons introduced by the long-term variations of the Earth's orbital elements at 125 kyr BP and 115 kyr BP (other boundary conditions, e.g. the sea surface temperature, being specified) (Royer et al., 1984). The model used for this experiment was a 10-level sigma-coordinates primitive equations model. The reason for the choice of these two particular periods is based on the following facts :

- these periods show the largest positive and negative deviations (from present conditions) for July insolation in the Northern Hemisphere during the last 200,000 years. At 125 kyr BP a large positive anomaly reaching more than 50 W/m^2 occurs in July in middle and high latitudes of the Northern Hemisphere, extending in the Southern Hemisphere in September-October and a large negative anomaly occurs in January in the Southern Hemisphere. At 115 kyr the situation is opposite with a negative anomaly in July and a positive one in January having about half the intensity of the 125 kyr anomaly. Such a large anomalies result directly from larger eccentricity values and passing at perihelion around mid-July and mid-January respectively. Comparing the 115 kyr insolation conditions with the 125 kyr, an excess reaching 16% in January and a similar deficit in July are found.

- these two periods mark approximately the beginning and the end of isotopic substage 5e found in geological data and which correspond to the optimum of the last interglacial. 125 kyr corresponds to high sea-level stands (few meters higher than at present) whereas 115 kyr marks the beginning of glacial inception and return to colder conditions.

The sea-surface temperatures and ice limits have been taken as observed today, based on a few geological evidence. Even if subsequent more detailed paleoclimatic records show that this hypothesis is not realistic enough, the experiment will nevertheless provides an estimation of the basic sensitivity of the atmospheric system to the insolation variations alone, other boundary conditions being constant.

The results of both simulations have been analysed by computing the difference between the 115 kyr and the 125 kyr monthly means. The map of surface temperature differences between the two simulations shows that even in the annual mean this response varies strongly according to geographical position. A region of higher temperatures over South-East Asia, over the Sahara across North America and over Northern Europe is simulated whereas colder areas are found from the Mediterranean to Siberia, on the West Coast of North America and over Labrador and Canada. This latter fact is of particular interest since it could support the theories of glacierization that have been put forward to explain the formation of the Laurentide ice cap from extensive snow cover. This decrease in annual temperature over this area is accompanied by an increase of soil water content reflecting a more positive balance of precipitation versus evaporation. This is too a favorable condition for the accumulation of an ice cover.

In conclusion to this experiment, it seems quite remarkable that the change of the insolation conditions alone from 125 kyr to 115 kyr has produced in the model a cooling over the Northern high latitudes and particularly over Eastern Canada. Such result seems to support the hypothesis that warm oceans and a deficit of summer insolation can be the prime conditions for the initiation of an ice sheet's growth in the Northern Hemipshere.

7. ASTRONOMICAL FUTURE PREDICTION

From most of the long-term climatic model results, there seems to be little doubt that the climatic system and the orbital elements are linked by a cause-and-effect relationship, although the precise linking mechanisms remains unknown. As a consequence, these climate models can not be used yet to make future long-term climate prediction. Meanwhile, however, regression model between insolations (or orbital elements) and paleoclimatic records can provide a highly accurate astronomic dating of past gross climate episodes. Such a model which bypasses the search for solar-terrestrial links in the climate system, if used carefully, can allow long-term future predictions.

The model that we have developped (Berger et al., 1980) is an autoregressive multivariate spectral model using geological and insolation data and based upon the climatological meaning and implication of the insolation signature concept. It simulates the dynamic evolution in time of the climatic response to the insolation forcing taking into account the memory of the climatic system itself. The climate, y, of one particular year, t, is considered as a function of both an objective selection of particular monthly insolations during year t and the climate y_{t-3} of the previous 3000 years (taken as the memory of the climate system). This is very important for the validity of any extrapolation into the future as one of the regressors, y_{t-3}, will have to be estimated. The calibration was made over the period 300 kyr BP to 50 kyr BP. The fit to past climates between present and 50 kyr BP and between 300 kyr and 400 kyr BP is almost perfect, which validates the regression.

Because y_{t-3} is one of the predictors, its extrapolation into the future requires testing of its stability and results allow future climate prediction over the next 60 kyr only. The general cooling trend that began some 6,000 years ago will continue over the next 60,000 years with the first cold peak arriving 4 kyr AP. The model foresees an improvement peaking at about 15 kyr AP, followed by a cold interval centered around 23 kyr AP. Major glaciation, comparable to the stage 4 of the last glacial cycle, is indicated at 60 kyr AP (Fig. 3).

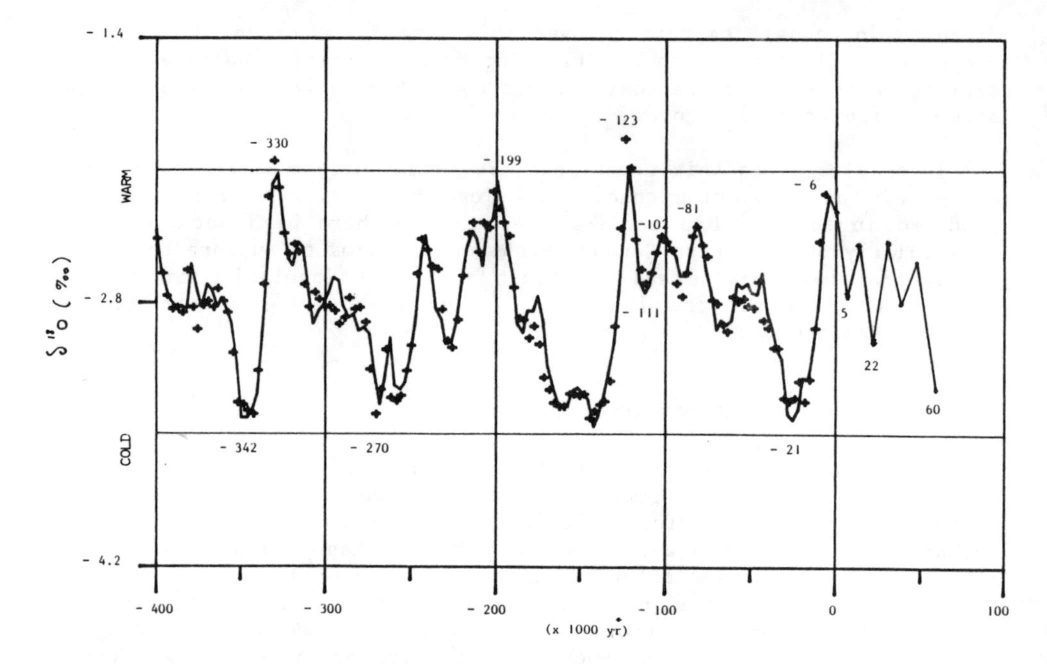

Figure 3. Long-term climatic variations over the past 400 kyr, and prediction for the next 60 kyr. Crosses represent the $\delta^{18}O$ deep-sea cores data from Hays et al. (1976). Full line is the climate simulated by the auto-regressive insolation model (Berger, 1980).

Such long-term future predictions must be used with caution when looking for future climate variations on smaller time-scales. For these we must evidently take into account the other causes, such as possible variations in the sun's output, in the amount of volcanic dust in the atmosphere, in the surface albedo in the atmospheric composition, and other non-linear feedbacks between the different parts of the climate system (ocean, cryosphere, lithosphere, biosphere), causes happening to reinforce or oppose powerfully the slower trend induced by the orbital variations.

Nevertheless, long-term predictions are far from being only of academic interest. Indeed, the effect of the 20th century CO_2 and other trace gases increase in the atmosphere, due to man's activities must be viewed in the frame of this glacial-interglacial cycle occuring over the past 150 kyr. Indeed, consumption of the bulk of the world's known fossil fuel reserves would plunge our planet into a super-interglacial following the Mitchell's concept (1977). This man-made perturbation of our present climate would be totally different if exerted in a different climatic background. As we are still in an interglacial phase, the CO_2-induced global mean temperature increase expected in the next century(ies) could give temperature levels several degrees higher than those experienced at any time in the last million years. Moreover, the slow long-term cooling towards the next "glacial" would have to wait until this warming had run its course, more than a thousand years from now, if ever, depending how the climate system responds to such an important perturbation.

8. CO_2 AND PALEOCLIMATES

Concerning the present carbon dioxide climate problem, another recent discovery by scientists working in Quaternary paleoclimatology is interesting : the atmospheric CO_2 concentration presents, indeed, variations with quasi-periodicities compatible with the astronomical theory. From measurements of carbon isotopic differences in marine sediments, Shackleton et al. (1983) conclude that, between present and 150 kyr BP, CO_2 concentration has oscillated in a range of 140 ppmv with a maximum of around 350 ppmv at 130 kyr BP and minima around 210 at 20 and 70 kyr BP.

Independent direct CO_2 measurements for the past 40 kyr and even for the last 150 kyr (Delmas et al., 1980; Neftel et al., 1982; Lorius et al., 1985) from the analysis of air bubbles entrapped in ice from Greenland and Antarctica are consistent with these results from indirect method by Shackleton et al. Moreover, ice core data show CO_2 concentration changes of 50 to 70 ppm occurring within a few hundred years during the last glaciation. The excellent correlation of these changes with the climatic changes (as given by $\delta^{18}O$) suggests a very close, direct relationship between CO_2 and climate which could play a role regarding the interhemispheric synchroneity of the glacial and interglacial periods and regarding the observed abrupt climatic changes during the last glaciation.

From these recent ice core measurements, it is deduced that the CO_2 variations occurred almost simultaneously or before the climatic variations as given by $\delta^{18}O$. On the other hand, from a reconstructed CO_2 curve until 340 Kyr, Pisias and Shackleton (1984) inferred from a simple climate model that CO_2 lags the orbital forcing (chosen as summer insolation at 65°N) but leads the ice volume, by an average of 2.5 kyr and that inclusion of CO_2 variations improved markedly the match between predicted and observed ice volume changes. They concluded that CO_2 changes are not a consequence of climatic changes but may represent part of the mechanisms by which orbital variations induce changes in climate.

The concentration of CO_2 in the atmosphere is controlled in a complicated manner by the geochemistry of the ocean (Broecker and Peng, 1984). Although modelers agree that the high latitude oceans are probably the key of CO_2 variations, they do not yet agree exactly on how that part of the ocean controls atmospheric carbon dioxide. In conclusion of this problem, until researchers have a better idea of what drove carbon dioxide and climate through such rapid variations in the past, they will remain uneasy about today's increasing atmospheric carbon dioxide and its expected greenhouse effect. For longer time prediction, following results explained above, a natural decrease in the atmospheric CO_2 concentration during the next 5 kyr must be expected, but which will most probably not exceed 50 ppmv.

9. CONCLUSION

In conclusion, recent models, both qualitative and quantitative, show that the orbital parameters have modulated the climate and will continue to do so assuming no human interferences. Considering only the long term climatic changes (consistent with astronomical frequencies), a simple autoregressive model concludes that the general cooling trend that began some 6,000 years ago will continue over the next 60,000 years.

As we have very accurate values for orbital elements and monthly insolations, we must continue to design both simple models, which would be able to reproduce the dynamic behaviour of climatic changes and variability, and more sophisticated ones which allow, in particular, to test the validity of the firsts for selected dates, particularly significant. Experiences already show that these models will have to include seasonal variability of insolation, climate-ocean-ice interactions, albedo-temperature-precipitations feedbacks, cryosphere/lithosphere dynamics and changes in atmospheric composition.

REFERENCES

Adem J., Berger A., Gaspar Ph., Pestiaux P., van Ypersele J.P., 1984. Preliminary results on the simulation of climate during the last deglaciation with a thermodynamic model. In : "Milankovitch and Climate", A. Berger, J. Imbrie, J. Hays, G. Kukla, B. Saltlzman (Eds), D. Reidel Publishing Company, Dordrecht, Holland, 527-537.

Berger A., 1975. Calcul de l'insolation par les intégrales elliptiques. Annales de la Société Scientifique de Bruxelles, 89, 69-91.

Berger A., 1976. Obliquity and precession for the last 5,000,000 years. Astr. Astrophys., 51, 127-135.

Berger A., 1977a. Long-term variations of the earth's orbital elements. Celestial Mech., 15, 53-74.

Berger A., 1977b. Support for the astronomical theory of climatic change. Nature, 269, 44-45.

Berger A., 1978. Long-term variations of daily insolation and Quaternary climatic changes. J. Atmospheric Sci., 35(2), 2362-2367.

Berger A., 1979. Spectrum of climatic variations and their causal mechanisms. Geophys. Surveys, 3(4), 351-402.

Berger A., 1979. Insolation signatures of Quaternary climatic changes. Il Nuovo Cimento, serie 1, 2(C), 63-87.

Berger A., 1980. The Milankovitch astronomical theory of paleoclimates, a review. Vistas in Astronomy, 24, 103-122.

Berger A., Guiot J., Kukla G., Pestiaux P., 1981. Long-term variations of monthly insolation as related to climatic changes. Geologischen Rundschau, Band 70, 748-758.

Berger A. (Ed.), 1981. Climatic Variations and Variability : Facts and Theories, NATO ASI, D. Reidel Publishing Company, Dordrecht, Holland, 795pp.

Berger A., Imbrie J., Hays J., Kukla G., and Saltzman B. (Eds), 1984. Milankovitch and Climate. Understanding the Response to Orbital Forcing. NATO ASI Series C vol. 126, D. Reidel Publishing Company, Dordrecht, Holland, 895pp.

Berger A., 1984. Accuracy and frequency stability of the Earth's orbital elements during the Quaternary. In : "Milankovitch and Climate", A. Berger, J. Imbrie, J. Hays, G. Kukla, B. Saltzman (Eds), D. Reidel Publishing Company, Dordrecht, Holland, 3-40.

Berger A., Pestiaux P., 1984. Accuracy and stability of the Quaternary terrestrial insolation. In : "Milankovitch and Climate", A. Berger, J. Imbrie, J. Hays, G. Kukla, B. Saltzman (Eds), D. Reidel Publishing Company, Dordrecht, Holland, 83-112.

Berger A., 1985. The astronomical theory of paleoclimates. World Climate Program Newsletter, 7, 1-3.

Berger A., Pestiaux P., 1985. Modelling the astronomical theory of paleoclimates in the time and frequency domain. An example of the relationship between long term and short term climate changes. In : "Current Issues in Climate Research", R. Fantechi and A. Ghazi (Eds), D. Reidel Publishing Company, Dordrecht, Holland, 77-95.

Broecker W.S., Peng T.H., 1984. The climate-chemistry connection. In : "Climate Processes and Climate Sensitivity", J.E. Hansen and T. Takashi (Eds), AGU, Geophys. Monogr. 29, 327-336.

Covey C., 1984. Earth's orbit and the ice ages. Scientific American, 42-50.

Crowley T.J., 1983. The geologic record of climatic change. Reviews of Geophysics and Space Physics, 21 n°4, 828-877.

Delmas R.J., Ascencio J.M., and Legrand M., 1980. Polar ice evidence that atmospheric CO_2 20,000 yr BP was 50% of present. Nature, 284, 155-157.

Duplessy J.C., Ruddiman F., 1984. La fonte des calottes glaciaires. La Recherche, 156, 806-818.

Hays J.D., Imbrie J., Shackleton N.J., 1976. Variations in the Earth's orbit : pacemaker of the ice ages. Science, 194, 1121-1132.

Imbrie J., and Imbrie K.P., 1979. Ice Ages. Solving the Mystery. Enslow Publishers, New Jersey, 224pp.

Imbrie J., and Imbrie J.Z., 1980. Modelling the climatic response to orbital variations. Science, 207, 943-953.

Imbrie J., and Kipp N.G., 1971. New micropaleontological method for quantitative paleoclimatology : application to a Late Pleistocene Caribbaen Core. In : "Late Canezoic Glacial Ages", K.K. Turekian (Ed.), Yale University Press, New Haven, 71-181.

Imbrie, J., Hays J., Martinson D.G., MacIntyre A., Mix A.C., Morley J.J., Pisias N.G., Prell W.L., Shackleton N.J., 1984. The orbital theory of Pleistocene climate : support from a revised chronology of the marine $\delta^{18}O$ record. In : "Milankovitch and Climate", A. Berger,

J Imbrie, J. Hays, G. Kukla, B. Saltzman (Eds), D. Reidel Publishing Company, Dordrecht, Holland, 269-305.

Kominz M.A., Heath, G.R., Ku T.L., Pisias N.G., 1979. Brunhes time scales and the interpretation of climatic change. Earth Planet. Sci. Lett., 45, 394-410.

Köppen W., Wegener A., 1924. Die Klimate der geologischen Vorzeit. Berlin, 256pp.

Kutzbach J.E., Guetter P.J., 1984. Sensitivity of late-glacial and Holocene climates to the combined effects of orbital parameter changes and lower boundary condition changes : "snapshot" simulations with a general circulation model for 18,9 and 6 ka BP. Annals of Glaciology, 5, 85-87.

Lorius C., Jouzel J., Ritz C., Merlivat L., Barkov N.I., Korotkevich Y.S., and Kotlyakov V.M., 1985. A 150,000-year climatic record from Antarctic ice. Nature, 316, 591-596.

Milankovitch M.M., 1920. Théorie Mathématique des Phénomènes Thermiques produits par la Radiation Solaire. Académie Yougoslave des Sciences et des Arts de Zagreb, Gauthier-Villars.

Milankovitch M.M., 1930. Mathematische Klimalehre und Astronomische Theorie der Klimaschwankungen. In : "Handbuch der Klimatologie", W. Köppen and R. Geiger (Eds), Borntraeger, Berlin, Band I. Teil A.

Milankovitch M.M., 1941. Kanon der Erdbestrahlung. Beograd. Köninglich Serbische Akademie. 484pp. (English translation by Israël Program for Scientific Translation and published for the U.S. Department of Commerce and the National Science Foundation).

Mitchell M.J.Jr., 1977. Carbon Dioxide and Future Climate. EDS, NOAA.

Morley J.J., Hays J.D., 1981. Towards a high-resolution, global, deep-sea chronology for the last 750,000 years. Earth and Planetary Science Letters, 53, 279-295.

Moore T.C., Pisias N.G., Dunn D.A., 1982. Carbonate time series of the Quaternary and late Miocene sediments in the Pacific Ocean : a spectral comparison. Marine Geol., 46, 217-233.

Neftel A., Oeschger H., Schwander J., Stauffer B., and Zumbrunn R., 1982. Ice core sample measurements give atmospheric CO_2 content during the past 40,000 yr. Nature, 295, 220-223.

North G.R., Cahalan R.F., Coakley J.A.Jr., 1981. Energy balance climate models. Rev. Geophysics and Space Phys., 19(1), 91-121.

Pisias N.G., Shackleton, N.J., 1984. Modelling the global climate response to orbital forcing and atmospheric carbon dioxide changes. Nature, 310, 757-759.

Royer J.F., Deque M., Pestiaux P., 1983. Orbital forcing of the inception of the Laurentide ice sheet. Nature, 304, 43-46.

Royer J.F., Deque M., Pestiaux P., 1984. A sensitivity experiment to astronomical forcing with a spectral GCM : simulation of the annual cycle at 125,000 BP and 115,000 BP. In : "Milankovitch and Climate", A. Berger, J. Imbrie, J. Hays, G. Kukla, B. Saltzman (Eds), D. Reidel Publishing Company, Dordrecht, Holland, 733-762.

Ruddiman W.F., Shackleton, N.J., McIntyre A., 1985. North Atlantic sea-surface temperatures for the last 1.2 M.Y., preprint.

Shackleton N.J., Opdyke, N.D., 1976. Oxygen-isotope and paleomagnetic stratigraphy of Pacific core V28-239 Late Pleistocene to Latest Pleistocene. Geological Society of America, Memoir 145, 449-464.

Shackleton N.J., Hall M.A., Line J., and Shuxi C., 1983. Carbon isotope data in core V19-30 confirm reduced carbon dioxide concentration in the ice age atmosphere. Nature, 306, 319-322.

Spitaler R., 1943. Die Bestralhungskurve der Eiszeit nach Milankovitch und Spitaler. Abh. deutsch. Akad. Wiss. Prag., 13, 18pp.

Start G.G., Prell W.L., 1984. Evidence for two Pleistocene climatic modes : data from DSDP site 502. In : "New Perspectives in Climate Modelling", Developments in Atmospheric Sciences, vol. 16, A. Berger and C. Nicolis (Eds), Elsevier Scientific Publ. Company, Holland, 3-22.

Wigley T.M.L., 1976. Spectral analysis and astronomical theory of climatic change. Nature, 264, 629-631.

METEOROLOGICAL DATA FOR EARTH ROTATION STUDIES

Richard D. Rosen
Atmospheric and Environmental Research, Inc.
840 Memorial Drive
Cambridge, MA 02139 USA

ABSTRACT. To assess the level of uncertainty in current calculations of the atmosphere's angular momentum, comparisons are presented of time series of this quantity produced by two operational weather centers. The two series generally agree quite well, although certain differences are evident and discussed here. The effect of neglecting middle and upper stratospheric winds in the production of these series is also considered and shown to be of some consequence on seasonal time scales.

1. INTRODUCTION

The increasing availability of global satellite observations during the mid-1970's led meteorologists to begin developing the means for assimilating these data into routine analyses of the state of the atmosphere from pole to pole. In September 1974, the U.S. National Meteorological Center (NMC) introduced its first operational global analysis system, designed to provide twice-daily analyses of the wind and other meteorological fields based on data from the satellites, rawinsonde balloons and aircraft.

Because of the disparate natures of the spatial resolution and error characteristics of the measurements from these different observing platforms, statistical techniques have had to be developed to blend the information contained in these data while accounting for their different characters. Moreover, the different observing times associated with each data type and the presence of large areas of the globe that are still poorly monitored have made the forecast model an integral part of the analysis process, by providing a "first guess" of the three-dimensional state of the atmosphere which can then be updated with observations as they become available. In this way, the forecast offers a method of maintaining some temporal continuity in the analyses and of filling in voids over data-sparse regions. The strong influence of the forecast in such data-sparse regions means that the quality of the analysis in these areas depends on the quality of the forecast model being used.

A. Cazenave (ed.), Earth Rotation: Solved and Unsolved Problems, 131–136.

The forecast model also affects the analyses more generally by requiring that the analyzed fields, which feed back on the forecasts by serving as the initial state for the forecast model equations, be "balanced" or "initialized" in some sense. The nature of this balance depends on the forecast model itself, and if balance is not achieved, then non-meteorological instabilities could grow in the model that would mar the quality of the forecasts. Further details regarding the techniques employed in modern forecast-analysis-initialization systems are beyond the scope here, but a discussion of the recent approaches followed at NMC is given by Dey and Morone (1985).

Analyzed fields of the zonal component of the wind, u, produced by NMC since 1 January 1976 have been used to compute up to twice-daily values of the angular momentum M of the atmosphere about the polar axis, relative to an earth-fixed frame:

$$M = \frac{2\pi a^3}{g} \iint [u] \cos^2 \phi \, d\phi \, dp ,$$

where $\underline{a}$ is the mean radius of the earth, $\underline{g}$ the acceleration due to gravity, $\underline{\phi}$ latitude, $\underline{p}$ pressure, and the brackets refer to a zonal mean. The assumptions involved in computing M and the relationship between M and earth rotation are discussed by Rosen and Salstein (1983). Barnes et al. (1983) have performed similar calculations as part of an extensive study using the atmospheric analyses produced by the European Centre for Medium Range Weather Forecasts (ECMWF) beginning 1 January 1981.

My purpose here is not to focus on comparisons between series of M and earth rotation values, which is dealt with in the two references just cited and in other contributions to this volume. Rather, I will consider two issues concerning the quality of the atmospheric series itself, namely the level of uncertainty in values of M and the effect of neglecting middle and upper stratospheric winds in the operational production of M values.

2. COMPARISON OF NMC AND ECMWF SERIES OF M

Although some errors in the zonal wind analyses exist over data-rich regions because of inaccuracies in the basic wind measurements, most of the errors occur in data-sparse areas of the globe where imperfect forecast models are relied upon to provide reasonable estimates. The complexity of modern forecast-analysis systems makes it difficult, however, to determine how much uncertainty can be quantitatively assigned to these estimates. Thus, error bars associated with the global values of M are not easily assessed through conventional error analysis techniques, even though the sources of the errors can be largely attributed to inadequacies in data coverage and forecast models.

An alternate approach to estimating the level of uncertainty in calculations of M is to compare time series of this quantity generated by two or more meteorological centers. Since the basic set of observations available to the operational weather centers is largely the same, differences in M which may exist are due to differences in forecast models and the manner in which the available observations are blended with the forecast "first guess". Comparing different M series thus provides a means for assessing the overall accuracy of current operational analysis systems. An example of such a comparison is given in Fig. 1, based on a study by Rosen et al. (1984). The values plotted are derived from the NMC and ECMWF analyses for 1981-1983 for atmospheric levels between 1000 and 50 mb, with care taken to ensure that identical data reduction procedures were followed in going from the analyzed [u] fields to M.

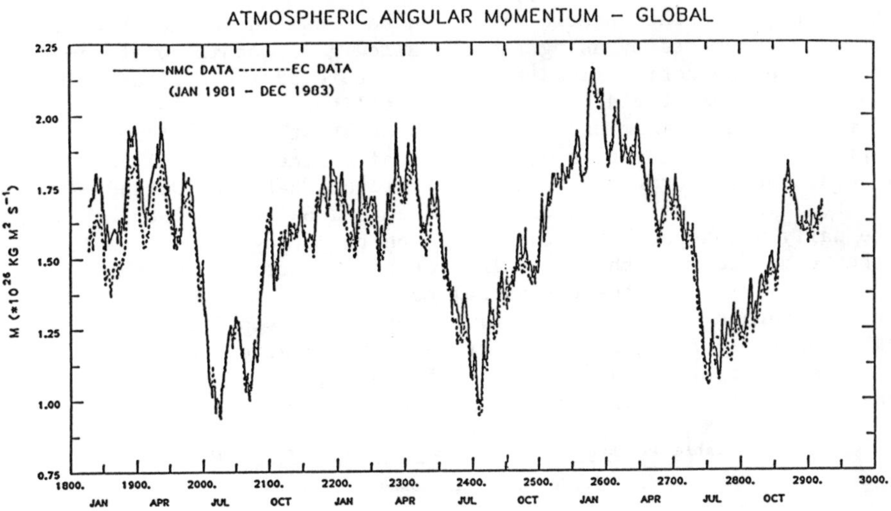

ATMOSPHERIC ANGULAR MOMENTUM – GLOBAL

Fig. 1. Daily values of M determined from zonal wind analyses to 50 mb by NMC and by the ECMWF.

The figure illustrates that the two M series generally agree quite well, although notable discrepancies are evident in the early part of the record. Overall for the whole period, the standard deviation of the difference between the two curves amounts to 0.0473×10^{26} kg m^2 s^{-1}, representing roughly 4% of the variance in either of the individual series. An analysis (not shown here) of the differences on more regional scales reveals that these differences become increasingly smaller as one proceeds from the tropics to higher latitudes in the Northern Hemisphere, whereas in the Southern Hemisphere relatively large differences exist in all latitude belts. This result is consistent with the poorer data coverage throughout the Southern Hemisphere relative to Northern Hemisphere middle and high latitudes. Also evident from plots of the NMC-ECMWF difference values is an annual component over much of the globe, the source of which is still unknown. During 1980-81, the difference in the global values of M between the two centers amounted to as much as 10% at the annual period.

3. CONTRIBUTION OF STRATOSPHERIC WINDS TO SEASONAL FLUCTUATIONS IN M

Because balloon soundings of atmospheric winds do not extend beyond the lower stratosphere, reliable operational analyses of global wind fields are available only to the 50 mb pressure level. Although these analyses thereby incorporate about 95% of the mass of the atmosphere, it is known (Belmont et al., 1974) that middle and upper stratospheric winds exhibit relatively large seasonal components, so their omission from the routine M calculations could have a noticeable impact. Indeed, the lack of agreement between M and earth rotation series at seasonal periods has been commented upon recently by Eubanks et al. (1985).

Estimates of stratospheric winds can be obtained in the absence of direct measurements by using the so-called geostrophic relationship to relate the motion field to the structure of the stratosphere's temperature field observed by satellites. Hirota et al. (1983) have recently made use of analyses of stratospheric temperatures produced by the British Meteorological Office to calculate daily values of [u] over the globe at six levels between 100 and 1 mb. These values were then used by Rosen and Salstein (1985) to compute daily values of M for the stratosphere during 1980-81, which were then combined with the NMC values of tropospheric M for the same period and analyzed for their seasonal components. A summary of the results is contained in Table 1, which compares seasonal terms for the atmospheric series with those for an earth rotation series (Δl.o.d.) derived by Morgan et al. (1985) through combining observations from VLBI, SLR and LLR techniques.

Table 1. Annual and Semiannual Components, 1980-81
(Amplitude in milliseconds, and phase)

	Annual	Semiannual
Δl.o.d.	0.37 (Feb 6-8)	0.27 (May 7-9)
TROPOSPHERE M (1000-100 mb)	0.43 (Feb 7)	0.21 (May 9)
STRATOSPHERE M (100-1 mb)	0.07 (Aug 24)	0.08 (May 24)
TROPOSPHERE + STRATOSPHERE M (1000-1 mb)	0.36 (Feb 4)	0.28 (May 13)

The table supports previous findings that series of tropospheric values of M and Δl.o.d. differ systematically on seasonal time scales, especially at the semiannual period. However, the inclusion of stratospheric data has considerably reduced the seasonal discrepancies, although some uncertainty still exists at the annual period because of the 10% difference in this term mentioned earlier between the NMC and ECMWF tropospheric M series. In any case, it is clear that future attempts at studying the causes for seasonal fluctuations in earth rotation should take into account winds in the upper portions of the atmosphere not routinely analyzed by the operational weather centers.

4. FUTURE PROSPECTS

Techniques for producing analyses of the state of the atmosphere are continually evolving, and the next several years should see increasingly accurate values of atmospheric angular momentum. Higher-speed computers, such as the Cray XMP-22 recently acquired by ECMWF, will allow analyses and forecasts to be produced at a finer resolution than is currently possible, thereby allowing circulation features on all scales to be portrayed more accurately.

A limiting factor on the accuracy of atmospheric analyses, of course, is the amount and quality of the basic observations themselves. In this regard, it is worth mentioning that plans are being formulated to develop a global wind monitoring system using satellite-borne lidar instruments, a so-called Windsat. Already approved for funding by NASA is the launch of a scatterometer designed to measure surface winds over the oceans. In addition to improving analyses over the oceans in general, data from the scatterometer ought to prove very useful in studying the surface friction torque which helps couple atmospheric motions to the solid earth.

In addition to the efforts at NMC and ECMWF, a number of other operational weather centers are developing the capability to produce global scale wind analyses. Comparisons among time series of M from a number of such centers in the future would offer a means for obtaining on-going estimates of the uncertainty in this quantity.

ACKNOWLEDGMENTS. I am grateful to my colleague D.A. Salstein for his many contributions to the work reported here. Our research has been supported by the Crustal Dynamics Project of NASA under contract NAS5-28195.

REFERENCES

Barnes, R.T.H., R. Hide, A.A. White and C.A. Wilson, 'Atmospheric angular momentum fluctuations, length-of-day changes and polar motion,' Proc. R. Soc. London, A387, 31-73, 1983.

Belmont, A.D., D.G. Dartt and G.D. Nastrom, 'Periodic variations in stratospheric zonal wind from 20 to 65 km at 80°N to 70°S,' Quart. J. R. Met. Soc., 100, 203-211, 1974.

Dey, C.H., and L.L. Morone, 'Evolution of the National Meteorological Center global data assimilation system: January 1982 - December 1983,' Mon. Wea. Rev., 113, 304-318, 1985.

Eubanks, T.M., J.A. Steppe, J.O. Dickey and P.S. Callahan, 'A spectral analysis of the earth's angular momentum budget,' J. Geophys. Res., in press, 1985.

Hirota, I., T. Hirooka and M. Shiotani, 'Upper stratospheric circulations in the two hemispheres observed by satellites,' Quart. J. R. Met. Soc., 109, 443-454, 1983.

Morgan, P.J., R.W. King and I.I. Shapiro, 'Length of day and atmospheric angular momentum: a comparison for 1981-1983,' submitted to J. Geophys. Res., 1985.

Rosen, R.D., and D.A. Salstein, 'Variations in atmospheric angular momentum on global and regional scales and the length of day,' J. Geophys. Res., 88, 5451-5470, 1983.

Rosen, R.D., and D.A. Salstein, 'Contribution of stratospheric winds to annual and semiannual fluctuations in atmospheric angular momentum and the length of day,' J. Geophys. Res., in press, 1985.

Rosen, R.D., D.A. Salstein and A.J. Miller, 'Momentum calculations with NMC and ECMWF wind analyses (and comparisons with "ground truth"),' in Proceedings of the Eighth Annual Climate Diagnostics Workshop, NOAA, pp. 268-277, 1984. [NTIS PB84-192418]

HIGH ACCURACY EARTH ROTATION AND ATMOSPHERIC ANGULAR MOMEMTUM

J. O. Dickey, T. M. Eubanks, J. A. Steppe
Jet Propulsion Laboratory
California Institute of Technology
Pasadena, CA 91109 U.S.A.

ABSTRACT. Recent advances in the measurement and interpretation of Earth rotation and polar motion are highlighted here, focusing on short period fluctuations and their relation with changes in the global atmospheric angular momentum (AAM). Particular emphasis and attention are placed on data acquired during the MERIT campaign (September 1983 - October 1984). The modern space geodetic techniques have achieved unprecented accuracy and precision; intercomparisons of the recent MERIT results indicate that Earth orientation is currently determined with an accuracy of ~ 5 cm. High-quality estimates of the atmospheric excitation of Earth rotation and polar motion are provided by the routine analysis of global weather data for operational weather forecasting. Fluctuations in Earth rotation over the time scale of a year or less are dominated by atmospheric effects; the agreement between changes in length of day (LOD) and AAM estimates is striking. In contrast, comparisons between geodetic and meteorological estimates of the excitation of the polar motion are as yet not as satisfactory, indicating the presence of challenging "unsolved problems." The presence of rapid high frequency fluctuations in the polar motion excitation pressure term is noted, these are marginally detectable by the modern geodetic techniques.

1. Introduction

In the absence of internal sources of energy or interactions with other astronomical bodies, the Earth would move as a rigid body with its various parts (the crust, mantle, inner and outer cores, atmosphere and oceans) rotating together at a constant fixed rate. In reality, the world is considerably more complicated, as is schematically illustrated by Figure 1. The rotation rate of the Earth's crust is not constant, but exhibits fluctuations on the order of 1 part in 10^8 over a broad range of time scales, and the rotation pole also moves with respect to both the fixed stars and the Earth's surface. Geodetic observations of Earth orientation changes thus provide insight into the geophysical processes mentioned in Figure 1, insight which is often difficult to obtain by other means. The

A. Cazenave (ed.), Earth Rotation: Solved and Unsolved Problems, 137–162.

atmosphere is responsible for many of these motions, especially those with seasonal or shorter periods, through exchanges of angular momentum with the solid Earth; analysis of these changes is both of meteorological interest and is essential to the correct interpretation of changes arising from other sources.

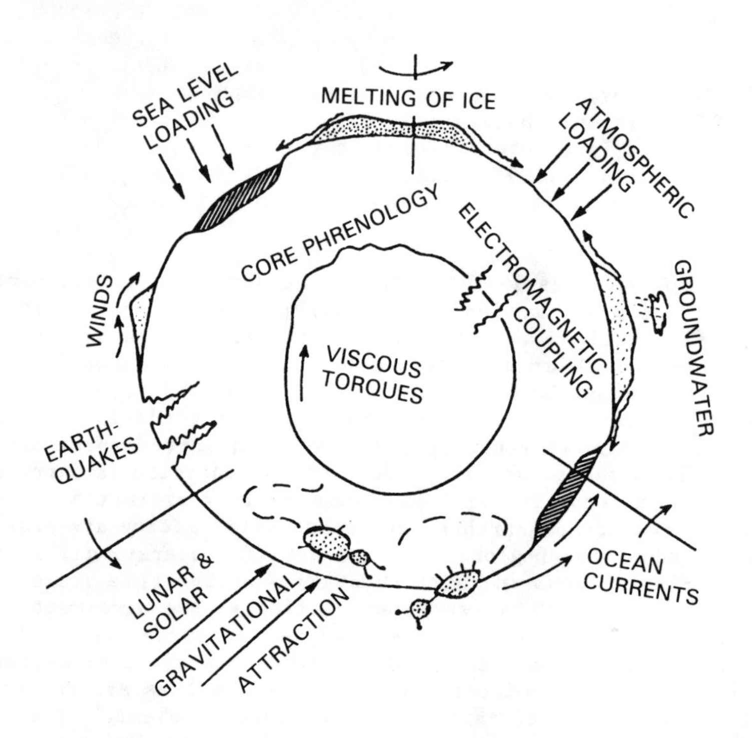

Figure 1. Schematic illustration of the forces that perturb Earth rotation. Topographic coupling of the mantle to the core has been referred to as "Core Phrenology" by R. Hide, while the beetles, following T. Gold, represent the effect of continental drift (after Lambeck, 1980a).

Observed variations in the Earth's rate of rotation - or changes in the length-of-day (LOD) - are attributed to a variety of sources (see Figure 1) and can be roughly divided into three categories: an overall linear increase from tidal dissipation, irregular large variations on a time scale of decades, and the smaller higher frequency changes. The linear increase in the LOD, determined from ancient eclipse records (see e.g., Morrison, this volume) is estimated to be about 1 to 2 msec per century. The LOD has large irregular, variations over times scales of 10 years to centuries with amplitudes of many milliseconds; these changes are too large to be of atmospheric origin and are thought to arise from fluid motions in the liquid outer core (Hide, 1977 and Lambeck, 1980b). Variations with periods of five years or less are driven primarily by exchanges of angular momentum with the atmosphere.

The Earth's rotation axis moves both with respect to a fixed inertial frame (i.e., precession and nutations), and with respect to the Earth's surface (i.e., polar motion or wobble). The polar motion consists mainly of oscillations at periods of one year (the annual cycle) and of 433 days (the Chandler wobble), with amplitudes of about 100 and 200 milliarcseconds, respectively (Lambeck, 1980b), together with a long-period drift of a few milliarcseconds per year. The yearly oscillation is meteorologically driven (Merriam, 1982 and Wahr, 1983), while the Chandler wobble is a free oscillation of the Earth whose source of excitation is currently uncertain (Wilson and Vincente, 1980; Smith and Dahlen, 1981 Wilson and Wahr, 1983).

The study of the Earth's orientation in space (this term as defined to include the effects of Earth rotation, polar motion, precession and nutation) attempts to understand the complex nature of changes in the Earth orientation, the excitation of these changes and their implications in a broad variety of areas. The rapid development of space geodesy has dramatically improved the accuracy of Earth orientation measurements and has created the potential for further profound advances which has led to a number of national and international programs to promote the collection and analysis of data from all techniques. MERIT (Monitor Earth Rotation and Intercompare the Techniques of Observation and Analysis), an international effort to evaluate the geodetic techniques sponsored by the International Union of Geodesy and Geophysics (IUGG) and the International Astronomical Union (IAU) (Wilkins, 1980), has, during its main campaign (September 1983-October 1984), yielded the most accurate Earth rotation data ever obtained. Daily (Eanes et al., 1984) and even sub-daily (Robertson et al., 1985) values of Earth rotation with accuracies well below the millisecond level have been reported during the intensive part of the MERIT campaign (April-June 1984); daily polar motion determinations were also obtained during this period (Eanes et al., 1984 and Tapley et al., 1985a). The availability of regular and accurate determinations of the angular momentum of the atmosphere, along with highly accurate geodetic Earth orientation results facilitates the study of the angular momentum budget of the Earth. Understanding the transfer of angular momentum between the atmosphere, solid Earth, and oceans is rapidly emerging as a problem of great scientific importance. The IUGG/IAG Special Study Group 5.98, "Atmospheric Excitation of the Earth's Rotation," strives for the improvement and promotion of further collaborative work in these areas and serves as an advocate for improved measurement techniques (Dickey, 1985a and b). These efforts have provided rich data sets which have and will continue to increase our understanding of Earth dynamics.

The excitation of Earth orientation changes has captured the attention of geophysicists for some time and has applications in a multitude of areas (geodesy, meteorology, astrometry, and solid Earth geophysics to mention a few). The first complete book length review of this subject was written by Munk and MacDonald (1960); Lambeck (1980b) reviewed and reported advances in this interdisciplinary field. Hide (1985) reviewed the short-term variations in Earth rotation and polar motion; Wahr (1985) stressed the geophysical implications of the Earth rotation measurements. A more detailed review covering a broad spectrum of Earth orientation variations is given by Dickey and Eubanks (1986).

This paper discusses recent advances in the measurement and the interpretation of Earth orientation data, focusing on short-period fluctuations and their relation with changes in the global atmospheric angular momentum (AAM). Many of the articles in this volume (e.g., Eubanks et al., and Rosen) are complementary in scope. The second section deals with Earth orientation measurements and discusses the recent intercomparisons that establish the accuracy of these data. The third section outlines the calculation of the atmospheric angular momentum (AAM) excitation functions, while the fourth section presents results from intercomparisons of Earth orientation measurements with the AAM excitation estimates and their implications. The summary divides the topics discussed into "solved and unsolved problems."

2. Earth Orientation: Measurements and Intercomparison

Current geodetic techniques for Earth orientation measurements can be classified into two subsets: classical (optical astrometry) and modern space techniques. Optical astrometry (e.g., Guinot, 1978 and Feissel, 1980) is based on measurements of the apparent angular positions of selected bright stars and astrometric measurements of the Earth orientation have been available for decades. The new space techniques are all based on measurements of electromagnetic signal delay or its time derivative (generally expressed as a Doppler shift). Delay and delay rate measurements are both more precise and less sensitive to systematic error than angular measurements. The modern measurement types include: satellite Doppler tracking (Kouba, 1983); Lunar Laser Ranging (LLR), (Langley et al., 1981a; Dickey, et al., 1985); Satellite Laser Ranging (SLR), (Tapley et al., 1985b; Smith, et al., 1985); and Very Long Baseline Interferometry (VLBI) (Eubanks et al., 1982; Ryan and Ma, 1983; Sovers et al., 1984; Robertson et al., 1985).

The MERIT campaign provided a rich data set of unprecedented accuracy; measurements from several techniques, as well as independent reductions from different analysis centers coupled with many intercomparisons performed by several independent centers are available for the period of the main campaign. The reader is referred to a three part series entitled _Reports on the MERIT-COTES Campaign on Earth Rotation and Reference Systems._ Part I (editor, G. Wilkins - Royal Greenwich Observatory, UK) is Proceedings of the Third MERIT Workshop and the Joint MERIT-COTES Working Group meetings which were held July 29 and 30, 1985, and following the scientific conference (August 3, 1985). Part II (editor, I. Mueller, Ohio State University, USA) is the Proceedings of the International Conference on Earth Rotation and the Terrestrial Reference Frame which was held July 31 -

August 2, 1985, at the Ohio State University. Part III (editor, M. Feissel - Bureau International de l'Heure, Paris) is a catalog of observational results. Volume 2 of Part II provided preliminary intercomparisons of all of the major results; the report of the intercomparison coordinator (King, 1985, Part I of this series) summarizes these results.

An understanding of the true errors in these measurements is essential to realizing the full geophysical potential of the new data. The best method to validate new techniques is through comparisons of independent measurement series. Fortunately, changes in orientation are dominated by rotations of the Earth as a whole, and it is thus possible to compare Earth orientation results from widely separated locations. When the new techniques were first introduced with their improved accuracies, it was unclear whether any of the new features observed in Universal Time were real changes in the Earth rotation. In an early intercomparison, Feissel and Gambis (1980) showed that the UT1 results from four different techniques were consistent and discovered rapid "50 day" fluctuations in UT1 as well. More recent geodetic intercomparisons demonstrated that for the MERIT main campaign, the new space techniques of VLBI and Laser Ranging are roughly a factor of 5 more accurate than the techniques of optical astrometry and Doppler satellite tracking (see e.g., Spieth et al., 1985; Steppe et al., 1985). The new techniques appear to be free of systematic errors; while the optical and Doppler results are corrupted by seasonal, as well as random effects. The current VLBI and SLR networks of stations are each able to determine polar motion and earth rotation (UT1) with an accuracy of about 5 cm; however, for Earth rotation, SLR requires the use of a long term stable source of UT1 (either VLBI or LLR) to maintain the stability at this level for periods of about 30 days or longer (King, 1985).

These preliminary analyses indicate a new milestone has been reached in monitoring Earth orientation; new insights have already been gained with these accurate and dense datasets. Additional advances are envisaged; these should provide a strong foundation to study the geophysical implications of Earth rotation and polar motion.

3. Atmospheric Angular Momentum

Changes in Earth orientation are caused by deformations of the solid Earth and by exchanges of angular momentum between the solid and fluid parts of the Earth, as well as by exchanges of angular momentum with extraterrestrial objects. Changes in Earth rotation and polar motion can be regarded as the response of a linear differential system to a three dimensional excitation vector. Meteorological excitation estimates are generally expressed in terms of the χ vector, the effective angular momentum function (EAMF) of Barnes et al., (1983). In this scheme the excitation of polar motion is described by χ_1 and χ_2, and that of earth rotation by χ_3. This function includes Love number corrections for rotational and surface loading deformation of the Earth (thus the term effective in EAMF), and can be evaluated accurately from meteorological data.

Estimates of the χ (more precisely, the three-component EAMF vector) are currently available based on the reduction and analysis of meteorological data for weather forecasting at three centers: European Centre for Medium Range Weather Forecasts (EC), Japanese Meteorological Agency (JMA) and the U.S. National Meteorological Center (NMC). A variety of atmospheric variables, including the local wind velocity vector, are estimated at each model grid point; the grid point data ("analysis fields") are then used as initial conditions in the numerical forecasts. The grid point data are updated several times daily by using a so-called forecast/assimilation, a weighted average of the forecast from the previous update and any new meteorological data near the grid point (see e.g., McPherson et al., 1979 and Rosen, this volume). These atmospheric models thus incorporate vast amounts of meteorological data and conveniently provide grid point estimates over the entire globe. The model's predictive power is used to estimate grid point values over regions with sparse data coverage; climatological averages may be used as well, if the model becomes unstable in some region due to absent or conflicting data. This reliance on model forecasts in regions (such as the South Pacific) with continually sparse coverage could introduce systematic errors in the resulting excitation estimates. The total angular momentum is estimated at twelve-hour or twenty-four- hour intervals (depending on the service used) from the appropriate integral of the grid point wind velocity and surface pressure analysis fields.

AAM changes can be divided into two categories: the changes in the net atmospheric rotation rate (the wind term), and those in the net atmospheric moment of inertia (the pressure term). Since the angular momentum changes are small compared to the total angular momentum of the Earth, these terms can be decoupled to first order and are generally calculated separately. The contributions of the Northern and Southern Hemispheres are also generally calculated separately. The wind term dominates axial AAM changes and is itself dominated by changes in the strength and location of the subtropical jet stream in each hemisphere (Rosen and Salstein, 1983). Allowance for the response of the oceans and the solid Earth to atmospheric surface pressure changes can be made via the "inverted barometer" and "Love number" corrections. The current inverted barometer model replaces any local changes in the surface pressure over the ocean with the average change in pressure over the entire ocean. This equilibrium model is thought to be valid for periods greater than a few months (Wahr, private communications) and tends to reduce the variability of the pressure components. Daily values of the EAMF have been calculated from the "initialized analysis global data base" of the EC starting in December 1979 (Barnes et al., 1983). Rosen and Salstein (1983 and private communication) have calculated twice-daily values of the EAMF from the U.S. National Meteorological Center (NMC) wind and surface pressure analyses following the formulation of Barnes et al. (1983). The time series component that corresponds to earth rotation (χ_3) begins in January 1976; the two components (χ_1 and χ_2) that relate to polar motion begin in July 1976. These functions are now calculated operationally (every 12 hours) at the NMC and are accessed and archived by the National Geodetic Survey (NGS). Naito and Yokoyama (1985) have computed the atmospheric angular momenta due to wind and pressure by using JMA global analysis data beginning in September 1983.

4. Earth Orientation and Atmospheric Results: Intercomparisons and Implications

4.1 Introduction

There are two basic ways to study the interchange of angular momentum within the earth: through estimates of the torques of one part of the Earth on the others or by studying the total Earth angular momentum balance (Lambeck 1980b, Section 7.4). Some meteorological studies of the atmospheric angular momentum budget take the torque approach, but the angular momentum approach is more common in geodetic analyses. The coupling between the air and the ground is by mountain (or pressure) torques and surface friction torques (Siderenkov, 1979). Mountain torques arise from an imbalance in the atmospheric surface pressure on either side of a mountain range (Smith, 1978; Oort and Bowman, 1974; Newton, 1971). Swinbank (1985) discussed the relative role of friction and mountain torques using the EC data for 1979. The evaluation of total atmospheric torque is quite difficult due to its dependence on the local conditions (Lambeck, 1980b; Smith, 1978; Garratt, 1977). Therefore, it is not generally possible to estimate global net torques as accurately as angular momentum changes. Wahr (1983) conducted an analysis of the torque approach and found reasonable agreement with astronomical observations for the annual excitation of the polar motion, but the torque approach was somewhat inferior to similiar estimates from angular momentum data in the analysis of seasonal changes in the LOD. The presentation here will follow the angular momentum balance approach.

4.2 Earth Rotation

The most outstanding characteristic of the general circulation of the Earth's atmosphere is its average "super-rotation" west to east relative to the solid Earth (at about 10 m/sec) with concentration of motion in jet streams (Hide, 1985). Changes in the atmospheric angular momentum (Fig. 2) contain a large seasonal cycle, dominated by annual and semiannual harmonics. The seasonal cycle of angular momentum causes seasonal LOD changes with an amplitude of about 0.5 milliseconds (msec), and UT1 changes with an amplitude of about 30 msec. The annual angular momentum cycles in the two Hemispheres are 180 degrees out of phase, and the Northern Hemisphere has a larger annual amplitude. The annual cycle is mostly due to changes in the mid-latitude westerlies (i.e., propagating eastward) including the subtropical jet streams; Rosen and Salstein (1983) show that most of this variance comes from the high velocity winds at or near the 250 millibar level. Superimposed on the seasonal cycles is the irregular

"50-day" oscillation, which varies in period roughly from 40 to 60 days. Most of the non-seasonal changes occur in regions from 10 degrees to 25 degrees South and 20 degrees to 35 degrees North (Rosen and Salstein, 1983).

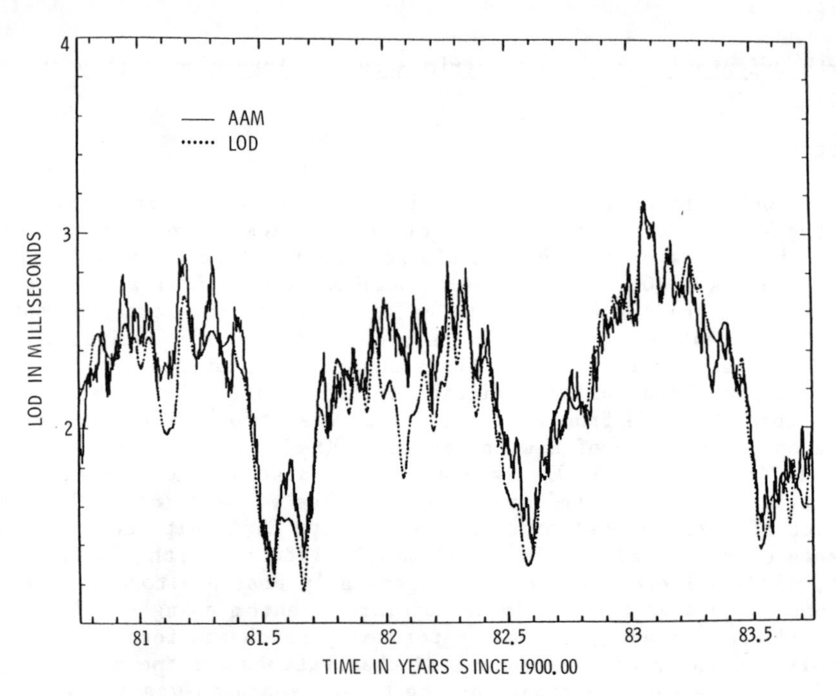

Figure 2. AAM changes as estimated by the NMC (solid line) together with Kalman smoothed estimates of LOD (dotted line).

Comparison of astronomical and geodetic LOD measurements with axial atmospheric angular momentum estimates demonstrates the significance of the atmospheric contribution to Earth rotation variations. Various studies (e.g., Rudloff, 1973; Lambeck and Cazenave, 1977; Lambeck and Hopgood 1981) have related LOD variation to changes in the atmospheric angular momentum on time scales ranging from months to a few years using mean monthly or longer period atmospheric data. Hide et al. (1980) evaluated the zonal component of angular momentum at 12-hour intervals using wind and pressure data collected by the Global Atmospheric Research Program (GARP) during the Special Observing Periods (two 8 week periods in 1979) of the First GARP Global Experiment (FGGE) and compared these with LOD data from the Bureau International de l'Heure (BIH). The comparison (Figure 3, especially for the second period) showed generally good agreement and indicated that on these short time scales; the angular momentum transfer between Earth and the atmosphere probably accounts for most of the observed variations, implying that core-mantle interactions are probably not significant on these time scales. The correlation of the atmospheric data with modern LOD results has been reported by several authors (e.g., Langley et al., 1981b, Carter et al., 1984, Eubanks et al., 1985b and Smith et al., 1986). Recent atmospheric and Earth rotation LOD results are shown in Figure 2 (Eubanks et al., 1985c). A number of studies (Eubanks et al., 1984 and 1985b; and Morgan et al. 1985) indicate that significant, but relatively small, imbalances in the angular momentum exist at the annual and semiannual periods, with the semiannual discrepancy being larger. Most of the annual imbalance can be attributed to the neglect of the upper atmosphere winds (Rosen and Salstein, 1985, and Rosen, this volume), but changes in the oceanic circulation may contribute as much as 0.2 to 0.3 msec to the unexplained seasonal changes in the LOD (Eubanks et al., 1984 and 1985b).

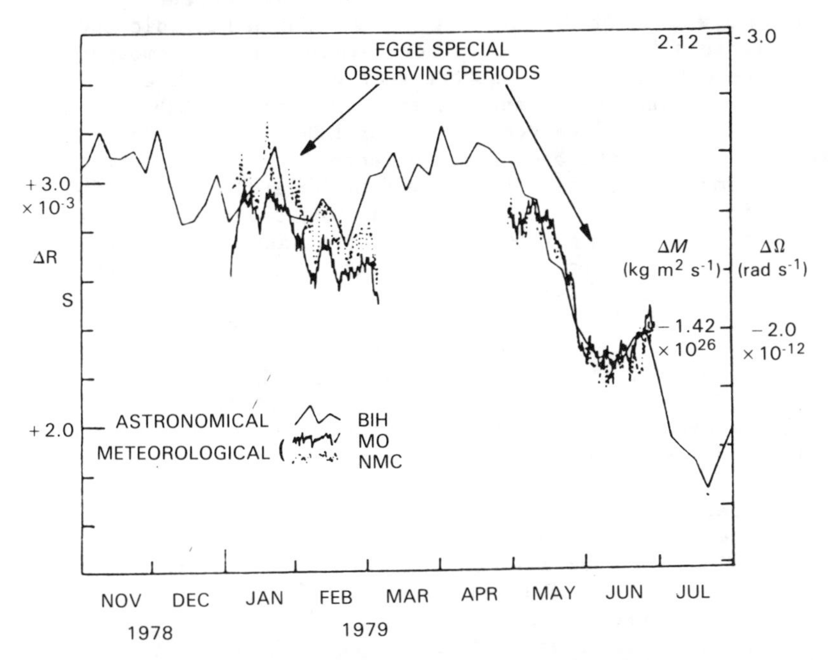

Figure 3. Comparison of the expected changes in the length-of-day derived from the meteorological results alongside observed changes from astronomical measurements. (Hide et al., 1980).

Oscillations in zonal winds and other meteorological quantities with periods of from 40 to 60 days were first discovered by Madden and Julian (1971; 1972) in wind data from the equatorial Pacific. Since that time, higher-frequency oscillations have also been seen (Miller, 1974), and the 40 to 60 day oscillation has been observed on a global scale (Krishnamurti and Subrahmanyam, 1982; Yasunari, 1981; Anderson and Rosen, 1983). The corresponding Earth rotation changes were first detected by Feissel and Gambis (1980) in four independent UT1 data types; Langley et al. (1981b) later reported this effect in both the

AAM and the LLR LOD. The peak amplitude seems to be about 0.25 msec of LOD, though the oscillation exhibits changes in both period and amplitude on a year-to-year basis. Morgan et al. (1985) showed that any nonmeteorological contribution to the 40-60 day oscillations is not significantly larger than the uncertainties in the observations (about 0.06 ms). Contributions from the winter hemisphere are generally stronger; Eubanks et al., (this volume) discuss these oscillations in more detail.

The period of the MERIT main campaign provides an excellent opportunity to examine the agreement between meteorological and geodetic excitation estimates. We conducted a systematic intercomparison of meteorological and geodetic excitation estimates during the MERIT main campaign. A session during the MERIT conference (International Conference on Earth Rotation and the Terrestrial Reference Frame) was devoted to the topic of "Short Periodic and Irregular Variations in Earth Orientation Parameters and Atmospheric Effects"; the reader is referred to the second volume of these proceedings for the papers presented there. A paper by Eubanks et al. (1985d) gives a more detailed account of our MERIT AAM results.

In order to provide the best Earth orientation and excitation estimates, a combined series was formed from several sources of geodetic measurements using the Kalman filtering technique. This filter, incorporating a statistical model, simultaneously smooths both the UT1 and the polar motion and estimates their excitations. It can accept one, two or three dimensional measurements of any components of Earth orientation (Eubanks et al., 1985a). The predictable UT1 fluctuations of tidal origin were removed from the UT1 data before smoothing (Yoder et al., 1981). The change in the LOD from the standard day of 86,400 seconds, after the removal of tidal oscillations, is denoted by the LOD*. The resultant combined data set is generally produced as an equally spaced series although values can be produced for any epoch.

The data types included are from the modern space techniques: LLR, SLR and VLBI. The IRIS (International Radio Interferometric Surveying) multi-baseline VLBI provided all three components of the Earth orientation. The Deep Space Network (DSN) contributed UT0 and variation of latitude results from single baseline VLBI on two baselines. Polar motion results from SLR were utilized from the University of Texas Center for Space Research Analysis. UT0 estimates from a JPL LLR reduction of both CERGA (France) and McDonald (Texas) data, as well as UT1 results from National Geodetic Survey reduction of the IRIS Westford - Wettzell baseline were included in this data set. The raw data are published in Section D of the BIH Annual Report and in Part III of the MERIT-COTES series.

Daily AAM estimates from the European Centre and the twice daily values from NMC were also included in our analysis. The NMC also provides corrections for the response of the sea level to changes in the surface pressure (the inverted barometer (IB) corrections). Overlapping EC data is available from the beginning of the MERIT campaign September 1, 1983 through October 1, 1984, and hence, the analysis is restricted to this period. Results from the Japanese

Meteorological Agency were not made available in time to be included in the analysis. The two meteorological centers use roughly the same raw data; however, the methods and techniques used to edit and combine the data with the previous forecast and to propagate it to the next forecast differ greatly between the two centers. Thus, an intercomparison of these series is included here as it provides some information about the accuracy of the atmospheric excitation estimates.

The overall agreement of the various series (both here and Section 4.3) is investigated by graphical presentation in the time domain and by examination of the Root Mean Square (RMS) of the differences between the series. Seasonal discrepancies were analyzed by the least squares estimation of seasonal signatures in the differenced data. A comparison between length of day as inferred by the two centers from the wind and pressure terms is shown in Figure 4. Note the dominance of the wind term; different vertical scales are required to adequately display the contribution of these terms. The close agreement between the two centers is seen clearly in the wind term (Figure 4a); in contrast, the European Centre pressure term has a downward step-like change in early 1984 (Figure 4b). The feature is presumably due to the EC model change on February 1, 1984 (Figure 4b); the major modification was apparently an increase in the resolution of the model's representation of mountainous regions (K. Whysall, personal communication). The step in the pressure term indicates a change in the distribution of atmospheric mass within the model, and does not reflect a change in the pressure term calculations. Figure 4c displays the NMC and EC data after the removal of a step in the EC χ_3 (pressure) of 0.375 $\pm$ 0.004 msec on February 1, with their difference shown below. The agreement in the two series is markedly improved; however, the difference clearly shows a trend indicating that the EC χ_3 pressure data recovered about 10% of the February 1 jump in the two weeks following. There were no obvious step changes on February 1, 1984, in any of the other EC data sets. A least squares solution was used to provide a quantitative estimate of the amplitude of a bias change on February 1, 1984 using the EC-NMC differences; all three pressure terms showed some evidence of a significant step function on that date, while the wind terms did not. In addition, there was a significant seasonal discrepancy in the χ_3 wind terms, but no evidence of such errors in the χ_3 pressure terms (Eubanks et al., 1985d). These types of phenomena indicate the need for close cooperation and communication between the atmospheric and geodetic communities. It is advisable to assess and to study the impact of any proposed model change on these excitation functions in advance of the modifications.

Figure 5 is the proverbial "one picture is worth a thousand words." Figure 5a is the comparison of the EC and the NMC χ_3 estimates(pressure plus wind); the RMS difference here is 0.067 msec. Figure 5b is the comparison of the χ_3 EC wind plus pressure term with the combined LOD series after the removal of a bias and a rate (to account for any nonatmospheric decade fluctuations). The RMS difference of the Figure 5b is 0.072 msec, only slightly higher than

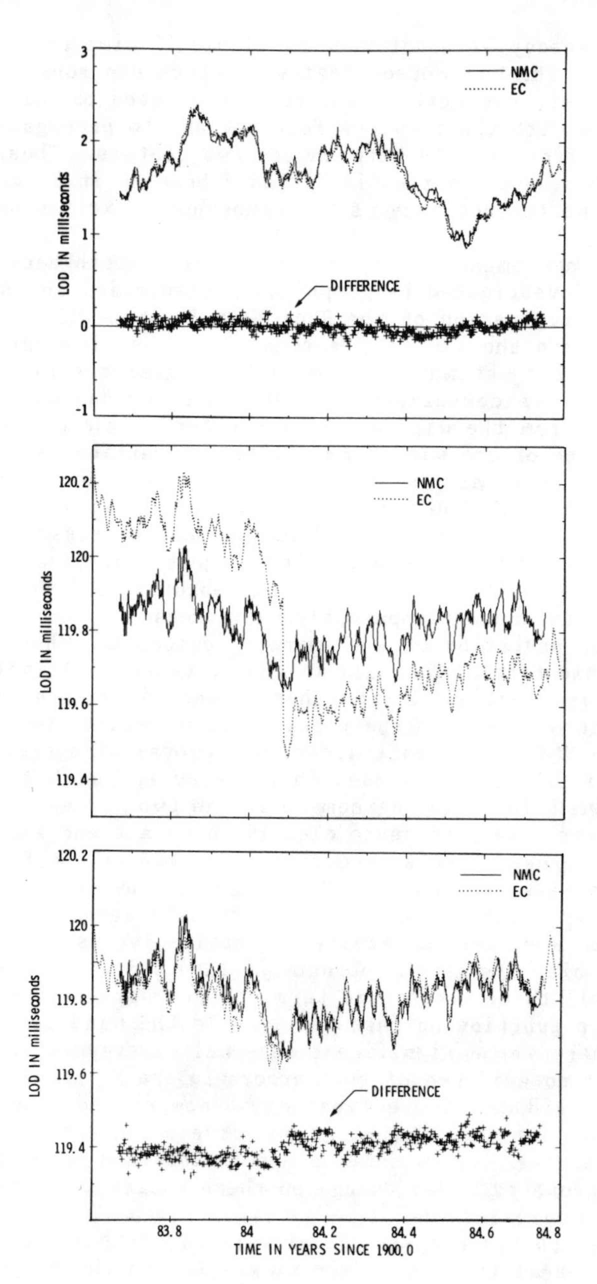

Figure 4a) The EC and NMC wind term for χ_3 (top), together with their difference (bottom); b) The EC and NMC χ_3 pressure terms. Notice the step in the EC term on February 1, 1984; c) The EC and NMC χ_3 pressure terms (top), together with their difference (bottom), after removal of the step from February 1 to 2, 1984

the 0.067 msec of Figure 5a. The comparison is striking. A systematic intercomparison was done comparing the combined length of day series and the various χ_3 components. These studies included comparison with NMC wind; NMC pressure and wind; NMC pressure, wind and inverted barometer; EC wind; and EC wind and pressure. RMS differences were computed for comparison between the LOD results and the various AAM sets before and after the removal of a rate. The subtraction of the best fitting linear trend is an ad hoc technique to remove the roughly linear divergence seen in the LOD during MERIT. The best fit was obtained with EC pressure plus wind (RMS difference = 0.141 (with no rate) and 0.072 (with rate term removed)). Use of the linear term clearly improves the agreement. Part of this discrepancy is caused by the neglect of the upper atmosphere in the routine AAM calculations, which introduces seasonal errors (see Rosen, this volume). Other possible sources of errors are unmodeled sources of angular momentum exchange (e.g., oceans), modeling errors in the various analysis and the data themselves.

4.3 Polar Motion

Studies of the axial angular momentum suggest that the atmosphere contributes to the wobble excitation. The polar motion or wobble (see Figure 6) consists mainly of oscillations at periods of one year (the annual cycle) and of 433 days (the Chandler wobble), together with a long period drift. The atmosphere and oceans, together with the effect of variable ground water storage, are clearly exciting the annual wobble. The source of the excitation of the Chandler wobble is uncertain; the two major candidates are the atmosphere and seismic deformation. Wahr (1983) found that the atmosphere and oceans appear to have had a noticeable effect on the Chandler wobble excitation during 1900-1973, mostly due to the effects of atmospheric pressure and mountain torques. The effects of friction stress, currents and upwellings are not a major contributor. The cumulative evidence presented by Wahr indicates that the atmosphere and oceans accounted for no more than 20-25 of the total observed wobble excitation during the 1900-73 period, and so were apparently not the major source of Chandler wobble excitation. Hide (1985) evaluated the atmospheric equatorial effective angular momentum functions for the period 1 December 1979 through 15 February 1984 from the EC analyses, comparing these results with polar motion estimates from the Bureau International de l'Heure (BIH). He concluded that the atmospheric excitation was sufficient to account for nearly all of the observed polar motion over this interval (3.6 Chandlerian periods or 0.5 beat periods) and that substantial excitation either by the fluid core of the Earth or by movements in the mantle associated with earthquakes was not required. It should be noted that there were several earthquakes over 7.2 in magnitude during the interval of Hide's study, but none exceeding 7.9. Analyses (Dickey et al., 1984 and Eubanks et al., 1986) of the NMC data set (Rosen and Salstein, private communication) compared with a Kalman smoothed polar motion series for the period July 1976 through August 1984 indicate that the atmosphere may be able

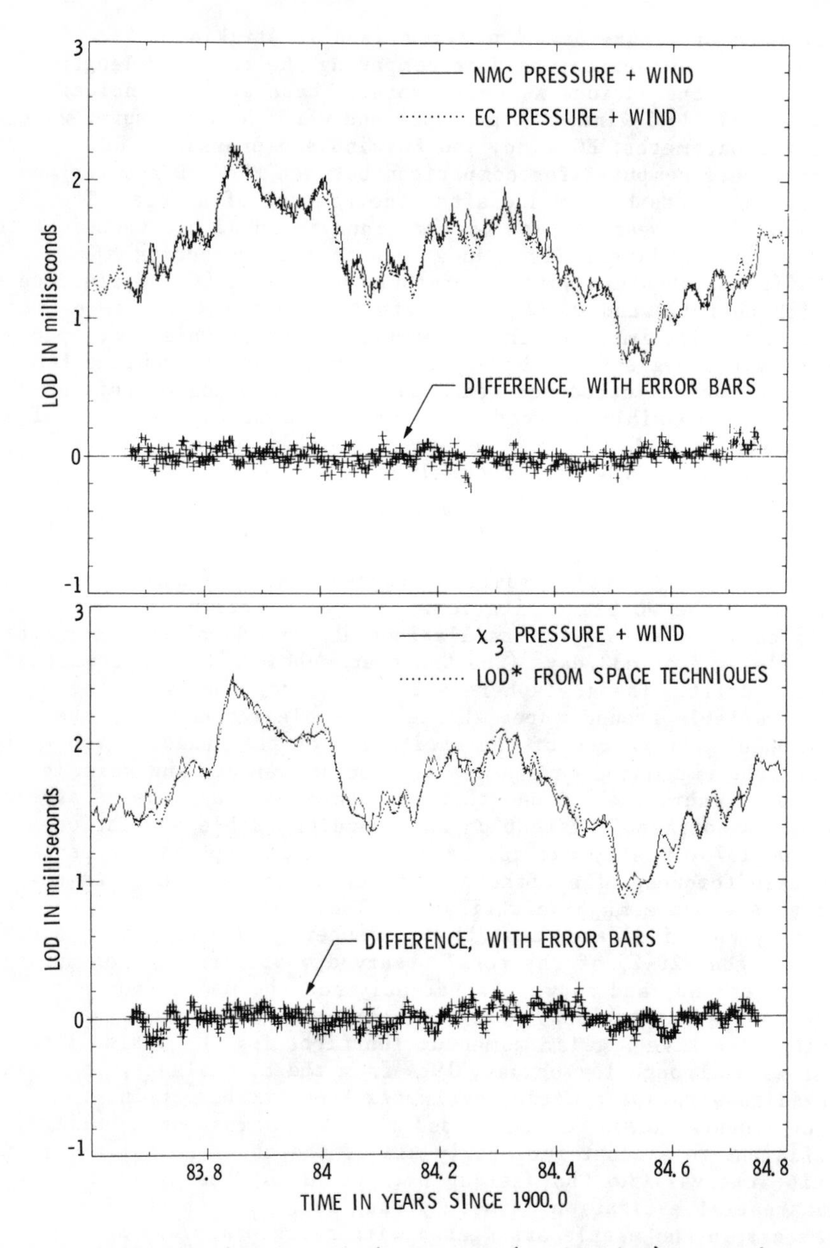

Figure 5a) The EC (dotted line) and NMC (solid line) χ_3 estimates (total of pressure and wind terms) (top) together with their difference (bottom); b) EC χ_3 pressure plus wind component (solid line), together with Kalman smoothed LOD estimates (dotted line) and their difference (bottom). The best fit bias and rate has been removed from the differences and the LOD together with all tidal LOD changes.

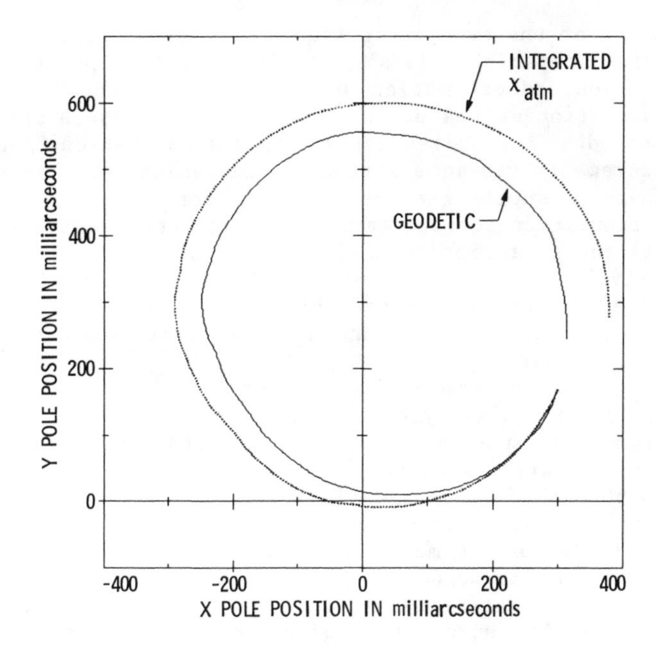

Figure 6. The Kalman smoothed Polar Motion series (solid line) together with the polar motion implied by the raw NMC pressure plus inverted barometer estimates.

to explain polar motion with periods less than one year. A study of International Latitude Service polar motion data for 1900-1980 and a common meteorological index for the Southern Oscillation (see Eubanks et al., this volume) found a statistically significant coherence between χ_1 data and the index, near the wobble period, and at near zero phase shift (Chao, 1985). There was no evidence of any coherence between χ_2 and this meteorological index.

Having indicated the recent dramatic improvement in the polar motion data quality during the MERIT campaign period (see Section 2), we focus the remainder of this discussion on recent polar motion results obtained during the MERIT campaign. The agreement between atmospheric and geodetic χ_1 and χ_2 series is less satisfactory than that obtained with the χ_3/ length of day intercomparisons indicating the presence of "unsolved problems." Unlike the χ_3 component, the equatorial components are dominated by the pressure term. The geodetic excitation series were obtained from the Kalman filter applied to polar motion data from the space techniques (see Section 4.2 for further details), while the atmospheric estimates were obtained from the EC and NMC analyses (see Sections 3 and 4.2 for further details). The period studied was again September 1, 1983 through October 1, 1984. Comparisons are possible in two sets of variables: polar motion coordinates (the conventional x and y) and the excitation function (χ_1 and χ_2). The observed polar motion changes are the convolution of the true excitation and response to the Earth; thus a polar motion comparison is a somewhat indirect approach requiring integration and estimation of initial conditions. Comparison of excitation estimates is a more direct approach and is the preferred method. It requires estimating the excitation from orientation measurements via some sort of a deconvolution. The Kalman filter is used here to smooth the combined data set as well as to deconvolve the orientation series; separate stochastic models for the excitation of UT1 and polar motion are utilized in this procedure (Eubanks et al., 1985a). Comparisons were made between the two available atmospheric series (EC and NMC), as well as between the meteorologically derived χ_1 and χ_2 and the excitation estimates from the combined geodetic series.

The χ_1 and χ_2 pressure term estimates from the NMC (solid lines) and EC (dotted lines) are displayed in Figures 7a and b; an arbitrary bias has been removed from each series. Both data sets contain seasonal structure, as well as short period oscillations that somewhat resemble the "40-60 day" oscillations in the LOD (see Section 4.2) although these oscillations have typical periods of 10 to 30 days, with typical amplitudes of ~50 mas. These oscillations would induce polar motion with an amplitude of few thousandths of an arcsecond, which is at the limit of detectability with the current geodetic data. The agreement between the two meteorological centers is quite evident correlation studies indicate a cross correlation coefficient of > 0.9; for these series even after the removal of the seasonal cycle. Analysis of the difference between the two series indicates that the discrepancy between the EC and NMC χ_1 and χ_2 series is not seasonal, but predominately at high frequencies. The RMS difference for χ_1 (χ_2) is 15.6 (16.8) milliarcseconds (mas) when a fit was made to biases only, and is reduced to 15.0 (15.6) mas when annual and semiannual sinusoids are included in the fit.

In contrast, the χ_1 and χ_2 wind terms appear to be dominated by noise; the difference between the two series (EC and NMC - shown in Figure 8a and b) is approximately the same size as either wind term. The cross correlation between the two series is less than 0.1 for both χ_1 and χ_2. The RMS scatter of the EC minus NMC χ_1 and χ_2 wind terms is almost 30% larger than the scatter of the raw wind data, indicating that the wind data from at least one center is dominated by error. The RMS difference between the EC and NMC χ_1 (χ_2) wind term is 39.3 (43.4) mas with a fit to biases only. Fits that include seasonal terms indicate significant seasonal errors in χ_2 data, where the RMS drops from 43.4 to 36.7 mas. The addition of the seasonal term reduces the scatter of the χ_2 wind term by 15%, while for both pressure terms and the χ_1 wind term, the drop is only at the 5% level. The RMS scatter between the two series is approximately two and a half times larger for the wind term than for the pressure term. From the point of view of the atmospheric modeling, the χ_1 and χ_2 wind terms are calculated from the summing of large numbers with large but almost equal positive and negative values, resulting in near cancellation. The end product is a series whose magnitude is comparable to its uncertainty. Our analysis confirms the noisiness of this term

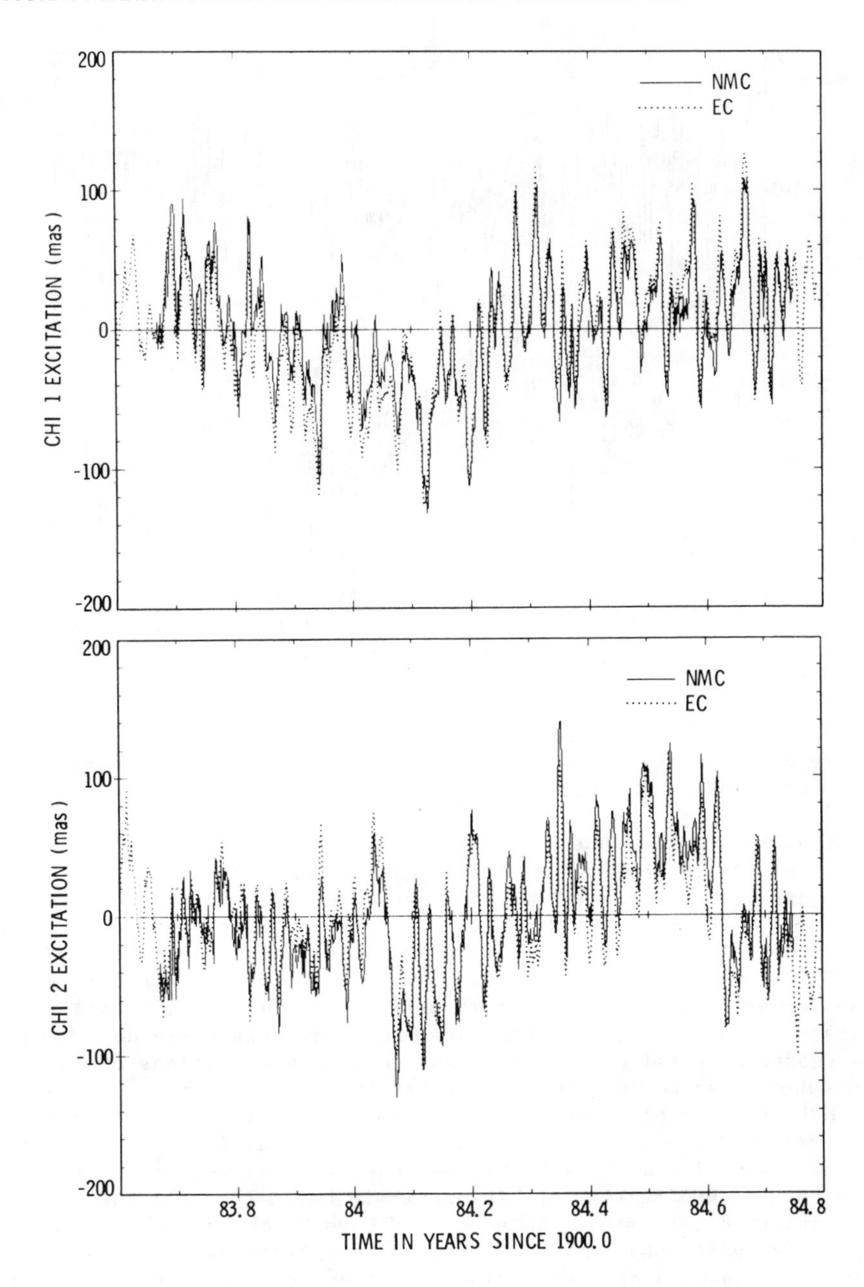

Figure 7. NMC and EC estimates of the raw χ_1 and χ_2 pressure component. An arbitrary bias was removed from both data sets.

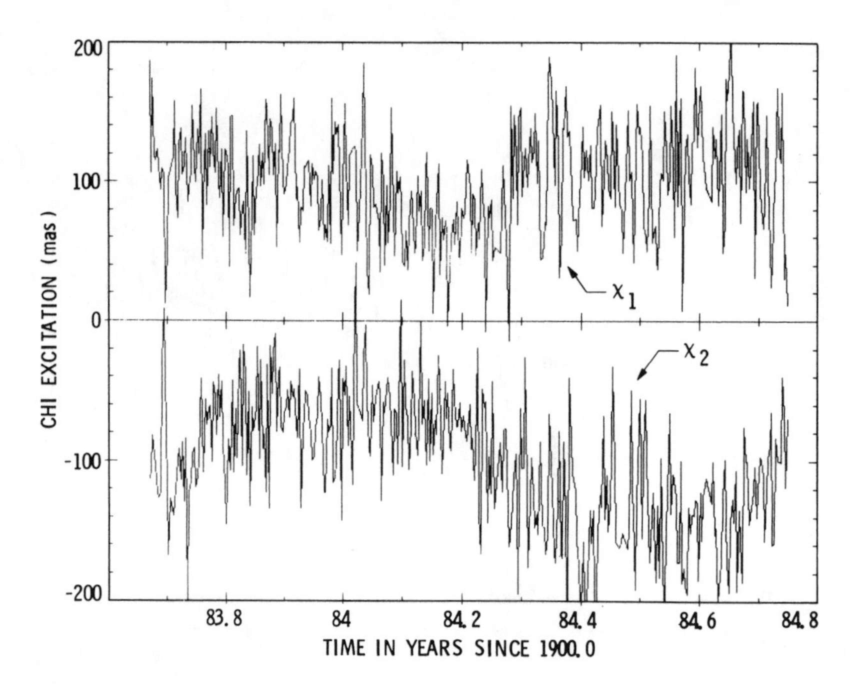

Figure 8. The difference between the EC and NMC wind components for χ_1 (top) and χ_2 (bottom).

relative to the pressure term.

The excitation estimates from the combined geodetic data are shown in Figure 9, together with the NMC pressure plus inverted barometer estimates. The corresponding comparison done in pole position with the appropriate convolution of the NMC results is displayed in Figure 6. The polar motion is dominated by the large prograde annual and Chandler wobbles making the short period fluctuations too small to discern on this scale. A systematic intercomparison was done between the geodetic χ_1 and χ_2 estimates and various combinations of the atmospheric terms in a manner similar to that used in the χ_3 analysis. The NMC pressure plus inverted barometer term gave the best agreement with RMS scatters of 23.4 mas for χ_1 and 27.5 mas for χ_2. The RMS scatters varied considerably between the various cases studied; some cases had an RMS scatter of approximately twice that the NMC pressure plus inverted barometer. As seen in Figure 9, the overall agreement for χ_2 is quite good with both curves showing similar seasonal structure, as well as some periods of detailed agreement such as in early 1984. The overall agreement for χ_1 is not as impressive and is somewhat poor; however, there is some period of agreement as in early 1984. The correlation between the χ_1 atmospheric and geodetic series

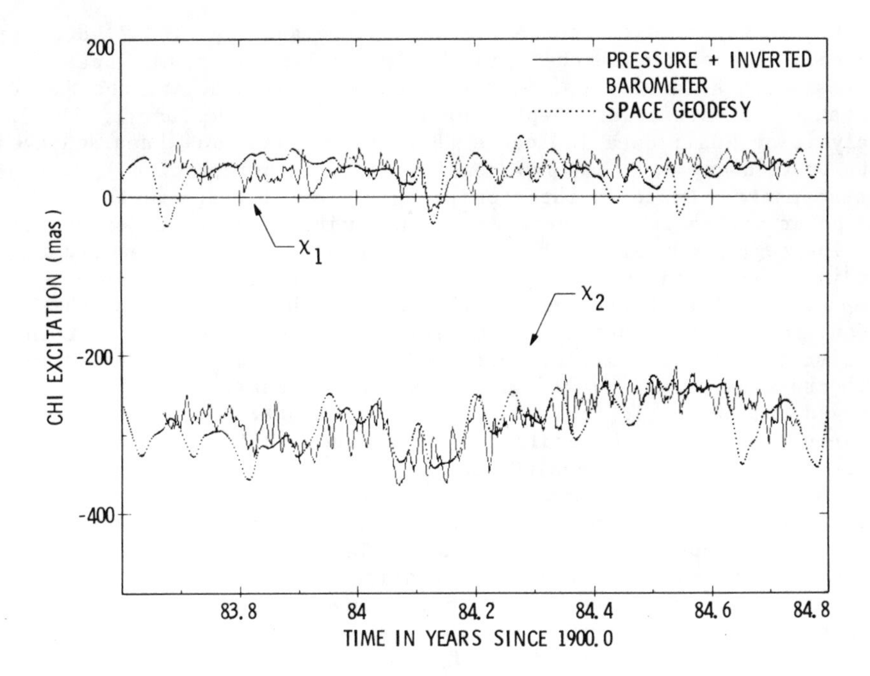

Figure 9. Raw NMC pressure plus inverted barometer estimates (solid lines) for χ_1 (top) and χ_2 (bottom), together with geodetic estimates (dotted lines) from the combined Kalman smoothing after the removal of an arbitrary bias between the series.

is increased by addition of the inverted barometer model. Comparison of Figures 7 and 9 indicates that the inclusion of the inverted barometer greatly reduced the size of the rapid oscillations. Further research is desirable to identify the limitations of the currently implemented inverted barometer model, as well as to explore these high frequency fluctuations in the χ_1 and χ_2 pressure term.

5. Summary

New technology has had a profound influence on the study of Earth rotation and polar motion. The advancement of space geodesy has greatly increased the accuracy and precision of Earth orientation measurements; the analysis of global weather data for operational forecasting provides high precision estimates of the atmospheric excitation of Earth rotation and polar motion. Here, we have highlighted recent advances in the measurement and interpretation of

Earth orientation data, concentrating on the short period fluctuations and their relation with the atmospheric angular momentum. Special emphasis and attention was placed on data acquired during the MERIT campaign (in particular, September 1 1983 through October 1, 1984). Analysis of MERIT data indicates that new heights have been achieved, both in accuracy and temporal density of Earth orientation measurements. Geodetic intercomparisons show that the Earth rotation and polar motion is currently determined with an accuracy of ~ 5 cm.

The comparison of Earth orientation and atmospheric results is leading to a better understanding of the coupling of various components of the Earth, especially between the solid Earth, the oceans and the atmosphere. Fluctuations in Earth rotation over the time scale of a year or less are dominated by atmospheric effects, with the wind term dominating. The agreement between the length of day and the meteorological χ_3 data is excellent. The RMS difference between the EC and NMC results is 0.067 msec, while the corresponding value between the combined LOD and the EC results (with a linear trend removed) is 0.072 msec, a variance difference of only 15%. This area represents largely a "solved problem" in Earth rotation. Some of the observed discrepancies are caused by neglect of the upper atmosphere in the routine meteorological computations; other probable causes are unmodeled sources of angular momentum exchange, modeling and data errors.

In contrast, the excitation of polar motion remains a largely "unsolved problem." The atmosphere and the oceans, together with variations in ground water, suffice to excite the annual wobble; however, the source of the Chandler wobble is unclear. The agreement between the geodetic and atmospheric χ_1 and χ_2 is less satisfactory than that obtained in the χ_3/LOD comparison. The best agreement is obtained with the NMC pressure plus the inverted barometer model, with an RMS difference for χ_1 and χ_2 at the 25 mas level. Data from both meteorological centers indicate the presence of rapid fluctuations in the pressure term, which are severely reduced in amplitude with the introduction of the inverted barometer hypothesis. These variations are marginally detectable by the modern space geodetic techniques. The χ_1 and χ_2 wind terms appear to be dominated by noise. Further research is required to explore these high frequency polar motion excitations and to identify the limitations of the current inverted barometer model. Possible improvements in the calculation of the wind term of the polar motion component should be sought.

Acknowledgements

This paper presents the results of one phase of research carried out at the Jet Propulsion Laboratory, California Institute of Technology, under contract with the National Aeronautics and Space Administration.

References

Anderson, J. A., Rosen, R. D., 'The Latitude-Height Structure of 40-50 day Variations in Atmospheric Angular Momentum,' Journal of Atmospheric Sciences, 40, 1584-1591, 1983.

Barnes, R. T. H., Hide, R., White, A. A., Wilson, C. A., 'Atmospheric Angular Momentum Fluctuations, Length of Day Changes and Polar Motion,' Proc. Roy. Soc. London, Series A, 387, 31-73, 1983.

Carter, W. E., Robertson, D. S., Pettey, J. E., Tapley, B. D., Schutz, B. E., Eanes, R. J. and Lufeng, Miao, 'Variations in the Rotation of the Earth,' Science, 224, 957-961, 1984.

Chao, B. F., 'Excitation of the Earth's Chandler Wobble by Southern Oscillation/El Nino, 1900-1980,' NASA Technical Memorandum 86231, 1985.

Dickey, J. O., Eubanks, T. M., Steppe, J. A., 'The Atmospheric Excitation of the Polar Motion,' Transactions of the American Geophysical Union, EØS, 65, 859, 1984.

Dickey, J. O., 'Coordinator's Report: Atmospheric Angular Momentum Studies,' Reports on the MERIT-COTES Campaign on Earth Rotation and Reference Systems, Part I, Proceedings of the Third MERIT Workshop and the Joint MERIT-COTES Working Group Meetings, ed. G. Wilkins, Royal Greenwich Observatory (UK), 1985a.

Dickey, J. O., 'Activity Report of the International Union of Geodesy and Geophysics/International Association of Geodesy (IUGG/IAG) Special Study Group 5-98, Atmospheric Excitation of the Earth's Rotation,' published in the CSTG (International Coordination of Space Techniques for Geodesy and Geodynamics) Bulletin, Dept. of Geodetic Science and Surveying, Ohio State University, Columbus, OH 43210, 1985b.

Dickey, J. O., Newhall, X X, and Williams, J. G., 'Earth Orientation from Lunar Laser Ranging and an Error Analysis of Polar Motion Services,' Special LAGEOS Issue of the Journal of Geophysical Research, 90, 9353-9363,1985.

Dickey, J. O. and Eubanks, T. M., 'The Application of Space Geodesy to Earth Orientation Studies,' an invited chapter in Space Geodesy and Geodynamics, Academic Press, editors, A J. Anderson and A. Cazenave, 1986.

Eanes, R. J., Schutz, B. E. and Tapley, B. D., 'Earth Rotation from LAGEOS: The 1984 CSR System,' EØS, Transactions, American Geophysical Union, 65, 16, 187-188, 1984.

Eubanks, T. M., Roth, M. G., Esposito, P. B., Steppe, J. A., Callahan, P. S., 'An Analysis of JPL TEMPO Earth Orientation Results,'

Proceedings of Symposium No. 5: Geodetic Applications of Radio Interferometry, NOAA Technical Report NOS 95 NGS 24, 63-70, 1982.

Eubanks, T. M., Dickey, J. O., and Steppe, J. A., 'Geophysical Significance of Systematic Errors in the Earth's Angular Momentum Budget,' Proceedings of the International Association of Geodesy (IAG) Symposia, International Union of Geodesy and Geophysics (IUGG) XVIIIth General Assembly (Hamburg, FRG; August 15-27, 1983) Vol. 2, 122-143, 1984.

Eubanks, T. M., Spieth, M. A., Steppe, J. A., 'The Accuracy of Radio Interferometric Estimates of Earth Rotation,' The Telecommunications and Data Acquisition Progress Report 42-80 229-235, NASA, Jet Propulsion Laboratory, Pasadena, CA, 1985a.

Eubanks, T. M., Steppe, J. A., Dickey, J. O. and Callahan, P. S., 'A Spectral Analysis of the Earth's Angular Momentum Budget,' Journal of Geophysical Research, 90, 5385-5404, 1985b.

Eubanks, T. M., Dickey, J. O., Steppe, J. A. 'The 1982-83 El Nino, the Southern Oscillation, and the Length of Day,' Tropical Ocean and Atmosphere Newsletter, Number 29, 21-22, 1985c.

Eubanks, T. M. , Steppe, J. A., and Dickey, J. O., 'The Atmospheric Excitation of Earth Orientation Changes During MERIT,' Report on the MERIT-COTES Campaign on Earth Rotation and Reference Systems, Part II: Proceedings of the International Conference on Earth Rotation and the Terrestrial Reference Frame, ed. I. Mueller, Ohio State University, (USA), Vol. 2, 434-440, 1985d.

Eubanks, T. M., Steppe, J. A., Dickey, J. O., Rosen, R. D., Salstein, D. A., 'Rapid Polar Motions Caused by the Atmosphere and Oceans,' in preparation, 1986.

Feissel, M., 'Determination of the Earth Rotation Parameters by the Bureau International de l'Heure, 1962-1979,' Bull. Geod., Vol. 54, 81-102, 1980.

Feissel, M., and Gambis, D., 'La mise en evidence de variations rapides de la duree du jour,' Comptes Rendus Hebdomadaires es Sceances de l'Academie des Sciences, Series B, Vol. 291, pp. 271-273, 1980.

Garratt, J. R., 'Review of Drag Coefficients Over Oceans and Continents,' Mon. Weather Rev., 105, 915-929, 1977.

Guinot, B., 'Rotation of the Earth and Polar Motion Services,' in Proc. of 9th GEOP Conference, An International Symposium on the Applications of Geodesy to Geodynamics, pp. 13-18, Dept. of Geodetic Science Rept. No. 280, Ohio State University, 1978.

Hide, R., Phil. Trans. R. Soc. Lond. A, 284, 547-554, 1977.

Hide, R., Birch, N. T., Morrison, L. V., Shea, D. J. and White, A. A., 'Atmospheric Angular Momentum Fluctuations and Changes in the Length of the Day,' Nature, 286, 114-7, 1980.

Hide, R., 'Rotation of the Atmospheres of the Earth and Planets,' Phil. Trans. R. Soc. Lond. A., 313, 107-121, 1985.

King, R. W., 'Coordinator's Report: Intercomparisons of Techniques,' Reports on the MERIT-COTES Campaign on Earth Rotation and Reference Systems, Part I, Proceedings of the Third MERIT Workshop and the Joint MERIT-COTES Working Group Meetings, ed. G. Wilkins, Royal Greenwich Observatory (UK), 1985a.

Kouba, J., 'A Review of Geodetic and Geodynamic Satellite Doppler Positioning,' Rev. Geophys. Space Phys., 21, 27-40, 1983.

Krishnamurti, T. N. and Subrahmanyam, D., 'The 30-50 Day Mode at 850 mb During MONEX,' J. Atmos. Sci., 39, 2088-2095, 1982.

Lambeck, K. and Cazenave, A., Phil. Trans. R. Soc. Lond. A, 284, 495-506, 1977.

Lambeck, K., 'Changes in Length of Day and Atmospheric Circulation,' Nature, 286, 104, 1980a.

Lambeck, K., The Earth's Variable Rotation, Cambridge University Press, 1980b.

Lambeck, K., Hopgood, P., 'The Earth's Rotation and Atmospheric Circulation from 1963 to 1973,' Geophys. J. R. Astr. Soc. 71, 67-84, 1981.

Langley, R. B., King, R. W., and Shapiro, I. I., 'Earth Rotation from Lunar Laser Ranging,' J. Geophys. Res., 86, 11,913-11,918, 1981a.

Langley, R. B., King, R. W., Shapiro, I. I., Rosen, R. D., and Salstein, D. A., 'Atmospheric Angular Momentum and the Length of the Day: A Common Fluctuation with a Period Near 50 Days,' Nature, 294, 730-3, 1981b.

Madden, R. A. and Julian, P. R., 'Detection of a 40-50 day oscillation in the Zonal wind in the Tropical Pacific,' J. Atmos. Sci., 28, 702-708, 1971.

Madden, R. A. and Julian, P. R., 'Description of Global-Scale Circulation Cells in the Tropics with a 40-50 day Period,' J. Atmos. Sci., 29, 1109-1123, 1972.

McPherson, R. D., Bergman, K. H., Kistler, R. E., Rash, G. E., Gordon, D. S., 'The NMC Operational Global Data Assimilation System,' Monthly Weather Review, 107, 1445-1461, November 1979.

Merriam, J. B., 'Meteorological Excitation of the Annual Polar Motion,' Geophys. J. R. Astr. Soc., 70, 41-56, 1982.

Miller, A. J., 'Periodic Variation of Atmospheric Circulation at 14-16 days,' J. Atmos. Sci., 31, 720-726, 1974.

Morgan, P. J., King, R. W. and Shapiro, I. I., 'Length of Day and Atmospheric Angular Momentum: A Comparison for 1981-1983,' Journal of Geophysical Research, 90, 12,645-12,652, 1985.

Munk, Walter H. and MacDonald, Gordan J. F., The Rotation of the Earth, a Geophysical Discussion, Cambridge University Press, London, 1960.

Naito, I. and Yokoyama, K., 'A Computation of Atmospheric Excitation Functions for the Earth's Rotation Based on JMA Global Analysis Data,' Reports on the MERIT-COTES Campaign on Earth Rotation and Reference Systems, Part II: Proceedings of the International Conference on Earth Rotation and the Terrestrial Reference Frame, ed. I. Mueller, Ohio state University (USA), Vol. 2, 434-440, 1985.

Newton, R. R., 'The Secular Acceleration of the Earth's Spin,' Geophys J. R. Astr. Soc., 80, 313-328, 1985.

Oort, A. H., and Bowman, H. D., 'A Study of the Mountain Torque and Its Interannual Variations in the Northern Hemisphere,' J. Atmos Sci., 31, 1974-1982, 1974.

Robertson, D. S., Carter, W. E., Campbell, J. and Schuh, H., 'Daily Earth Rotation Determinations from IRIS Very Long Baseline Interferometry,' Nature, 316, 424, 1985.

Rosen, R. D. and Salstein, D. A., 'Variations in Atmospheric Angular Momentum on Global and Regional Scales and the Length of Day,' Journal of Geophysical Research, 88, No. C9, 5451-5470, 1983.

Rosen, R. D. Salstein, D. A., Eubanks, T. M., Dickey, J. O., Steppe, J. A., 'An El Nino Signal in Atmospheric Angular Momentum and Earth Rotation,' Science, 225, 411-414, 1984.

Rosen, R. D. and Salstein, D. A., 'Contribution of Stratospheric Winds to Annual and Semi-annual Fluctuations in Atmospheric Angular Momentum and the Length of Day,' Journal of Geophysical Research, 90, 8033-8041, 1985.

Rudloff, W., annln Meteorol. neue folge, no. 6, 221-223, 1973.

Ryan, R. C., Ma, C. 'Results from the NASA Crustal Dynamics Project,' Bureau International de l'Heure Annual Report for 1982, Paris, 1983.

Sidorenkov, N. S., 'An Investigation of the Role of the Atmosphere in the Excitation of Multiyear Variations in the Earth's Rotation Rate,' Astron. Zh., 56, 187-199, 1979.

Smith, D. E., Christodoulides, D. C., Kolenkiewicz, R., Dunn, P. J., Klosko, S. M., Torrence, M. H., Fricke, S. and Blackwell, S., 'A Global Geodetic Reference Frame from LAGEOS Ranging (SL5.1AP),' Journal of Geophysical Research, 90, B11, 9221-9235, 1985.

Smith, D. E., Christodoulidis, D. C., Wyatt, G. H., Dunn, P., Klosko, S., Torrence, M., 'Variations in Length-of-Day, and the Atmospheric Angular Momentum,' to be published in the Journal of Geophysical Research, 1986.

Smith, M. L., 'Wobble and Nutation of the Earth,' Geophys. J. R. Astr. Soc. 50, 103-140, 1977.

Smith, M. L. and Dahlen, F. A., 'The period and Q of the Chandler Wobble,' Geophys. J. R. Astro. Soc., 64, 223-281, 1981.

Smith, R. B., 'A Measurement of Mountain Drag,' J. Atmos Sci., 35, 1644-1654, 1978.

Sovers, O. J., Thomas, J. B., Fanselow, J. L., Cohen, E. J., Purcell Jr., G. H., Rogstad, D. H., Skjerve, L. J. and Spitzmesser, D. J., 'Radio Interferometric Determination of Intercontinental Baselines and Earth Orientation Utilizing Deep Space Network Antennas: 1971 to 1980,' Journal of Geophysical Research, 89, B9, 7597-7607, 1984.

Spieth, M. A., Eubanks, T. M. and Steppe, J. A., 'Intercomparison of UT1 Measurements During the MERIT Campaign Period,' Reports on the MERIT-COTES Campaign on Earth Rotation and Reference Systems, Part II: Proceedings of the International Conference on Earth Rotation and the Terrestrial Reference Frame, ed. I. Mueller, Ohio State University, (USA), Vol. 2, 609-622, 1985.

Steppe, J. A., Eubanks, T. M. and Spieth, M. A., 'Intercomparison of Polar Motion Measurements During the MERIT Period,' Reports on the MERIT-COTES Campaign on Earth Rotation and Reference Systems, Part II: Proceedings of the International Conference on Earth Rotation and the Terrestrial Reference Frame, ed. I. Mueller, Ohio State University (USA), Vol 2, 622-637, 1985.

Swinbank, R., 'The Global Atmospheric Angular Momentum Balance Inferred from Analysis During the FGGE,' Quart. J. R. Met. Soc., 14, 977-992, 1985.

Tapley, B. D., Eanes, R. J. and Schutz, B. E., 'UT/CSR Analysis of Earth Rotation from LAGEOS SLR Data,' Reports on the MERIT-COTES Campaign on Earth Rotation and Reference Systems, Part II: Proceedings of the International Conference on Earth Rotation and the Terrestrial Reference Frame, ed. I. Mueller, Ohio State University (USA), Vol. 1, 111-126, 1985a.

Tapley, B. D., Schutz, B. E. and Eanes, R. J., 'Station Coordinates, Baselines and Earth Rotation from Lageos Laser Ranging: 1976-1984,' LAGEOS special issue of the Journal of Geophysical Research, 90, 9235-9249, 1985b.

Wahr, J. M., 'The effects of the Atmosphere and Oceans on the Earth's Wobble and on the Seasonal Variations in the Length of Day - I. Theory,' Geophys. J. R. Astr. Soc., 70, 349-372, 1982.

Wahr, J. M., 'The Effects of the Atmosphere and Oceans on the Earth's Wobble and on the Seasonal Variations in the Length of Day - II. Results,' Geophys. J. R. Astr. Soc., 74, 451-487, 1983.

Wahr, J. M., 'The Earth's Rotation Rate,' American Scientist, 73, 41-46, 1985.

Wilkins, G. A., edt., PROJECT MERIT, Joint Working Group on the Rotation of the Earth, IAU/IUGG, 1980.

Wilson, C. R. and Vincente, R. O., 'An Analysis of the Homogeneous ILS Polar Motion Series,' Geophys. J. R. Astr. Soc., 62, 605-616, 1980.

Yasunari, T. 'Structure of an Indian Summer Monsoon System with Around 40-Day Period,' J. Met. Soc. Japan, 59, 336-354, 1981.

Yoder, C. F., Williams, J. G., Parke, M. E., 'Tidal Variations of Earth Rotation,' Journal of Geophysical Research, 86, B2, pp. 881-891, 1981.

THE EL-NINO, THE SOUTHERN OSCILLATION AND THE EARTH ROTATION

T.M. Eubanks, J.A. Steppe, J.O. Dickey
Jet Propulsion Laboratory
California Institute of Technology
Pasadena, California, 91109

ABSTRACT. Unpredictable changes in the Length of Day (LOD) at seasonal and higher frequencies are mostly driven by exchanges of axial angular momentum between the atmosphere and the solid Earth. The corresponding changes in the axial angular momentum of the atmosphere can be divided into thermally forced seasonal oscillations and longer and shorter period free oscillations. All of these variations seem to be largely due to zonal wind velocity changes, and existing LOD measurements thus provide a global zonal wind record of climatological interest (Salstein and Rosen, 1985). We present a description of our research into the relationship between LOD variations and the Southern Oscillation, one of the most important and best studied free oscillations, and with the associated El Nino phenomenon. There is a statistically significant correlation observed between interannual LOD changes and a widely used meteorological index for the Southern Oscillation during the period 1962-1985. There also seems to be a weak relationship between El Nino events and the rapid 40 to 60 day fluctuations in the LOD. We discuss the meteorological significance of these findings.

1. INTRODUCTION

Changes in the Length of Day (LOD) at seasonal and higher frequencies are primarily caused by exchanges of angular momentum with the atmosphere (see e.g., Eubanks, et al. 1985a, Morgan et al., 1985). The atmosphere cannot be responsible for the large LOD changes seen over periods of decades and centuries, and its role in any of the interannual LOD changes was, until recently, controversial. A number of authors have linked interannual LOD changes to the stratospheric Quasi-Biennial Oscillation (QBO), with a period of ~ 2 years. Lambeck (1980, Chapter 7) presents a review of this work, and reports detection of a nearly periodic oscillation, with a period equal to 24 to 26 months, and a total LOD amplitude of ~ 0.08 milliseconds. This work, although promising, still leaves most of the observed interannual variations unexplained. The LOD excursions of many milliseconds observed over time scales of decades to centuries are too large to plausibly be due to any

A. Cazenave (ed.), Earth Rotation: Solved and Unsolved Problems, 163–186.

source other than the Earth's liquid core (Lambeck 1980), and are correlated with geomagnetic field variations (Le Mouel et al., 1981), and so any observed interannual fluctuations are often assumed to be due to core-mantle torques. Since the atmosphere causes the strong (amplitude ~ 0.4 millisecond) annual LOD variations, and since the core causes the stronger decade period fluctuations, it seems reasonable to expect that the atmosphere and the core both contribute to LOD fluctuations with periods from 1 to 10 years. Explication of the role of the atmosphere in this frequency band is thus not only of inherent interest, but also should help to isolate the effects of the liquid core on the LOD.

A major increase in the understanding of interannual LOD variations began when Stephanic (1982) proposed a connection between the LOD and the Southern Oscillation, a long period (2 to 10 year) oscillation in the atmospheric pressure, rainfall and sea surface temperature in the equatorial Pacific. This paper showed some coherence between the LOD and equatorial Pacific air temperatures (whose variation is one facet of the Southern Oscillation), and proposed that the coherence was caused by thermally driven zonal winds in the tropics. The thermal wind relationship is not valid in the tropics, as Stephanic admits, and the proposed mechanism was not generally viewed as convincing. Concurrently, interest in the Southern Oscillation, and in the related El Nino phenomena, was building amongst meteorologists and oceanographers (see, for example, Horel and Wallace 1981). This interest was further stimulated by the unusually strong El Nino of 1982-83, which was associated with unusual weather in many parts of the world, and was very well observed (Philander, 1983, Rasmusson and Wallace, 1983). There were also unusually large and well observed changes in the LOD and the Atmospheric Angular Momentum (AAM) during January and February, 1983 (Rosen et al, 1984), which coincided almost exactly with the most intense period of the 1982-83 El Nino and with large changes in the atmospheric pressure over the tropical Pacific. The apparent connection between the El Nino and the duration of the day in early 1983 suggested that there may have been similar rapid LOD changes during previous El Ninos and, also, that the interannual changes in the LOD may indeed be related to the Southern Oscillation. Several studies (Chao, 1984 and Eubanks et al., 1985b) have now shown significant correlation between long period LOD changes and one meteorological index for the Southern Oscillation, although the relation with the shorter period 40 to 60 day LOD fluctuations remains unclear. Gross and Chao (1985) and Chao (1985), in a further analysis, found some coherence between a Southern Oscillation index and one component of the excitation of the polar motion near the Chandler wobble period of ~ 14 months. This work, although stimulating, fails to consider the retrograde and prograde circular components of excitation. Since the Chandler wobble is much more sensitive to prograde (eastward propagating) than to retrograde (westward) planetary pressure waves at its resonance frequency, any coherence at that frequency should be predominately with the prograde component of excitation, not in one component of linear polarization as found by Chao (1985).

This paper describes an analysis of changes in the duration of the day, the Southern Oscillation and the related El Nino phenomena during the period 1962-1985. The Earth rotation data set is a combination of Earth orientation measurements from a number of different sources, to provide the best possible determinations of the LOD over the entire period. Meteorological estimates of the atmospheric angular momentum are available for the period after 1976 and make it possible to directly examine the role of the atmosphere since that time. These data sets are described further in Section 2. As is typical in climatology the Southern Oscillation must be represented by an index, and we use the most common Southern Oscillation Index (or SOI) in our analysis. This index, and the underlying oscillation, are described further in Section 3.

One problem in relating the free oscillations of the atmosphere to geodetic measurements is their irregular nature. In contrast to the astronomically forced variations, which have precisely fixed frequencies, the free oscillations are aperiodic, and apparent periods and amplitudes often change rapidly (see, e.g., Feissel and Nitschelm, 1985). The atmosphere is a nonlinear system, and many of the assumptions used in classical statistical analysis may be inappropriate or misleading in climatology. This is illustrated well by the 40 to 60 day oscillations in the LOD, which are clearly visible in graphs of recent LOD variations (see Figure 1 a), but which have power spectra almost indistinguishable from random walks (Eubanks et al., 1985a). These considerations suggest that statistical analysis, although useful, be employed and interpreted with caution in climatology. We use both graphical presentation and linear correlation coefficients in Section 4 to explore the relationship between the LOD, the Atmospheric Angular Momentum, and an index for the Southern Oscillation. The LOD event in early 1983 is examined in some detail. Axial AAM measurements are only available for one complete cycle of the Southern Oscillation, and most of our analysis was conducted with the longer LOD series. Section 4.4 explores the relationship between interannual LOD fluctuations and the available axial AAM measurements. Section 5 contains our conclusions.

2. AXIAL ANGULAR MOMENTUM MEASUREMENTS

2.1 Measurements of the Earth's Rotation and the Length of Day

Current geodetic techniques measure changes in the angular orientation of the Earth with respect to a quasi-inertial reference frame based on extraterrestrial objects, and these measurements are used to infer the rotation rate. Changes in the Earth orientation can be described in terms of small motions of the pole of rotation and by the total amount of rotation about that pole. The UT1 is the angle of rotation about the pole, and can be directly inferred from geodetic measurements. Estimation of the Earth rotation rate requires a reference clock more stable than the Earth itself, and so accurate determinations of the duration of the day required development of atomic clocks. The actual quantity measured is the difference between the UT1 and the International Atomic Time scale IAT (or its predecessors), and is thus denoted by UT1 - IAT.

The duration of the day is simply related to the time derivative of the measured quantities by

$$\frac{LOD}{86400\ sec} = 1.0 - \frac{d(UT1 - IAT)}{d\ IAT} \qquad (1)$$

The estimation of the LOD from noisy UT1 measurements by Kalman filtering will be discussed in Section 2.2

We combined UT1 measurements from a variety of efforts using modern and classical techniques to provide the best possible estimates of changes in the LOD. Such data are available for the period 1962 to 1985, although the modern data is only available since 1970 (for Laser ranging to the Moon) and 1978 (for radio interferometry). The resulting series thus improves in quality during the analysis period, but the interesting rotation rate fluctuations appear to be well above the measurement noise throughout. Many of our measurements are of the local variation of longitude (or UT0) of an observatory or interferometric baseline, and polar motion and variation of latitude measurements were included to improve the determination of the LOD. Spieth et al. (1985) discuss the accuracy of the various UT1 data sets for a period in 1983 and 1984.

Optical astrometry was the first method to measure changes in the Earth's rotation rate. Such data exist from (or even before) the start of Atomic Time scales in 1955, but the early data are not homogeneous. Recently Li (1985) and the Bureau International de l'Heure (BIH) have conducted a re-reduction of optical data from the beginning of 1962 through 1982, and we used this data set in our analysis. This reduction provides five day averages of Earth orientation changes based on data from ~ 80 observatories world-wide. Even after the careful adjustment of systematic error by Li (1985), we still found it necessary to remove a UT1 rate of 0.47 msec/year from these data before combination with the more modern measurements. This data set provides UT1 and polar motion estimates every 5 days, with a UT1 error of ~ 1.4 milliseconds, based on intercomparison with data from the modern techniques.

Lunar Laser Ranging (LLR) uses dynamical models of the lunar motion to provide a nearly inertial reference frame suitable for monitoring the Earth's rotation. Currently, this technique provides measurements of the UT0 (variation of longitude) for each participating observatory. Measurements of the UT0 of the McDonald Observatory, Texas, are available since 1970, with typical accuracies of ~ 0.5 milliseconds (Dickey et al. 1985). A similar technique, Satellite Laser Ranging (SLR) was used to provide polar motion estimates for the period since 1976 (Tapley et al. 1985).

Very Long Baseline radio Interferometry (VLBI) can provide highly accurate UT1 estimates based on time averages of a few hours to a day, but such data are only available for the last sixth of the analysis period. We used two independent VLBI data sets in our research. The International Radio Interferometric Surveying (IRIS) project, and its predecessor, the U.S. POLARIS project, have conducted single and multiple baseline VLBI observations on a regular basis since mid 1980 (Robertson et al. 1985). The IRIS project commenced multi-baseline Europe to America VLBI experiments on a 5 day schedule at the beginning of 1984. Much of the pre-1984 Polaris data comes from observations from a single baseline (Massachusetts to Texas) interferometer; we used only the baseline transverse component of orientation from this interferometer in our combined smoothing. The data are derived from the IRIS/POLARIS reduction of the U.S. National Geodetic Survey (series 2/1/85). Since mid-1980 the U.S. Deep Space Network (DSN) has conducted single baseline VLBI experiments, typically of three hour duration, over Spain California and Australia California baselines, generally on a weekly basis. The effects of errors in the IAU 1980 nutation theory were removed from the short duration data (Eubanks et al. 1985e) before this analysis. Additional 18 to 24 hour duration dual frequency observing sessions have been conducted over the same baselines since late 1978. These data were reduced at JPL (Eubanks et al. 1985f).

2.2 Kalman Filtering of Earth Rotation Measurements

Astronomical techniques are generally not directly sensitive to the rotation rate of the Earth. The duration of the day must be estimated from the resulting UT1 - IAT measurements, typically through some combination of differencing and smoothing. Rational choice of the amount and nature of smoothing requires knowledge of the signal as well as of the measurement noise. The unpredictable high frequency fluctuations in the UT1 are mostly driven by changes in the axial angular momentum of the atmosphere, and meteorologically derived angular momentum estimates indicate that the resulting changes in the UT1 should have a power spectrum proportional to the frequency^{-4} at periods $\lesssim$ 100 days (Eubanks et al. 1985a). We used a Kalman filter incorporating this statistical model both to smooth the combined data set and to estimate the LOD. This filter was implemented at JPL to provide the DSN with accurate smoothings and predictions of earth orientation changes for spacecraft navigation. The filter simultaneously smooths both the UT1 and the Polar Motion and estimates their excitations and can accept one, two or three dimensional measurements of any components of earth orientation. The UT1 models and filter are described further in Eubanks et al. (1985d). The predictable UT1 fluctuations of tidal origin were removed from the UT1 data before smoothing (Yoder et al. 1981). We denote the change in the LOD from a standard day of 86400 seconds, after removal of all tidal oscillations, by the LOD*. The Kalman filter can provide smoothed values at any epoch; generally equally spaced series are produced with a value for each day at midnight, UTC.

2.3 Measurements of the Axial Atmospheric Angular Momentum

Changes in the atmospheric circulation cause variations in the axial angular momentum of the atmosphere. In the absence of other sources or sinks of angular momentum, conservation of angular momentum induces opposing changes in the Earth's rotation rate, and thus proportional changes in the LOD*. A number of studies have shown that meteorological estimates of the axial AAM can account for most of the observed seasonal and higher frequency fluctuations in the LOD* (see, e.g., Eubanks et al 1985a). Such estimates are now regularly available from atmospheric models used in operational weather forecasting by several organizations. These models provide a variety of atmospheric parameters at a set of grid points through a weighted average of recent meteorological data and the results of the last forecast. The AAM can then be easily derived from integrations of the gridded wind and pressure data. Our analysis uses twice daily AAM measurements derived from the National Meteorological Center (NMC) operational weather analysis for the period between July, 1976, through December, 1984 (Rosen and Salstein 1983). The accuracy of the AAM data is discussed in Eubanks et al. (1983 and 1985c). The NMC data was sampled at midnight UTC daily, and occasional short gaps were filled by linear interpolation.

The total angular momentum of the atmosphere can be described by a three vector, with M_3 describing the axial AAM component. To first order this component can be further decomposed into wind and pressure terms, calculated from those fields separately. The resulting integrals are

$$M_3(W) = \frac{2\pi R^3}{g} \int_{50\ \mathrm{mbar}}^{1000\ \mathrm{mbar}} \int_{-\pi/2}^{\pi/2} [u] \cos^2\phi \, d\phi \, dp \qquad (2)$$

and

$$M_3(P) = \frac{0.70\, R^4}{g} \int_{0}^{2\pi} \int_{-\pi/2}^{\pi/2} p_s \cos^3\phi \, d\phi \, d\theta \qquad (3)$$

where R is the mean radius of the Earth, p is the air pressure, p_s is the local surface pressure, g is the acceleration due to gravity (assumed constant), [u] is the zonally averaged west to east wind velocity and ϕ and θ are the latitude and East longitude, respectively. The numerical factor of 0.7 in (3) accounts for the compensating deformation of the earth in response to atmospheric loading (Barnes et al., 1983). M is in units of effective angular momentum, and was converted to seconds of LOD through the conversion constant ($1.668\ 10^{-29}\ \mathrm{sec}^2/\mathrm{kg\ m}^2$) in Eubanks et al. (1985a). We follow the notation of Barnes et al. (1983) and denote the axial angular momentum of the atmosphere, converted into LOD units, by χ_3.

The χ_3 wind component is simply a weighted average of the global zonal winds. This component dominates the axial AAM changes, and is itself dominated (at least at high frequencies) by changes in the strength and location of the subtropical jet stream in each hemisphere (Rosen and Salstein, 1983). The NMC also provides a correction for the response of the sea level to changes in the surface pressure. This model replaces any local changes in the surface pressure over the ocean with the average change in pressure over the entire ocean. The equilibrium model is thought to be acceptable at low frequencies but is probably inadequate at periods much shorter than seasonal (J.M. Wahr, personal communication). This correction tends to reduce further the small pressure component in the axial AAM.

3. THE SOUTHERN OSCILLATION AND ITS INDICIES

The Southern Oscillation denotes a long period (~ 6 years) irregular variation of atmospheric pressure over the South Pacific. This oscillation is the cause of the most prominent year to year variations in the climate of the South Pacific basin (Rasmusson and Wallace 1983), and is also associated with wide-spread changes in the atmospheric and oceanic circulation, in sea surface temperatures, and with the El Nino phenomenon (Philander 1983). It is thought that the oscillation, and the related El Nino events, are due to a non-linear air-sea interaction, with changes in the sea surface temperature causing changes in the surface winds, which then move and modify ocean surface waters, providing a feed-back loop. The persistence of the oscillation is probably a consequence of the thermal inertia of the oceans, which provides a means of storing heat from one year to the next. A recent model (Wyrtki 1985) proposes that the Southern Oscillation is driven by the steady accumulation of warm water in the western part of the equatorial Pacific, caused by the prevailing westward propagating winds. In this model, El Nino events begin when relaxation of the prevailing winds cause a surge of warm water to move from the West to the East Pacific and then to dissipate at high latitudes. The El Nino events thus serve to remove the excess warm equatorial water, setting the stage for the oscillation to begin again. Cane (1983) (among others) showed the passage of an equatorial wave in the sea surface with ~ 20 cm amplitude, moving from the West to the East Pacific during all of 1982. This wave of warm water reached Peru in January, 1983, and was associated with many of the unusual climate changes in the tropics.

The Southern Oscillation in surface air pressure has the structure of a standing wave, with antinodes near the eastern and western boundaries of the South Pacific, ~ 90° of longitude apart (Horel and Wallace 1981). This has led to the development of various Southern Oscillation Indicies (SOI), based on the difference between sea level surface pressure in the East and West Pacific. The most suitable index, given available pressure measurements, seems to be the seasonally adjusted pressure difference between Tahiti, in the east, and Darwin, Australia, in the west (Chen 1982). We use a modified version of the Tahiti - Darwin index provided by the NMC Climate Analysis Center

(Quiroz 1983). The NMC index is normalized by dividing each month's value by the long term standard deviation for that month, although this is not an important change for the Tahiti - Darwin SOI (Chen 1982). We removed this normalization, pending further study of seasonal variations in the LOD, and changed the sign to make the resulting series positively correlated with the LOD*. We denote the resulting Modified Darwin - Tahiti SOI the MSOI. Such data are available since 1935. Trenberth (1983) has shown that such indicies contain high frequency noise and should be smoothed. We smoothed our anomaly series with 5 month running mean, which is consistent with the amount of noise observed in these data, and with the smoothing used by the Climate Analysis Center (Quiroz, 1983). It is worth pointing out that axial AAM changes come from changes in zonally averaged winds and pressures (see Equations 2 and 3) and that zonal winds arise from meridional pressure gradients. The Darwin and Tahiti sea level pressure indicies represent a longitudinal pressure gradient, and are thus not directly related to either component of the axial AAM. Any correlation between the axial AAM (or the LOD*) and the SOI or MSOI thus must arise dynamically.

El Nino events are associated with low surface air pressure on the eastern side of the south Pacific basin and with high pressure on the western side, and thus occur during one part of the Southern Oscillation. The 1982-83 El Nino lasted roughly from the spring or summer of 1982 until the spring of 1983 and was one of the strongest on record, with the Tahiti - Darwin SOI reaching a record low in January, 1983 (Rasmusson and Wallace 1983). This El Nino event, although somewhat anomalous, is the best observed of any so far, and thus provides insight into conditions during past El Nino events. It is now clear that El Nino events are associated with unusual weather throughout the globe (Gill and Rasmusson 1983), and it is hoped that studies of the oscillation may lead to a means of long range weather prediction. Relating LOD changes to the Southern Oscillation could provide further insight into the relation between the subtropical jet streams and El Nino events.

4. ATMOSPHERIC OSCILLATIONS AND THE LOD

4.1. The Earth's Axial Angular Momentum Budget

Figure 1 a shows graphically the agreement between the LOD* and the total axial angular momentum (wind plus pressure plus inverted barometer correction) on time scales of weeks to years. Clearly visible are the rapid variations, with apparent periods of 60 to 40 days or less, as well as the thermally forced seasonal oscillations. This LOD* series was obtained using rotation data only from the modern techniques of VLBI and LLR, and can be taken as being virtually free of systematic error (Spieth et al. 1985). The difference between the LOD* and the AAM, at the bottom of the figure, thus represents unexplained variations in the LOD*. (The large differences in early 1981 are due to gaps in the UT1 data set). Although the LOD* data agree very well with

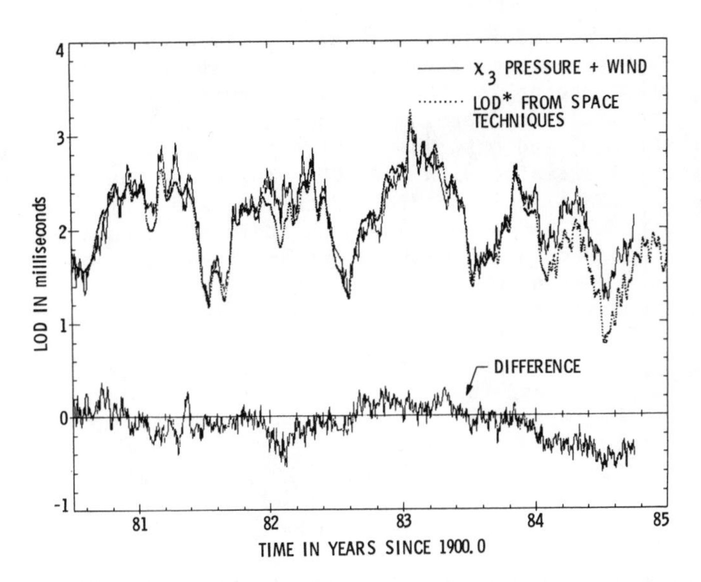

Figure 1 a. NMC AAM estimates (solid line) together with Kalman smoothed LOD* estimates from the space techniques (dotted line) and the difference (LOD* - AAM). An arbitrary bias has been removed from the AAM data set and the difference.

the rapid fluctuations in the AAM, there is a slow divergence between the two time series. This divergence is mostly due to LOD* changes, and is generally assumed not to be of atmospheric origin. Our research (see Section 4.4) suggests, however, that the atmosphere (or oceans) may account for part of the LOD* trend seen in Figure 1 a.

The unusual maxima in the LOD* and the AAM in January and February, 1983, are clearly visible in Figure 1 a. The simultaneous occurance of these events with the record low in the Tahiti-Darwin SOI led Carter et al. (1984) and Rosen et al. (1984) to hypothesize a connection between rapid changes in the LOD* and the EL Nino phenomena. These LOD* changes represent an absorption and then loss by the atmosphere of ~ $3 * 10^{25}$ kg m^2 sec^{-1} of eastward angular momentum during a period of less than a month, and thus required earth-atmosphere torques of more than $2 * 10^{19}$ kg m^2 sec^{-2} (or 20 Hadley Units). Rosen et al. (1984) speculated that these torques could be directly driven by the surface pressure changes associated with the Southern Oscillation, but Eubanks et al. (1985b) showed that this is probably not the case, since this hypothesis predicts a correlation between the SOI and the time derivative of the AAM, which is not observed. Rosen et al. (1984) found that most of the anomalous increase in the AAM was caused by Northern Hemisphere winds, between 0° and 40° North, and it thus seems reasonable to suspect that the LOD* burst in early 1983 is associated with changes in the velocity or location of the subtropical jet stream in the Northern Hemisphere.

The rapid burst in the AAM and LOD* in early 1983 was clearly an

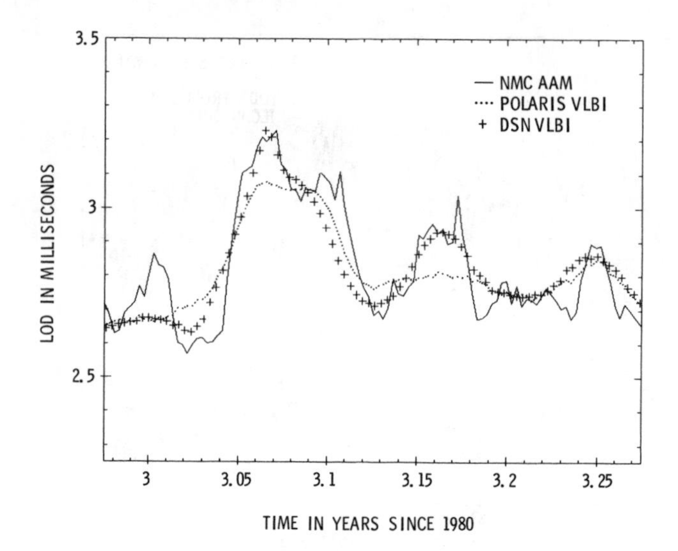

Figure 1 b. NMC AAM estimates (solid line) together with Kalman smoothed LOD* estimates from the POLARIS project (dotted lines) and from the DSN (crosses) during early 1983.

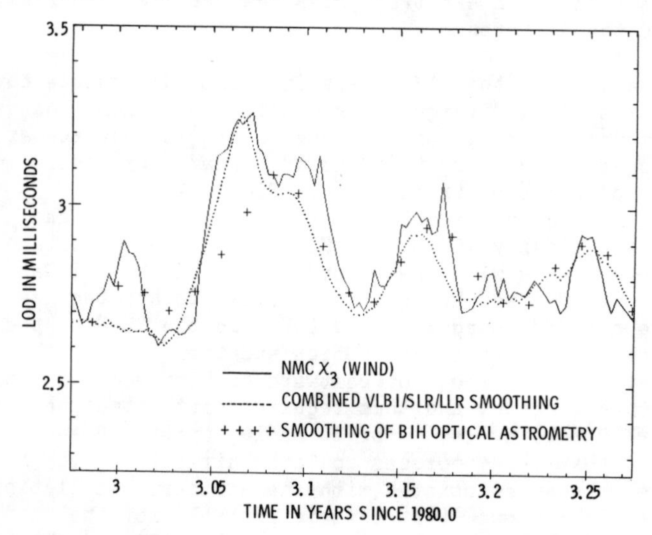

Figure 1 c. NMC AAM estimates (solid line) together with combined Kalman smoothed LOD* estimates from the space techniques (dotted lines) and from optical astrometry as reduced by the BIH (crosses) during early 1983.

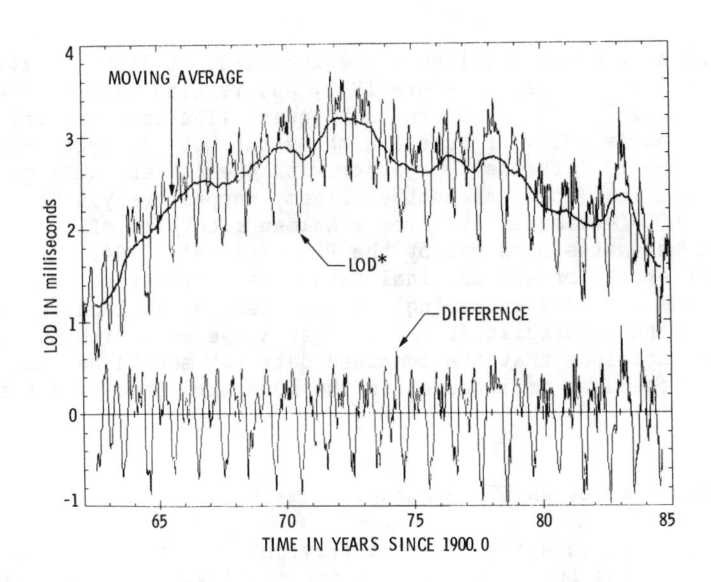

Figure 2. The LOD* estimates from a combined Kalman smoothing of all techniques (top), together with a 365 day rectangular moving average (dark line) and the difference (bottom).

unusual event, producing the largest values yet recorded for the axial angular momentum of the atmosphere (Rosen et al, 1984). This burst is ~ 4 standard deviations from the seasonal variation for its time of year; the largest such deviation reported for either the LOD or the AAM for any time of year (Eubanks et al., 1985 b). The Earth's rotation during early 1983 was monitored by at least 4 different observing networks, and it is possible to describe the nature of the burst in unprecedented detail. Figure 1 b presents an enlarged view of LOD* and AAM changes during early 1983. (Note that each horizontal division represents only 3.6 days). The AAM data set is the same as in Figure 1 a, while the LOD* has been estimated independently using VLBI data from the DSN (crosses) and POLARIS (dotted line). The SLR polar motion data were used to improve the smoothing of the DSN data. The DSN had, fortuitously, scheduled 4 observing sessions, including two of ~ 24 hours duration, within a few days of the peak in the AAM data. The quality of these data, together with the unusually large and rapid changes in Earth rotation, reveal the Earth's response to atmospheric torques at unusually high frequencies. The peak in the NMC axial AAM data is at the leading edge of the burst, and this peak itself has two maxima, on January 23 and 26. The peak in the smoothed DSN VLBI data is on January 24, and there is no evidence of phase lags or leads between these time series. The POLARIS network had not attained full operation by early 1983, and these data were thus not able to resolve the fine structure in the burst. The main AAM event was preceded and followed by smaller variations, and these are reasonably well matched by the LOD* data sets.

Optical astrometry provides the sole source of Earth rotation measurements for the period before 1970, and is also important during gaps in the data from more modern techniques. (The last significant gaps were in early 1981.) Figure 1 c shows the LOD* derived from optical astrometry (crosses), together with the AAM and LOD* data sets used in Figure 1 a (solid and dotted lines, respectively). The astrometric LOD* estimates are from a Kalman smoothing of five day average UT1 estimates produced by the BIH (Feissel, 1984), and are shown at the epochs of the original data. The astrometric data (perhaps because of the five day averaging) do not resolve the leading edge structure of the main burst, but otherwise agree well with the VLBI and AAM data. We conclude that the combined data set should be able to resolve most features of interest in the LOD* throughout the analysis period.

4.2. Construction of an LOD Seasonal Anomaly Index

The existing Earth rotation data set provides a historical record of past changes in the LOD*, and thus in the axial AAM. Figure 2 shows, at the top, the full combined LOD* smoothing. Note that the LOD* gradually increased until ~ 1972, and then commenced a period of decline that has lasted until the present. This overall trend in the LOD* seems too large to reasonably be due to atmospheric circulation changes (it is considerably larger than the known features of atmospheric origin), and is probably part of the decade fluctuations originating in the core. There are also smaller and more rapid interannual variations, most easily visible in the 365 day moving average in Figure 2, that could plausibly originate in torques either at the core-mantle boundary or at the surface.

We constructed an LOD* Anomaly Index (LAI) to isolate the LOD* changes possibly resulting from free atmospheric oscillations for intercomparison with other meteorological indicies. The combined LOD* data set was smoothed with a 365 day rectangular moving average (heavy line in Figure 2), selected to totally reject all harmonics of one year. The difference between the LOD* data set and this moving average (bottom of Figure 2) thus contains all yearly harmonics, as well as the irregular rapid oscillations. We then constructed an average seasonal cycle for the LOD* dividing the calendar year into 73 bins, and averaging together the moving average residuals located in the same bin during successive years (Eubanks et al. 1985 b). This seasonal average thus contains only the yearly harmonics of the LOD* data. Since we are interested in the departures from the seasonal cycle, we subtracted the seasonal average from the original data set to form seasonal anomalies. We fit a quadratic polynomial to these LOD* anomalies, and used the residuals from this fit for our LAI. The polynomial fit is a purely ad hoc procedure to remove the roughly quadratic trend seen in Figure 2 without removing the interannual variations of possible atmospheric origin. A similar anomaly index was constructed from the axial AAM data, but without removal of a polynomial trend.

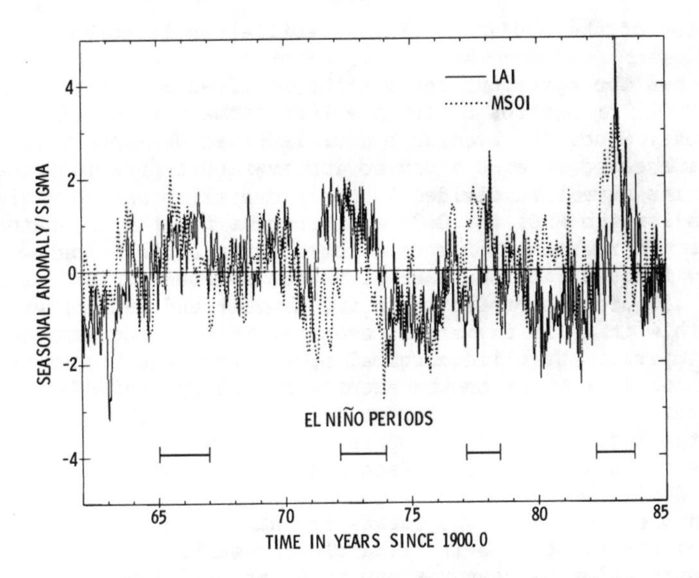

Figure 3 a. Monthly mean values of the LOD* Anomaly Index (solid line) together with monthly mean values of the modified (Darwin - Tahiti) Southern Oscillation Index. Both series have been scaled to have a RMS value of unity. The approximate dates of recent El Nino events have been marked at the bottom on this and Figures 3 b and 5.

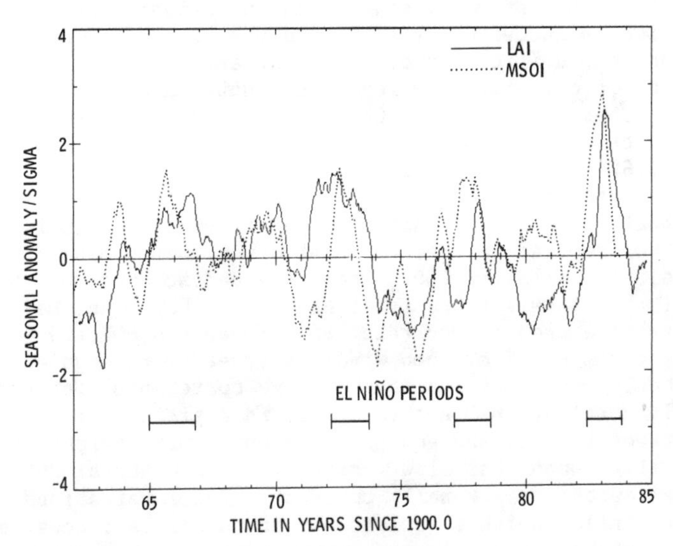

Figure 3 b. The same indicies from Figure 3 a. after smoothing by a 150 day rectangular moving average.

4.3 Comparison of the LOD and Southern Oscillation Indicies

Figure 3 a shows the agreement between the Modified Southern Oscillation Index (MSOI) of Section 3, and the LOD* Anomaly Index (LAI) discussed above. Since the available MSOI is based on monthly averages, the LAI was subjected to a 30.4 day moving average before intercomparison. Both series were then divided by their overall standard deviation of 1.80 mbars for the MSOI and 0.25 milliseconds for the LAI before presentation. There seems to be some correlation between the two series at low frequencies, together with uncorrelated "noise" at high frequencies. Figure 3 b presents the LOD anomaly and MSOI after smoothing both with a 150 day moving average. This moving average was selected to separate the climatological signal from the higher frequency noise and is similar to the smoothing used by the NMC Climate Analysis Center (Quiroz 1983). Figure 3a and 3b, together with Figure 5, indicate the approximate dates of recent El Nino events, defined by the existence of Darwin-Tahiti surface pressure anomalies of more than one standard deviation.

Although the atmosphere and oceans are non-linear systems, much of the progress achieved to date in climatology results from studies of the linear correlation between various time series. Figure 4 a presents autocorrelation estimates for the two series in Figure 3 a. Both series have roughly similar autocorrelations, and appear to behave like exponentially correlated random variables with 1 / e correlation times of ~ 200 days. The crosscorrelation between the LAI and the MSOI is given in Figure 4 b for both the monthly and the five month average series. There is a statistically significant correlation between the LAI and MSOI which is increased by smoothing the data. The maximum crosscorrelation is ~ 0.5 for the smoothed series, for an LAI lag of ~ 100 days, which is comparable to lags found between South Pacific pressure series (Chen 1982), and by Chao (1984). By contrast, the crosscorrelation between the monthly mean pressure anomalies at Darwin and Tahiti is -0.67 (Horel and Wallace 1981), and so perfect LAI-MSOI correlation is not expected.

We think that it is likely that the Southern Oscillation is associated with widespread changes in zonal winds, which cause the low frequency correlation with the LOD. Section 4.4 describes our attempts to validate this hypothesis using axial AAM data. Pazan and Meyers (1982) demonstrated a negative correlation between the SOI and eastward winds in the tropical Pacific. Such winds carry eastward angular momentum and thus should contribute to the negative correlation between the LAI and the SOI. Horel and Wallace (1981) found a similar negative correlation between the SOI and equatorial geopotential heights at 200 millibars, and also found that slow variations in equatorial 200 millibar geopotential heights are zonally coherent. Zonal variations in equatorial geopotential heights, unless common to all latitudes, should cause zonal wind changes and this work thus indicates that 200 millibar winds away from the equator should also contribute to the MSOI-LAI correlation.

Away from the equator, zonal winds and temperatures approximately satisfy the thermal wind relationship (see, e.g., Holton 1979), in which the vertical zonal wind shear is proportional to the meridional temperature gradient. Such thermal wind shears also depend on the sine of the latitude, and thus increase strongly away from the equator. Given the general decrease in temperature with height in the troposphere the total thermal wind also typically increases with altitude up to the base of the stratosphere, at ~ 100 to 200 millibars. As discussed in Section 3, the Southern Oscillation is associated with discharges of heat from the equator towards the poles. Variations in zonal thermal winds caused by discharges of heat from the equatorial Pacific during El Nino events could thus cause some or all of the associated changes in the duration of the day. This is essentially the hypothesis advanced by Stephanic (1982), except for the recognition that heat is transported away from the tropics and it is not necessary that the induced thermal winds be near the equator. It is also possible that some of the observed low frequency variations arise from instabilities at midlatitudes, which could be enhanced by increases in the total heat transport without being coherent with those changes (Legras and Ghil 1985).

There is an apparently significant negative correlation between the MSOI and the LAI at an LOD* lead of ~ 900 days (Figure 4 b), roughly half the typical recurrence time of the Southern Oscillation, but only 11 % of the length of the analysis period. (The crosscorrelation was calculated using a so-called biased correlation estimator (Priestley 1981, Equation 9.5.3), which has a roughly constant error for small lags.) This apparent correlation is especially interesting because it cannot be explained by peaks in the autocorrelations of the two series (see Figure 4 a). Such a relationship, if proven, would mean that the Earth's rotation might be able to predict El Nino events, and thus could be of importance to long range weather forecasting. Further research is needed to explore the significance of this apparent correlation.

Lau (1985) and Lau and Chan (1985) speculate that the El Nino events may be triggered by the 40 to 60 day oscillation in equatorial convection and zonal winds, through a close connection with the westerly (eastward propagating) tropical Pacific surface wind bursts described by Luther et al. (1983) and Lukas et al. (1984). The tropical surface wind bursts undoubtedly play an important role in the El Nino phenomenon through directly forcing the transfer of warm water from the West to the East Pacific, one important stage of an El Nino event (Lukas et al. 1984). It is thus tempting to associate these surface wind bursts with the 40 to 60 day oscillations of the LOD*, but the bursts are intermittent while the rapid LOD* oscillations are almost always present (Eubanks et al. 1985a and Feissel and Nitschelm 1985). Figure 5, which contains the difference between the raw and smoothed LOD anomalies in Figures 3 a and b, is a record of rapid changes in the LAI. The extreme nature of the rapid burst in 1983 can be clearly seen, as can the almost continuously present smaller oscillations. This figure also indicates, at the bottom, the El Nino events for the period, and, at the top, the strong westerly bursts in equatorial Pacific sea level winds, from Luther et al. (1983) and Lukas et al. (1984). (A strong burst is defined as having zonal surface winds > 8 m

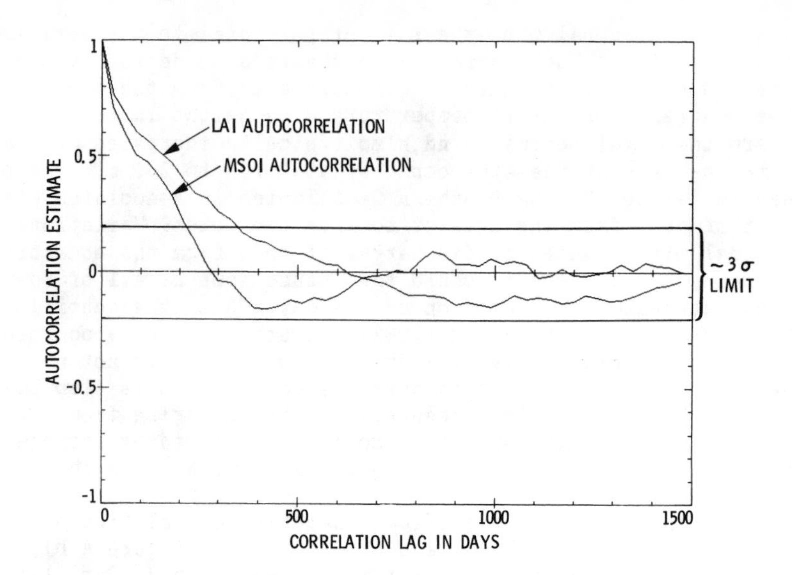

Figure 4 a. The auto-correlation of the raw monthly anomaly indicies in Figure 3 a. Approximate "3 sigma" confidence limits are indicated on the right.

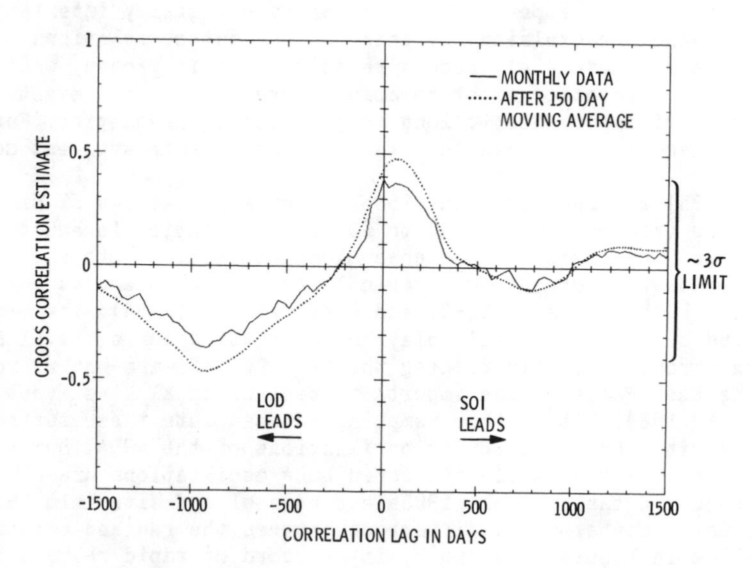

Figure 4 b. The cross-correlation of the anomaly indicies in Figure 3 a (solid line), and of the smoothed indicies in Figure 3 b (dotted line). Approximate "3 sigma" confidence limits for the smoothed data are indicated on the right.

/ sec eastwards at longitude 170° E after the low pass filtering in Luther et al. (1983)). Some of the positive LAI bursts seem to be associated with westerly wind bursts, such as in late 1977 and in 1982-83. On the whole, though, the correlation between EL Nino events, westerly wind bursts and LOD* bursts does not seem striking.

As discussed in Section 4.1, there are a number of indications that the LOD* burst in early 1983 was related to the 1982-1983 El Nino event. The LOD* burst might be directly caused by thermal winds created by the heat discharged from the Pacific, or the excess heat transport might decrease the stability of the northern hemisphere jet stream, causing large internal variations. The 1982-1983 El Nino event was associated with the passage of a surge of warm water across the equatorial Pacific during most of 1982. This surge reached the coast of South America in late September, 1982, and equatorial sea surface temperatures rose ~ 2° C / month at the coast through the rest of 1982. Lukas et al. (1984) show that the associated eastward wind bursts in the tropical Pacific start around August, 1982, and peak and end in November of that year, still two months before the LOD* peak on January 24, 1983. The lag between the beginning of the dispersion of the warm equatorial water and the LOD* burst probably represents the time for the excess heat to go from the equator to the subtropical jet streams, where warm air from the tropics meets cold air from the polar regions and where the thermal gradients are thus largest. The implied meridional propagation velocity is ~ 10 to 20° of latitude / month, or ~ 0.5 to 1 meter / second. Similar meridional propagation velocities have been observed in other studies of the 40 to 60 day oscillation, both empirically (Krishnamurti and Gadgil 1985), and in numerical models (Goswami and Shukla 1984). The eastward propagating thermal winds created by an increase in the equator to pole temperature gradient carry positive axial angular momentum, and therefore should cause an increase in the axial AAM and LOD*, exactly as observed. The winter hemisphere has the coldest polar air and thus the largest thermal gradients, and thermal winds tend to be stronger in winter. The axial AAM burst in January, 1983, was mostly caused by zonal wind changes in the northern (winter) hemisphere at midlatitudes (Rosen et al. 1984), again consistent with the thermal wind hypothesis. The 1982 equatorial wind bursts peak in November, which is unusual, since most such bursts occur near the New Year. This suggests a possible dependence of the strength of the resulting LOD burst on the time of year the propagating disturbance reaches the mid latitude jet streams. Dependence of response upon season could thus explain the lack of strong LOD* bursts during other El Nino events.

It seems likely that the rapid oscillations in the LOD* have a number of different causes. In models at least, instabilities can cause periodic or aperiodic flow variations even given constant heat transport forcing (Legras and Ghil 1985). The 40 to 60 day oscillations in the axial AAM and the LOD* could be caused by rapid changes in heat transport unrelated to the Southern Oscillation, since there is a 40 to 60 day free oscillation in surface convection and air temperatures in the tropics (Goswami and Shukla 1984). In any case, there does not appear to be any consistent connection between El Nino

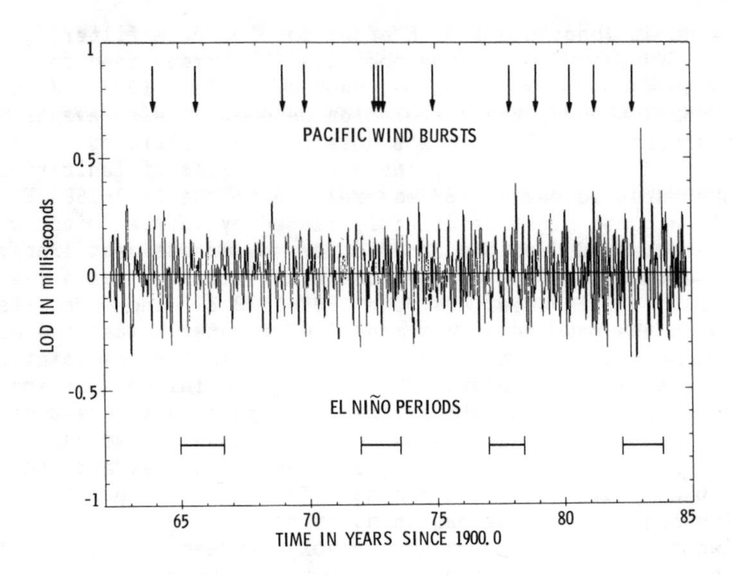

Figure 5. The difference between the raw monthly LOD* anomaly index in Figure 3 a and the moving average in Figure 3 b. At the top are the equatorial wind bursts of Luther et al. (1983).

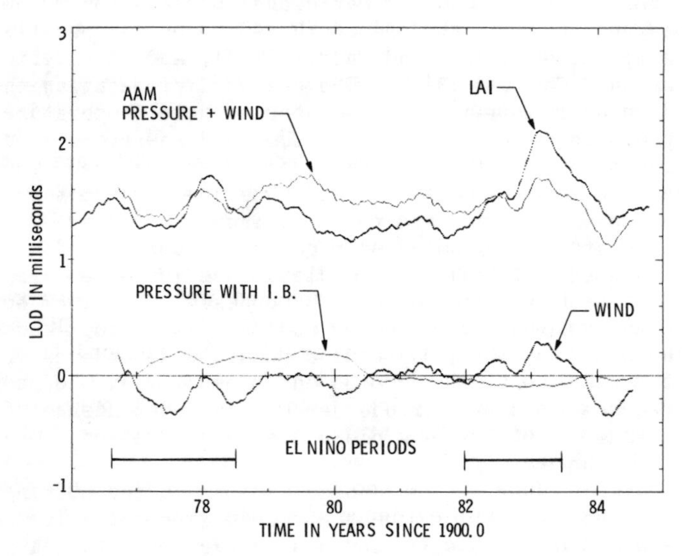

Figure 6. The LOD* seasonal anomaly index (heavy line, top), together with smoothed seasonal anomalies from the combined NMC AAM data set (light line, top), and the seasonal anomalies of the NMC estimates of the wind (dark line, bottom) and pressure (light line, bottom) contributions to the axial AAM.

events and the rapid LOD* variations (Figure 5). It thus seems probable that the observed high frequency LOD* changes consist of occasional rapid bursts caused by EL Nino generated temperature gradients, together with more or less continuous oscillations from another source.

4.4 Atmospheric Excitation of Long Period LOD Changes

Numerous studies have shown the close correlation between LOD* changes and meteorological estimates of the axial AAM at seasonal and higher frequencies, and it is natural to expect the AAM data to shed some light on the interannual changes in the LOD*. We compared the LAI with similar anomaly indicies constructed from the NMC χ_3 estimates, which start in 1976. As mentioned in Section 4.2, polynomial trends were not removed from the meteorological indicies. Figure 6 shows, at bottom, the anomaly indicies from the pressure plus inverted barometer (light line) and wind (heavy line) components of the NMC χ_3, and, at the top, the total χ_3 index (light line) together with the LAI (dark line). (The two curves at the top were arbitrarily displaced for graphical purposes and all of these curves have been smoothed with a 150 day moving average.) Most of the observed interannual AAM variations are caused by variations in zonal winds. The observed AAM anomalies are clearly correlated with the LAI, but have a considerably reduced amplitude. Note also the more rapid oscillations in the AAM wind anomalies, which agree well with LAI features. It seems obvious that further study of the QBO in the LOD* requires subtraction of the larger, but slower, effects of the Southern Oscillation.

Figure 6 indicates that the observed axial AAM may not account for all of the interannual changes in the LOD*. We tested this possibility by using the smoothed LAI in a linear regression for an admittance multiplying the smoothed NMC axial AAM anomalies. The least squares solution also included a quadratic polynomial trend in the LAI for the period of overlap (mid-1976 to mid-1984). The resulting admittance estimate is 1.43 $\pm$ 0.27, after adjusting the formal error to account for the scatter in the postfit residuals as well as the correlation introduced by smoothing. Use of the smoothed AAM seasonal anomalies does explain much of the variation seen in the LAI. The RMS scatter of the smoothed LAI during the period of overlap was reduced from 0.195 msec (after subtraction of the best fit polynomial trend) to 0.134 msec (after sutraction of the smoothed NMC anomalies and a polynomial trend). Introduction of the admittance parameter reduces the postfit residuals still further, to 0.126 msec. The axial AAM can thus explain roughly half the variance in the smoothed LAI, and there is some (statistically inconclusive) evidence that the smoothed NMC anomalies under report interannual changes in the axial AAM.

Figure 1 a provides further evidence that the interannual LAI changes correlated with the MSOI are not accounted for by the NMC axial AAM measurements. The LOD*-AAM difference at the bottom of that figure rises throughout most of 1982 and falls during 1983 and 1984. These interannual changes are clearly coincident with the on-going El Nino event, and our research has shown that the corresponding interannual

LOD* changes are correlated with Pacific sea level pressure indicies, which implies that the coincidence is not accidental. There is no known physical mechanism to relate the Southern Oscillation with torques at the core-mantle boundary, and so any related angular momentum is presumably exchanged through torques at the Earth's surface. This suggests either interannual errors in the NMC axial AAM estimates, or exchanges of angular momentum with some other source. The NMC integration excludes zonal winds above the 50 millibar level (Equation 3). Rosen and Salstein (1985) show that these levels contribute ~ 0.06 milliseconds to the annual LOD variation, small compared to the total annual variation, and so the stratosphere likely contributes some, but not all, of the missing angular momentum. The Southern oscillation is certainly accompanied by changes in ocean circulation, but changes in the Antarctic Circumpolar Current transport (with the most axial angular momentum of any ocean current) of ~ 100 * 10^9 kg / sec (or roughly the total transport) would be necessary to explain the differences observed in Figure 1 a, and this seems unreasonably large (Wearn and Baker 1980). Most of the changes in sea level associated with El Nino events seem to be due to thermal expansion, and thus would not affect the Earth rotation (A.E. Gill, personal communication). It seems most likely that any missing angular momentum results from attenuation of very slow changes in wind fields by atmospheric models designed for 5 day forecasts. Such errors might be totally masked in forecast statistics by the day to day and week to week variations, and their detection could lead to improvements in forecast models.

5. CONCLUSIONS

It now seems fairly clear that there is a relationship between the Southern Oscillation and variations in the duration of the day, although the exact mechanism is still unclear and much work remains to be done. Data from the new technique of radio interferometry make it possible to examine the rapid changes in the duration of the day in early 1983 with unprecedented clarity, and such events should be even better observed in the future. Although the LOD burst in early 1983 occurred at the peak of the 1982-83 El Nino event, and only a few months after bursts in the tropical Pacific surface winds, there does not seem to be a consistent relation between other such events during the analysis period. It is quite possible that the relation between El Nino events and the duration of the day depends upon the phase of the seasonal cycle, and further research is needed to disentangle these effects. By contrast, interannual changes in the LOD* are linearly related to a Southern Oscillation sea level pressure index. The most likely physical mechanism for this relationship seems to be thermal winds caused by the changes in the equator to pole heat transports associated with El Nino events and the Southern Oscillation. The available measurements of the axial angular momentum of the atmosphere explain some, but probably not all, of the interannual variations of the LOD*; further research is needed to decide if this possible discrepancy is due to errors in the weather forecasting models, or to other sources of angular momentum.

ACKNOWLEDGEMENTS. The work described in this paper was carried out by the Jet Propulsion Laboratory, California Institute of Technology, under contract with the National Aeronautics and Space Administration. T. M. Eubanks wishes to thank the hospitality of the BIH and the Observatoire de Paris during his visit there.

REFERENCES

Anderson, J.R. and Rosen, R.D., 1983, 'The Latitude-Height Structure of 40-50 Day Variations in Atmospheric Angular Momentum', J. Atmos. Sci., 40, 1584-1591.

Barnes, R.T.H., Hide, R., White, A.A., Wilson, C.A., 1983, 'Atmospheric Angular Momentum Fluctuations Correlated With Length Of Day Changes and Polar Motion', Proc. Roy. Soc. London, Series A, 387, 31-73.

Cane, M., 1983, 'Oceanographic events during El Nino', Science, 222, 1189-1195.

Carter, W. E., Robertson, D. S., Pettey, J. E., Tapley, B. D., Schutz, B. E., Eanes, R. J., and Lufeng, M., 1984, 'Variations in the Rotation of the Earth,' Science, 224, 957-961.

Chao, B.F., 1984, 'Interannual Length-of-Day Variations with relation to the Southern Oscillation / El Nino', Geophys. Res. Lett., 11, 541-544.

Chao, B.F., 1985, 'Excitation of the Earth's Chandler wobble by Southern Oscillation / El Nino, 1900-1979', NASA Technical Memorandum 86231.

Chen, W.Y., 1982, 'Assesment of Southern Oscillation Sea-Level Pressure indices', Mon. Weather Rev., 110, 800-807.

Dickey, J.O., Newhall, X X, and Williams, J.G., 1985, 'Earth rotation from Lunar laser ranging', Bureau International de l'Heure, Annual Report for 1984, D69-D72.

Eubanks, T.M., Dickey, J.O., and Steppe, J.A., 1983, 'The Geophysical Significance of Systematic Errors in the Earth's Angular Momentum Budget,' Proc. IAG Symp., IUGG XVIIIth general assembly, vol. 2, 122-143, Ohio State University, Columbus.

Eubanks, T.M., Steppe, J.A., Dickey, J.O., and Callahan, P.S., 1985a, 'A Spectral Analysis of the Earth's Angular Momentum Budget,' J. Geophys. Res., 90, B7, 5385-5404.

Eubanks, T.M., Dickey, J.O., and Steppe, J.A., 1985b, 'The 1982-83 El Nino, the Southern Oscillation, and Changes in the Length of Day,' Trop. Ocean and Atmos. Newsletter, 29, 21-23.

Eubanks, T.M., J.A. Steppe, and J.O. Dickey, 1985c, 'The Atmospheric Excitation of Earth Orientation Changes During MERIT,' 1985c, Proc. Inter. Conf. Earth Rotation and Terrestrial Reference Frame, Vol. 2, 469-483, Ohio State University, Columbus.

Eubanks, T.M., Steppe, J.A., and Spieth, M.A., 1985d, 'The accuracy of radio interferometric estimates of Earth rotation', DSN TDA Progress Report, 42-80, 229-235.

Eubanks, T.M., J.A. Steppe, and O.J. Sovers, 1985e, 'An analysis and intercomparison of VLBI nutation estimates', Proc. International Conf. on Earth Rotation and the Terrestrial Reference Frame, Vol. 1, 326-340, Ohio State University, Columbus, Ohio.

Eubanks, T.M., Steppe, J.A., and Spieth, M.A., 1985f, 'Earth orientation results from DSN VLBI at JPL', Bureau International de l'Heure, Annual Report for 1984, D19-D23.

Feissel, M., and Nitschelm, C., 1985, 'Time-dependent aspects of the atmospheric driven fluctuations in the duration of the day', Ann. Geophys., 3, 181-186.

Feissel, M., 1984, 'Earth rotation from optical astrometry', Bureau International de l'Heure, Annual Report for 1983, D3-D8.

Gill, A.E., and Rasmusson, E.M., 1983, 'The 1982-83 climate anomaly in the equatorial Pacific', Nature, 306, 229-234.

Goswami, B.N., and Shukla, J., 1984, 'Quasi-Periodic Oscillations in a Symmetric General Circulation Model', J. Atmos. Sci., 41, 20-37.

Gross, R.S., and Chao, B.F., 1985, 'Excitation Study of the Lageos-derived Chandler wobble', J. Geophys. Res., 90, 9369-9380.

Holton, J.R., 1979, An introduction to dynamic meteorology, Academic Press, New York.

Horel, J.D., and Wallace, J.M., 1981, 'Planetary-scale atmospheric phenomena associated with the Southern Oscillation', Mon. Weather Rev., 109, 813-828.

Krishnamurti, T.N. and Gadgil, S., 1985, 'On the structure of the 30 to 50 day mode over the globe during FGGE', Tellus, 37A, 336-360.

Lambeck, K., 1980, The Earth's Variable Rotation, Cambridge University Press.

Lau, K.M., 1985, 'Elements of a stochastic-dynamical theory of the long-term variability of the El Nino/Southern Oscillation', J. Atmos. Sci., 42, 1552-1558.

Lau, K.M., and Chan, P.H., 1985, 'Aspects of the 40-50 day oscillation during the northern winter as inferred from outgoing longwave radiation', Mon. Weather Rev., 113, 1889-1909.

Legras, B. and Ghil, M., 1985, 'Persistent Anomalies, Blocking and Variations in Atmospheric Predictability', J. Atmos. Sci., 42, 433-471.

Le Mouel, J.L., Madden, T.R., Ducruix, J., and Courtillot, V., 1981, 'Decade fluctuations in geomagnetic westward drift and Earth rotation', Nature, 290, 763-765.

Li, Z., 1985, 'Earth Rotation from Optical Astrometry, 1962.0-1982.0', Bureau International de l'Heure, Annual Report for 1984, D31-D63.

Lukas, R., Hayes, S.P., and Wyrtki, K., 1984, 'Equatorial Sea Level Response During the 1982-1983 El Nino', J. Geophys. Res., 89, 10425-10430.

Luther, D.S., Harrison, D.E., and Knox, R.A., 1983, 'Zonal Winds in the Central Equatorial Pacific and El Nino', Science, 222, 327-330.

Morgan, P.J., King, R.W., and Shapiro, I.I., 1985, 'Length of day and atmospheric angular momentum : a comparison for 1981-1983', J. Geophys. Res., 90, 12645-12652.

Pazan, S.E., and Meyers, G., 1982, 'Interannual fluctuations of the tropical Pacific wind field and the Southern Oscillation', Mon. Weather Rev., 110, 587-600.

Philander, S.G.H. 1983, 'El Nino Southern Oscillation phenomena', Nature Vol. 302, 295-301.

Priestley, M.B., 1981, Spectral Analysis and Time Series, Academic Press, London.

Quiroz, R.S., 1983, 'The Climate of the 'El Nino' Winter of 1982-83; Mon. Weather Rev., 111, 1685-1706.

Rasmusson, E.U., Wallace, J.M., 1983, 'Meteorological Aspects of the El Nino/Southern Oscillation', Science, 222, 1195-1202.

Robertson, D.S., Carter, W.E., Campbell, J. and Schuh, H. 1985, 'Daily Earth rotation determinations from IRIS very long baseline interferometry,' Nature, 316, 424-427.

Rosen, R.D. and Salstein, D.A., 1983, 'Variations in Atmospheric Angular Momentum on Global and Regional Scales and the Length of Day', J. Geophys. Res., 88, 5451-5470.

Rosen, R.D., Salstein, D.A., Eubanks, T.M., Dickey, J.O., and Steppe, J.A., 1984, 'An El Nino signal in atmospheric angular momentum and Earth rotation, Science, 225, 411-414.

Salstein, D.A. and Rosen, R.D., 1985, 'Earth Rotation as a proxy index of global wind fluctuations,' Third Conference on Climate Variations, American Meteorological Society, Boston.

Spieth, M.A., Eubanks, T.M., and Steppe, J.A. 1985, 'Intercomparison of UT1 Measurements during the MERIT campaign period,' Proc. Inter. Conf. Earth Rotation and Terrestrial Reference Frame, Vol. 2, 609-621, Ohio State University, Columbus.

Stephanic, M., 1982, 'Interannual atmospheric angular momentum variability 1963-1973 and the southern oscillation', J. Geophys. Res., 87, 428-432.

Tapley, B.D., Schutz, B.E., and Eanes, R.J., 1985, 'Station coordinates, baselines, and Earth rotation from Lageos laser ranging: 1976-1984', J. Geophys. Res., 90, 9235-9248.

Trenberth, K.E., 1983, Signal Versus Noise in the Southern Oscillation', Trop. Ocean and Atmos. Newsletter, 20, 1-3.

Wearn, R.B. and Baker, D.J., 1980, 'Bottom pressure measurements across the Antarctic Circumpolar current and their relation to the wind', Deep Sea Res., 27A, 875-888.

Wyrtki, K., 1985, 'Water Displacements in the Pacific and the Genesis of El Nino Cycles', J. Geophys. Res., 90, 7129-7132.

Yoder, C.F., Williams, J.G., Parke, M.E., 1981, 'Tidal Variations of Earth Rotation', J. Geophys. Res., 86, 881-891.

THE ATMOSPHERIC CIRCULATION, THE EARTH'S ROTATION AND SOLAR ACTIVITY.

D. Djurovic

Department of Astronomy, University of Belgrade
Studentski trg 16, Belgrade

Nowadays we dispose of some new observational proofs in favour of the hypothesis that solar activity contributes to the variations of the angular velocity of the Earth's rotation (ω).

Since the "Vanguard" and "Explorer" satellite experiments we know that the density variations in the upper atmosphere are dependent of the extreme ultraviolet solar radiation, which is closely correlated to the Sunspot activity (Jacchia 1967). For exemple in 1963, near the minimum of sunspots, it has been discovered by "Vanguard-2" that over the dark hemisphere and at the altitude $h \approx 600$ km the air density was about 100 times less than over light hemisphere near the last sunspots maximum of 1959.

From satellites observations it has also been discovered that the geomagnetic storms are followed by the atmospheric density variations.

Such variations, induced by solar radiation, were observed at altitudes over 100 km. Therefore, the contribution of solar radiation, considered directly in relation with the tropospheric circulation and, consequently, with the Earth's rotation, is not revealed by the observations. On the other side, it has nevertheless been demonstrated that solar activity causes the variations of the geomagnetic field and the upper atmospheric density (Jacchia, 1967; Keating and Levin, 1972).

The cyclic variations of solar activity represent a phenomenon of permanent duration. Still nowadays we are not sure whether they may have a cumulative influence on the tropospheric density and cause a weak, still nonobservable, zonal circulation which may be masked, for exemple, by the large convective motions near the ground. Therefore, solar activity, as one possible origin of ω variations, strictly speaking, is not proved, nor excluded.

During some ten years ago it was proved that the contribution of the global atmospheric circulation on ω may be important, particularly in the range of short-periodic and irregular fluctuations (Lambeck

A. Cazenave (ed.), Earth Rotation: Solved and Unsolved Problems, 187–192.

and Cazenave, 1974; Lambeck and Hopgood, 1981; Okazaki, 1977, 1979; Hide et al., 1980; Djurovic, 1983; etc ...). An interesting contribution to the above mentioned conclusion was the discovery of the 55-day quasi-cyclic variation, common to the zonal component of the AAM and ω, made by Feissel and Gambis (1980), and confirmed later by Langley et al. (1981) and by Djurovic (1983). This last variation will be the central point of our considerations in this paper.

In Fig. 1, published in Djurovic (1983), we reproduce the smoothed residuals of UT1-TAI (a), obtained after the elimination on the 3rd order polynomial and 5 sinusoïdal terms (biannual, annual, semi-annual, 4-mth term), and the smoothed residuals of AAM (b), obtained after elimination of the annual and semi-annual term. The method of smoothing is the method of Whitaker-Robinson-Vondrak.

Despite the small amplitudes of the two quasi-cyclic variations, we see that they have approximately the same period and the same phase. The phase difference estimated to 17° ± 13° could be explained by the inaccuracy of the measurements.

Nevertheless and from the numerical evaluation (see, for instance, Djurovic 1983) we conclude that the 55-day fluctuation of UT1-TAI is induced by the 55-day fluctuation of AAM.

All the previous mentioned conclusions, being known our aim will be the investigation of the relationship between the solar activity and the 55-day variation of AAM and UT1-TAI. For this purpose we have analysed the spectra of the geomagnetic index A_p and the Wolf number W.

The data used in this paper are daily W and A_p for the period 1965-1978, published in "Die Sterne" and "Observations solaires".

Let ΔW represents the residuals of the observed W after the elimination of all sinusoïdal terms in the range of frequencies $f \leq 2c/yr$. This elimination has been made by the triple Labrouste transform of type (Labrouste et Labrouste, 1943).

In Fig. 2 the graphs a, b, c represent the spectra of ΔW computed over 4-yr period of observations (on right hand side is given the middle epoch of these subintervals), the graphs d, e, f - the spectra of ΔW computed over 3-yr period of observations.

Beside the 29-day peak, which is due to the synodic Sun's rotation, we note a second one just over 75 days. Both peaks also exist in the spectra of the geomagnetic index A_p (Fig. 3). In Fig. 3, the spectra of A_p for the period 1966-1974 and for 3-yr subintervals are plotted.

Therefore, the observational data analysis reveals the weak 75-day cyclic variation of the solar activity and the geomagnetic field. The question which could be imposed now is : whether 55-day variation of UT1-TAI and AAM has a causal connection with the 75-day variation of W and A_p ? This question, as we will see, is quite realistic.

Let ΔW_T represents the transformed values of ΔW, computed by the Labrouste's method of type $\pi(T) = T_p\ T_2\ T_r\ T_s$. This method does not transform the time function into the frequency function, like spectral analysis methods. It allows to estimate how the period and the amplitude of the signal behave in function of time.

The residuals ΔW_T are presented in Fig. 4. The first remark is that the 75-day variation is not strictly cyclic. Its amplitude varies between A = 5 and A = 23. Moreover the period of this term, which is more important for our concern, is also variable and is ranging from a minimum of 51 days to a maximum of 102 days. Therefore, the 55-day and the 75-day variations may represent only one, which is common to the solar activity, the atmospheric angular momentum, the geomagnetic field and the Earth's rotation. The precision and the accuracy of data which we have had at our disposal have not been sufficient for the more exact period and phase determination, and, consequently, for the gain of the more convincing arguments in favour of our hypothesis.

However, the above presented results, considered together with the other ones on the same subject (see, for example, Djurovic and Stajic, 1985) get larger value.

Having in mind the number of papers, published on the spectral structure of W and A_p, the suggestive question is : how is it possible that the 75-day cyclic variation was remaining undetected ?

Of course the amplitude of 75-day component is very small. But at this fact we append a second one : its frequency is unstable. Therefore, the energy of the weak 75-day signal in the spectra of W and A_p could be dissipated on the larger range of frequences or on the few successive peaks and sunk in the noise.

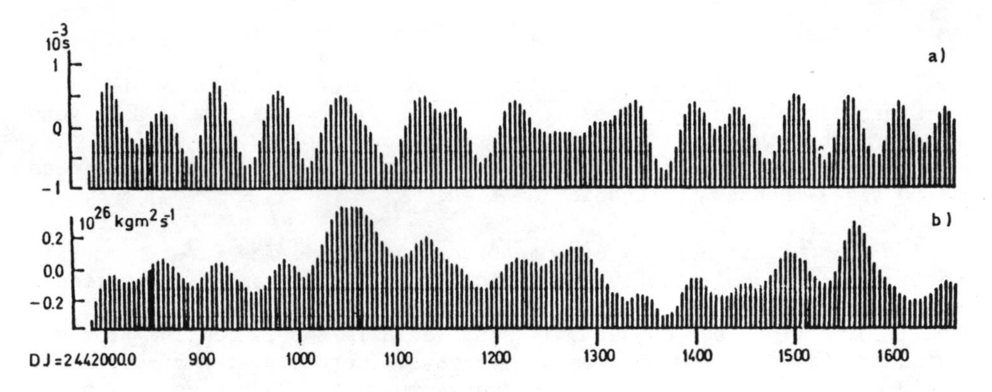

Fig. 1 - Residuals of fitting of
a) UT1 - TAI
b) AAM

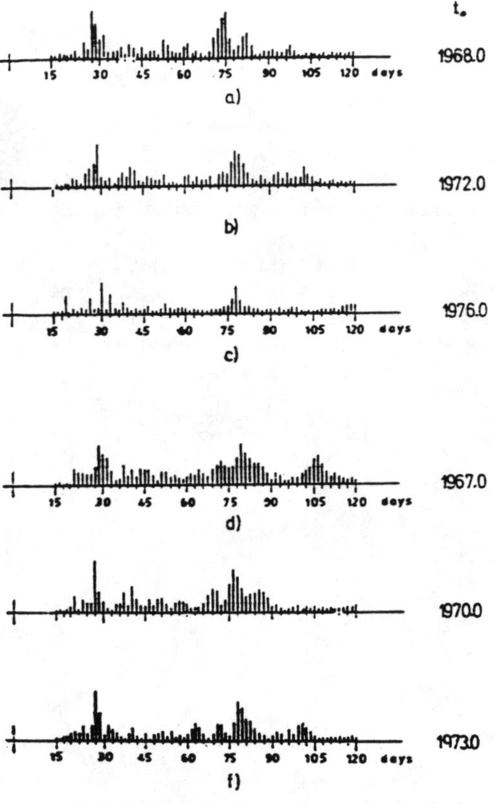

Fig. 2 - a-b-c, Spectra of Wolf number over period of 4 yrs.
d-e-f, Spectra of Wolf number from which sinusoïdal terms were removed ($f \leq 2$ cycles/yr).

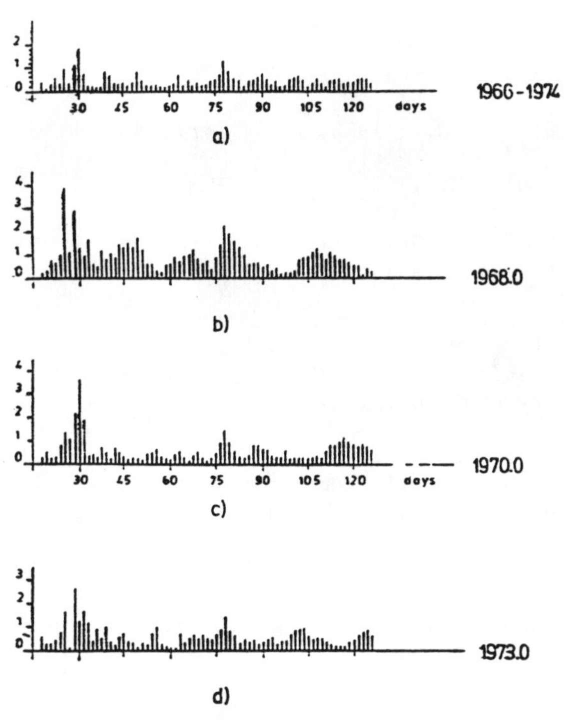

Fig. 3. Spectra of the geomagnetic index over successive periods of 3yrs.

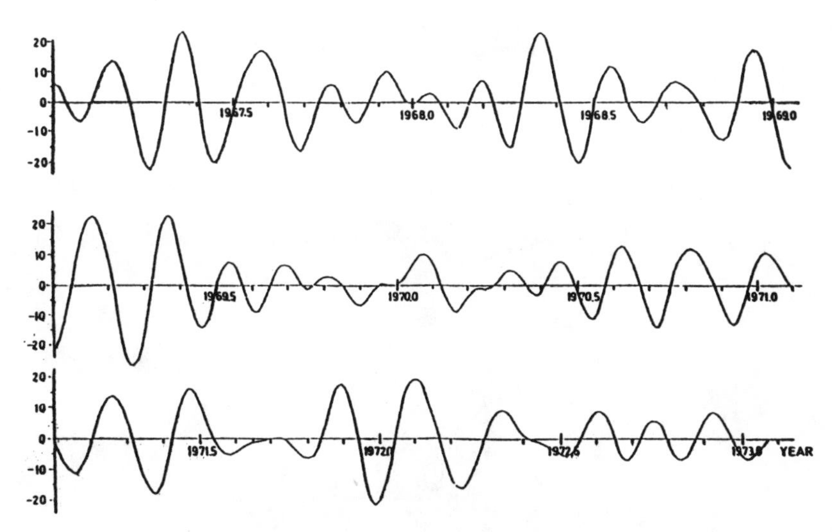

Fig. 4. Representation of ΔW_T which are the transformed values of ΔW, computed by the Labrouste's method of type $\pi(T) = T_p T_2 T_r T_s$, ΔW being the residuals of the Wolf numbers after elimination of all sinusoidal terms in the range of frequencies $f \leq 2$ c/yr.

REFERENCES

Djurovic, D. : 1983, *Astron. Astrophys.* 118, (26-28).

Djurovic, D., Stajic, D. : 1985, *Publ. Dept. Astron. Belgrade*, 13.

Feissel, M., Gambis, D. : 1980, *Compt. Rend. Accad. Sci. Paris*, B 271.

Hide, R., Birch, N.T., Morrison, L.V., Schea, D.J., White, A.A. : 1980, *Nature* 286, 5769.

Jacchia, L.G. : 1967, *Solar-Terrestrial Physics* 5.

Keating, G.M., Levin, G.S. : 1972, "Solar Activity Observations and Predictions", *ed. Mc Intosh and M. Dryer*.

Labrouste, H., Labrouste, Y. : 1943, "Analyse des graphiques résultant de la superposition de sinusoïdes", *Mémoires de l'Accad. Sci. Paris*, 64, 1940.

Lambeck, K., Cazenave, A. : 1974, *Geophys. I. Roy. Astron. Soc.* 38, 49.

Lambeck, K., Hopgood, P. : 1981, *Geophys. I. Roy. Astron. Soc.* 64, 67.

Langley, R.B., King, R.V., Shapiro, I.I., Rosen, R.D., Salstein, D.A. : 1981, *Nature*, Lond. 294.

Okazaki, S. : 1977, *Publ. Astron. Soc. Japan* 29, 619.

Okazaki, S. : 1979, *Publ. Astron. Soc. Japan* 31, 613.

OCEAN-ATMOSPHERE COUPLING AND SHORT TERM FLUCTUATIONS OF EARTH ROTATION

J.A. GONELLA
Museum National d'Histoire Naturelle
75005 Paris
France.

ABSTRACT. Several authors have shown that measurements of fluctuations in length-of-day (l.o.d.) and angular momentum of the atmosphere are strongly correlated ; these measurements support the assumption that the entire planet is a closed dynamical system. Accelerations of angular momentum of the atmosphere occur simultaneously with decelerations of the rotation of the solid earth, and vice versa, with change of l.o.d. of about a millisecond on annual period. In this paper, only the minor role of the world ocean is discussed. The Southern Ocean which is wind driven, could give, as an upper bound, a contribution of a few percents of the total change of l.o.d.. In the oceans, phase velocities of barotropic waves are ranging between 2000 to 20000 km/day ; these large free oceanic waves could act on l.o.d. changes only in the frequency band, ranging from 0.1 to 1 cycle per day. On annual and seasonal scales, it is emphasised that global water balance, within the water cycle (evaporation - condensation - precipitation), could have a larger effect than the oceanic circulation on the earth rotation. In the future, more attention must be focused on the influence of the latitudinal exchange of water substance between ocean, atmosphere and continents (ice sheets included) on l.o.d. changes.

RESUME."Le couplage Océan-Atmosphère et les fluctuations de courte période de la rotation terrestre". Plusieurs auteurs ont montré la forte corrélation existant entre les fluctuations de la vitesse de la rotation terrestre et le moment cinétique de l'atmosphère par rapport à l'axe des pôles (moment angulaire). Bien que le processus de transfert ne soit pas encore compris, l'amplitude et la cohérence de ces mesures justifient l'hypothèse que la planète, c'est à dire la terre solide et son atmosphère, constitue un système dynamique fermé. Toute accélération du moment angulaire de l'atmosphère se traduit par une décélération correspondante de la rotation de la terre solide et vice versa, avec une amplitude de l'oscillation de la durée du jour de l'ordre de ± 0,5 milliseconde par an. Dans cet article, nous examinons le rôle possible de l'océan mondial. Seul l'Océan Austral (ou

A. Cazenave (ed.), Earth Rotation: Solved and Unsolved Problems, 193–201.

Antarctique) dont la circulation circumpolaire est essentiellement entretenue par le vent pourrait expliquer quelques pourcent de l'amplitude des fluctuations de la durée du jour. Dans les océans, les vitesses de phase des ondes barotropes varient de 2000 à 20000 km par jour ; la propagation de ces ondes longues ne peut induire des perturbations des paramètres de la rotation terrestre que dans le domaine de fréquence allant de 0,1 à 1 cycle par jour. Aux échelles saisonnières et annuelles, l'influence des fluctuations du bilan hydrique global (évaporation, condensation, précipitation) ne peut être négligée. Leurs effets sur la rotation terrestre pourraient être plus importants que ceux liés aux variations de la circulation océanique. Dans le futur, une plus grande attention devrait être portée à l'étude du cycle de l'eau (notamment les flux méridiens) à l'échelle de la planète pour mieux comprendre les fluctuations de la longueur du jour.

1 - INTRODUCTION

How the ocean-atmosphere coupling can act on the earth rotation parameters ? This interesting question still remains unsolved . Interactions between ocean and atmosphere are complex but, a priori, the two major phenomena which can have some effects are :

- the mechanical input from the atmosphere to the ocean by the wind stress (momentum transfer) and by the atmospheric pressure ; both act on the dynamics of the ocean and on its angular momentum ;
- the water mass exchange through the fluctuating cycle : excess of evaporation from oceans, condensation in the atmosphere, excess of precipitation on continents and ice sheets ; this cycle has a direct influence on the moment of inertia of the earth and consequently on the variation of the length-of-day (l.o.d.).

Oceanic currents data are not yet available on the global scale and the oceanic angular momentum cannot be computed as accurately as the atmospheric one (Barnes, Hide, White & Wilson, 1983 ; Rosen & Salstein, 1983 and 1985). Nevertheless the understanding of the ocean dynamics could allow us to appreciate its possible effect on the earth-rotation. Any kind of forcing, acting on the world ocean (tidal forces, atmospheric pressure, wind stress, earthquake,...) generates oceanic waves which propagate kinetic and potential energy. Only large oceanic gravity waves (i.e. tides, Kelvin waves...) and planetary waves (which are also called Rossby waves) may have some influence on earth rotation parameters.

This paper is made up as follows. First,a short intercomparison is given between the magnitude of the atmospheric and oceanic mean angular momenta. Then, a brief description follows on the characteristics of long free oceanic waves and the frequency range on which they could affect the earth rotation parameters. Because the ocean-atmosphere coupling is not only dynamic,the thermodynamic effect will be explored at a global scale in term of water mass exchange within the cycle : evaporation, condensation, precipitation.

The discussion will bring attention on the role of the global water balance on the variations of the earth rotation parameters at annual and seasonal scale.

2- OCEANIC & ATMOSPHERIC ANGULAR MOMENTA: THE SOUTHERN OCEAN ZONAL FLUX

It is well known that one order of magnitude exists between atmospheric (Ma) and oceanic (Mo) angular momenta (Munk and MacDonald, 1960 ; Lambeck, 1980). The ratio of these angular momenta (Mo/Ma) is the product of the ratios of the masses (mo/ma) and mean velocities (vo/va) modulated by the percentage of free zonal circulation for both fluids all around the globe.

The mass ratio can be estimated by the ratio of mean depth of oceans on a globe without continents (2500 m) by the equivalent water height of the atmospheric pressure (10m). In the oceans, the averaged value of currents from surface to bottom is of order of 1 cm/s ; in the atmosphere, the averaged wind velocity is around 10m/s(1000 cm/s). Furthermore the atmosphere can go free all around the globe. For oceans, only the Southern one (16 % of the earth surface ; 25 % of the ocean volume) has a free zonal circulation ; we will take a compromise at 20 % for the computation. Because of the fluid continuity equation, the angular momentum for the others oceans with meridional boundaries can be neglected. All these estimates can be summarised :

- mass ratio mo/ma = 2500/10 = 250
- velocity ratio vo/va = 1/1000
- zonal circulation ratio %o/%a = 20/100 = 2/10
- angular momentum ratio Mo/Ma = 250(1/1000)(2/10)= 5%.

The oceanic angular momentum could represent only a few percents of the atmospheric one. This rough approximation represents an upper bound in agreement with results from previous works (T.Eubanks, 1985). A more recent one (Rosen & Salstein, 1985) conclude "that there seem little need to invoke processes other than atmospheric motions to explain seasonal, nontidal changes in l.o.d."; that means that the zonal oceanic flux has no effect at all on l.o.d. changes!

Nevertheless, in a closed system, if the mean westerly value of Ma = 1.4×10^{26} kg.m^2/s (Rosen and Salstein, 1983) was transfered to the solid earth, the length of day would be shortened by 2.3 milliseconds. On the other hand, if we consider the values measured by differents authors for the Antarctic Circumpolar Current (AACC) flux, we have :

- throught the Drake Passage: 124 SV(Sverdrups) by Nowlin, Whitworth and Pillsbury,(1977), value reviewed more accurately to 123(± rms=10.5)SV by Whitworth and Peterson (1985) ;
- in the Crozet-Kerguelen Area: 118 SV by Gamberoni, Geronimi, Jeannin and Murail (1982).

A fluctuation of oceanic zonal transport of 1 SV at 45° latitude could be expressed in l.o.d changes by one microsecond. So the real mean ratio <Mo/Ma> is about 120/2300= 5,2 %, close to our estimate (5%).

Then again, fluctuations of Ma expressed in change of l.o.d. have amplitudes of order of ±0.5 milliseconds (Rosen and Salstein, 1983) ; with the same reasoning, we can estimate that fluctuations of the Antarctic Circumpolar Current will be 25 Sv : measurements throught the Drake Passage show fluctuations of order of 30 Sv (Pillsbury, Whitworth, Nowlin and Sciremammano, 1979 ; Wearn and Baker, 1980 ; Whitworth and Peterson, 1985).

Modeling the Antarctic Circumpolar Current in the future could be beneficial for understanding its effect on earth rotation fluctuations; but the angular momentum transfer process between the solid earth and the oceans (70% of the earth surface) is still an unsolved problem. Therefore we have to distinguish the role of wind driven currents - related to wind stress - and the role of oceanic free waves -related to long wave propagation . A brief review is given in the following paragraph on long oceanic free waves at a planeraty scale and on their possible influence on l.o.d. changes.

3 - CHARACTERISTICS OF LONG OCEANIC FREE WAVES AT A PLANETARY SCALE

The world ocean is stratified in density and oceanographers distinguish "barotropic" from "baroclinic" modes for waves in the ocean. The term "barotropic" means that the pressure is constant on surfaces of constant water density, the word "baroclinic" meaning that pressure is not constant on these density surfaces. It is now conventional to call "barotropic mode" the surface gravity wave even if this is not exactly true for the real ocean, but only for a fluid of uniform density : the phase speed "c" of long surface gravity waves is related to the gravity g and the depth H of the ocean: $c = \sqrt{(gH)}$. With $g=10m/s^2$, a typical value for c is 200m/s in a 4000m deep ocean.

3.1. The "barotropic" mode.

Let us consider the "homogeneous" ocean and a surface wave of the form : $n = A \cos(kx+ly-\sigma t)$ where n is the displacement of the sea level from the rest position, K(k,l) the horizontal wave number vector, x the coordinate in longitude, y the coordinate in latitude, σ the frequency and t the time.

3.1.1. Long gravity waves.

Tidal forces, atmospheric systems (wind-stress and atmospheric pressure fields) generate important effects on the sea level at coastal boundaries (i.e. tides, storm surges, Kelvin waves...). Although storm surges could have tremendous consequences at the coast ,their influences occur on an area of several thousands of km^2, negligible at a global scale; only oceanic tides could bring significant contributions in l.o.d. changes. The sea-level response depends on the ratio of speeds of progress of equilibrium tides around the earth compared with the long gravity wave speed "c".The fortnightly tide would be close to equilibrium, but that is not true for diurnal and semi-diurnal tides which move around the earth once per day, giving a speed of 350 m/s, nearly twice faster than the long-wave speed c (circa 200 m/s or 20 000 km/day). Nevertheless free waves play an important role in the description of the tides : for instance the Kelvin wave parts are a major element of the tides in the Pacific. Tidal components of earth-rotation parameters are well documented and are discussed with solid tides (P. Melchior, 1985).

3.1.2. Planetary waves.

Planetary or Rossby waves are quasi-geostrophic and are controled by the "beta" effect (β=df/dy ; f, Colioris parameter). Their phase speed comes from the dispersion equation :
$C = \sigma/k = -\beta/(k^2+l^2+f^2/c^2)$ showing that all planetary waves have westward propagation. The ratio R of the kinetic energy density over potential energy density (averaged over a wave lengh) is given by $R = (Ka)^2$ where K is the horizontal wave number vector already defined ($K^2=k^2+l^2$) and "a" the Rossby radius ($a = c/f$). Typical values for oceans are : H = 4000 m, c = 200 m/s, a = 2000 km. It follows that we can distinguish :

- $R \ll 1$: barotropic planetary waves would have most of their energy in the potential form for scale larger than the Rossby radius "a", with westward propagation: $C \approx -\beta a^2 = -80$ m/s = -8000 km/day. Because of the scale dimension, these waves which are non dispersive, are difficult to generate in the real ocean on our earth planet ; the wave length has to be of the order of $L=2\pi a$ = 12500 km which is the dimension of the Pacific Ocean.
- $R \gg 1$: atmospheric systems which have north-south scales of circa δy =1000 km (smaller than the oceanic barotropic Rossby radius a = 2000 km) can produce in the oceans short barotropic waves which are dispersive and whose energy is mainly kinetic ; they propagate westward with phase speeds of order of $C \approx -\beta/l^2$ = - 20 m/s = 2000 km/day.

The Southern Ocean is the only one which could be concerned with a strong atmospheric imput all around the world on 20000 km ; the barotropic adjustement time would be about : T = (20000 km)/(2000 km/day) = 10 days. But fluctuations of the Antarctic Circumpolar Current which are about 30 Sverdrups, are mainly related to the variations of the total wind-stress all over the Southern Ocean (Wearn & Baker, 1980). These authors set up a time lag of 7 days, over which fluctuations of total wind stress have a maximum of correlation with changes in oceanic transport. This time lag seems too short to be attributed to barotropic adjustement. We will come back on this time lag in the discussion.

In summary, for oceanic barotropic modes, ocean dynamics give large wave phase velocities ranging from 2000 to 20000 km/day ; if any influence could be exerted on the earth rotation parameters by oceanic barotropic waves, this would be in the frequency band ranging from 1 to 1/10 cycle-per-day with a significant phase lag. The actual sampling of earth rotation parameters (1 every 5 days for long series) does not allows the study of this frequency domain ; but, except for tidal frequencies, we cannot expect any discernible influence from these oceanic waves on earth rotation parameters.

3.2. The "baroclinic" modes.

The linearized theory of ocean dynamics gives for the long "baroclinic" waves similar result as the barotropic ones ; the depth "H" of the homogeneous ocean has to be replaced by an equivalent depth "He" for each baroclinic mode. For the first baroclinic mode (the more important in the oceans), typical values for phase velocities (c) are 2 to 3 m/s, corresponding to an equivalent depth He of 0.5 to 1 meter.

The baroclinic Rossby radius (a) would be of 30 km. Westward phase speeds (C) of long baroclinic planetary waves are a function of latitude : about 2.5 cm/s at 30° latitude, this speed reaches 50 cm/s at the equator for the first mode. That explains why baroclinic adjustments in the ocean are made on a time scale of several years at middle and high latitudes and of a few months in the equatorial belt.

Although these baroclinic modes could be responsible of strong currents in surface layers above the main and seasonal thermoclines, the vertical integrated current is zero and has no effect on the oceanic angular momentum variations. Moreover, in baroclinic mode, the bottom pressure does not change and these oceanic waves could not bring any contribution in the loading of the earth by the oceans.

4 - OCEAN-ATMOSPHERE COUPLING ; THE GLOBAL WATER MASS EXCHANGE.

The ocean-atmosphere coupling is governed not only by momentum transfers, but also by thermodynamic processes : the water mass exchange, through the cycle evaporation -condensation-precipitation, concerns a mean global evaporation of about 1000 mm per year compensated by an equivalent amount of global precipitation (Baumgartner & Reichel, 1975). The scheme of the global water cycle is quite simple ; evaporation and precipatation over the global ocean are more important than over the land. Nevertheless, over the oceans, the total evaporation exceeds total precipitations ; the difference is compensated by the run-off from the continents where, on the average, precipitations exceed the evaporation. The mean run-off is estimated as 1.2 SV = $0.4x10^{14}$ m^3/year, that means (on annual scale) an excess of evaporation over precipitations of 10 cm over the oceans (360 millions Km^2) and an excess of precipitations over evaporation of 24 cm over land (150 millions km^2).

According to F.Bryan and A.Oort(1984), the global water balance, computed from aerological data over the period, from May 1963 to April 1973, has a strong seasonal cycle. During the northern fall (Sept., Oct., Dec.,), global precipitations exceed the total evaporation ; the situation is reverse during the others 9 months. But the annual total run off, estimated by Bryan and Oort shows a discrepency by a factor 5 from Baumgartner and Reichel'results which are in good agreement with 1979 data (Fist GARP Global Experiment: FGGE) . Nevertheless, if we consider that ratio values of seasonal fluctuations/annual mean run-off are correct in Bryan and Oort paper, the oceans would loose about 1.5 cm of water from January to August. According to Trenberth (1981), this amplitude would be of 0.4 cm only.

A discussion on errors introduced on l.o.d. changes by accounting for the static contribution of water before precipitation, but not after, can be found in the appendix to the Eubanks et al. (1985). Unfortunately, these authors could not include this effect in their analysis because spherical harmonics (J_0 and J_2) of the atmospheric surface pressure field are not currently performed by national data centers.

Nevertheless, because the dynamics of the global atmosphere is closely related to the air-sea interaction within the process of evaporation and precipitation, we can notice the following relationships between l.o.d. and strong convection phenomena in the atmosphere :

- in the annual cycle, the earth rotation velocity is maximum in July-August when the total amount of water loosed by the oceans is maximum ; on the other hand, the maximum of water is present in the solid earth (oceans included) in winter when the l.o.d. is maximum.
- strong intertropical convection over Indian and Pacific Oceans are the motor of the 30-60 days atmospheric oscillations which are closely related to about 0.2 milliseconds fluctuations of the earth-rotation (Whyshall and Hide, 1985 ; Weickmann, 1983).

The study of these connexion could be a topic for further research on the role of the water cycle on the process of the angular momentum transfer between the solid earth and the atmosphere. A global uniform evaporation of 1 cm of water which turns to be condensated on polar sheets (e.g. on the 14 millions km^2 of Antarctica) could shorten the l.o.d. by about 0.18 milliseconds. This process could have more influence than the mean flux of th AACC (120 SV $\approx$ 0.12 milliseconds on l.o.d. change).

5 - DISCUSSION AND CONCLUSION

We have reviewed the possible influence of the oceans on the earth rotation.

5.1. Phase velocities of free oceanic barotropic waves (long gravity and Rossby waves) are ranging between 2000 to 20000 km/day : these waves could act on l.o.d.changes in frequency band ranging from 0.1 to 1 cycle per day. Only oceanic waves resulting from tidal forces can have their signature in l.o.d. changes in this frequency band.

5.2. On seasonal and interannual scales, two major contributions of the oceans could be the dynamics of the Southern Ocean and the global water mass balance fluctuations :

The dynamics of the Southern Ocean could contribute for 5 % of l.o.d. changes at seasonal and annual scale ; but, because the Southern Ocean is wind driven (friction effect), the time lag between the total wind stress and the oceanic transport would be about 7 days. No such time lag has been observed in cross correlation between l.o.d. change and atmospheric momentum fluctuations : the oceanic friction could not be the main process for momentum transfer from the atmosphere to the earth and vice versa. Wind stress on the dry land or atmospheric pressure gradients across mountain ranges could act on l.o.d. changes with a shorter time lag; these two mecanisms must certainly have a greater influence than the oceanic friction.

Spatial sampling deficiencies and bad accuracy of aerological data do not allow to compute reliable quantitative estimates of parameters required in the global water balance ; peculiarly, it is quite hard to measure the interannual and seasonnal variations of total water in the atmosphere. But from rough estimations, it is clear that the latitudinal exchange of water between ocean and atmosphere cannot be neglected and could have more influence on l.o.d. changes than the zonal flux fluctuations of the Southern Ocean. In the future, the water mass balance on seasonal and interannual scale has to be explored more accuratly as a detectable geophysical cause in the earth rotation parameters.

ACKNOWLEDGEMENTS

Dr O.Talagrand from the Laboratoire de Météorologie Dynamique (Paris) and two anonymous reviewers made helpful comments on an earlier version.

REFERENCES

BARNES R.T.H., R. HIDES, A.A. WHITE and C.A. WILSON, 1983 - "Atmospheric Angular Momentum Fluctuations, Length-of-day changes and Polar Motions". Proc.R.Soc.Lond., Vol.A 387, pp 31-73.

BAUMGARTNER A. and E. REICHEL, 1975 - "The World Water Balance". 127 p., Elsevier, New-York.

BRYAN F.,and A.OORT, 1984-"Seasonal Variation of the Global Water Balance Based on Aerological Data". J.Geophys.Res, Vol.80, n° D7, pp 11 717-11 730.

EUBANKS T., 1985 - "El Nino Type Oceanic Effects on Earth Rotation"; in Proc. Nato Adv. Res. Workshop, Bonas, June 1985, France.

EUBANKS T.M., J.A. STEPPE, J.O. DICKEY and P.S. CALLAHAN, 1985 -"A Spectral Analysis of the Earth's Angular Momemtum Budget".J.Geophys. Res, vol 90, N°B7, pp.5385-5404.

GAMBERONI L., J. GERONIMI, P.F. JEANNIN and J.F. MURAIL, 1982 -"Study of Frontal Zones in the Crozet-Kerguelen Region".Oceanologica Acta, vol. 5, 3, pp. 289-299.

LAMBECK K., 1980 -"The Earth's Variable Rotation, Geophysical Causes and Consequences". Cambridge University Press, Cambridge, 449 p.

MELCHIOR P., 1985 -"Earth Rotation and Solid Tides"; in Proc. Nato Adv. Res. Workshop, Bonas, June, 1985, France.

MUNK H.H. and G.J.F. MACDONALD, 1960 - "The rotation of the Earth". Cambridge University Press, Cambridge, 329p.

NOWLIN W.D., J.r., T. WHITWORTH III and R.D. PILLISBURY, 1977 - "Structure and Transport of the Antarctic Circumpolar Currents at Drake Passage from Short-term Measurements". J.Phys.Oceanogr., Vol.7, pp. 788-802.

PILLSBURY R.D., T. WHITWORTH III, W.D. NOWLIN Jr. and F. SCIREMAMMANO Jr., 1979-"Currents and temperatures as observed in Drake Passage during 1975". J.Phys.Oceanogr., Vol.9, pp 469-482.

ROSEN R.D. and D.A. SALSTEIN, 1983 - "Variations in Atmospheric Angular Momentum on Global and Regional Scales and the Lenght-of-Day". J.Geophys.Res., Vol.88, n°C9, pp. 5451-5470.

ROSEN R.D and D.A. SALSTEIN, 1985-"Contribution of stratospheric winds to Annual and Semiannual Fluctuations in Atmospheric Angular Momentum and the length of Day". J.Geophys.Res., Vol.90, N° D5, pp. 8033-8041.

TRENBERTH K.V., 1981 -"Seasonal Variations in Global Sea Level Pressure and the Total Mass of the Atmosphere". J.Geophys.Res., Vol.86, n°C6, pp. 5238-5246.

WEARN R.B.,Jr. and D.J.BAKER,Jr., 1980 -"Bottom Pressure Measurements across the Antarctic Circumpolar Current and their Relation to the Wind". Deep Sea Res., Vol.27A, pp. 875-888.

WEICKMANN K.M., 1983 -"Intraseasonal circulation and outgoing long wave radiation modes during Northern Hemisphere Winter". Monthly Weather Review, vol. 111, pp. 1838-1858.

WHYSALL K. and R.HIDE, 1984-"A seven week fluctuation of the general circulation of the earth's atmosphere : a brief survey"- U.K. Meteorological Office, Tech. Memo. n° 0218402.

THE DAMPING OF THE CHANDLER WOBBLE AND THE POLE TIDE

S. R. Dickman
Department of Geological Sciences
State University of New York
Binghamton, NY 13901
USA

ABSTRACT. The damping of the Chandler wobble is an old yet actively researched problem still awaiting resolution in a number of respects. Due to both observational and theoretical limitations the actual damping rate is uncertain even now; a decay time of 45-70 years or more seems probable. The two most likely candidates for dissipating wobble energy are mantle anelasticity and a dynamic response of the oceans to wobble. Because of the frequency dependence of anelastic mechanisms, study of the oceanic response may be useful for constraining low-frequency anelasticity.

The nature of the pole tide, as the oceanic response is known, has been investigated both observationally and theoretically. Data analysis reveals clear evidence of non-equilibrium pole tide characteristics in shallow seas; the products of inertia implied by such characteristics yield significant shallow-sea dissipation of wobble energy. Theoretical explanations of those characteristics are presently being revised, and so far confirm the importance of the shallow seas. Data in the open ocean is sparse; at a few locations it implies statistically significant enhancements great enough to completely damp the wobble. Open-ocean theory so far fails to support the data, but has yet to be extended to realistic oceans.

1. INTRODUCTION: A SURVEY

The damping of the Chandler wobble is an exciting topic to research because it relates to a great variety of global geophysical problems. It is challenging because it demands expertise in so many fields. And, it is pleasurable because (as we will see today) there is controversy and room for innovative hypotheses in every one of its aspects.

To gain an appreciation of the role of the pole tide in various aspects of the damping problem, I thought it best to begin with a general survey. Following the survey we will look at the pole tide in shallow seas, then in the deep ocean.

The four major candidates with the ability to dissipate Chandler wobble energy are core viscosity, electromagnetic coupling between core

A. Cazenave (ed.), Earth Rotation: Solved and Unsolved Problems, 203–228.

and mantle, mantle anelasticity, and a non-equilibrium oceanic response to wobble. Because the outer core fluid is viscous, relative motion between core and mantle during wobble would be resisted and the wobble would gradually decay. But extrapolation from laboratory experiments (Gans, 1972), atomic structure theories (for pure iron; Bukowinski & Knopoff, 1976), and observational constraints from earth's nutations (e.g. Toomre, 1974), especially the recent determination by Herring et al. (1985), all indicate that the core's viscosity is too small to significantly affect wobble. Similarly, Rochester & Smylie (1965) showed that electromagnetic core-mantle coupling is too weak by several orders of magnitude to damp the Chandler wobble. Even if lower-mantle electrical conductivity is much greater than had been supposed, as suggested later by Braginskiy (Braginskiy & Nicholaychik, 1973; Braginskiy & Fishman, 1976) and others, such coupling would still be too weak (see Lambeck, 1980).

The two candidates upon which I will mainly focus in this talk are mantle anelasticity, which dissipates energy as the mantle flexes in response to wobble, and the "pole tide", as the oceanic response to wobble is known; if the pole tide is dynamic, ocean currents associated with the non-equilibrium response to wobble would necessarily dissipate wobble energy.

There are a number of difficulties we must face concerning the subject of wobble damping in general, and these candidates in particular. Most fundamentally, we are not even certain <u>how much</u> Chandler energy is actually dissipated per cycle of wobble. Figure 1 shows the Chandler wobble x and y components, extracted from the more homogeneous ILS data for 1900-1978 after elimination of other signals.

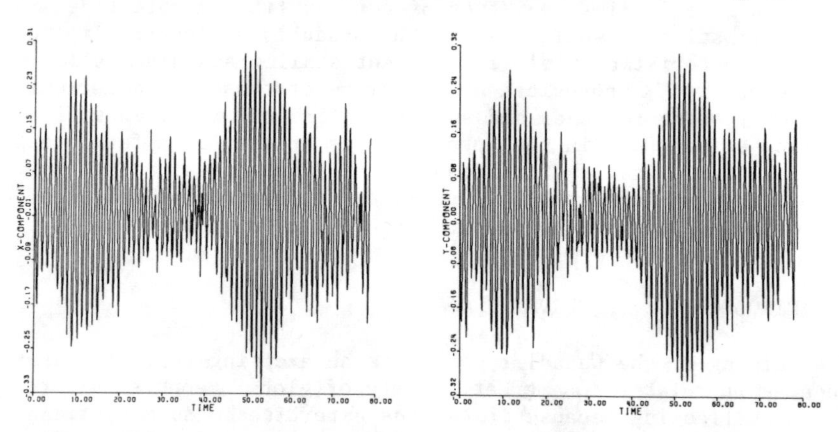

Figure 1. X- and y-components of the Chandler wobble, 1900-1978, as revealed by the more homogeneous International Latitude Service monthly data after removal of a linear trend and the annual and Markowitz wobbles. Data are in units of arc sec.

It is not known whether the decaying portions of these time series represent the natural decay time of wobble or a forced effect. The decaying portions are part of a clear amplitude modulation which is statistically significant, so we must consider physically reasonable alternative explanations. One possibility is that the varying amplitude represents a combination of natural decay plus non-random excitation of wobble; our problem in this case is that the excitation source of the Chandler wobble is yet unresolved.

Perhaps the most suggestive connection was proposed in 1977 by Kanamori (1977; see also O'Connell & Dziewonski, 1976; Anderson, 1974). He compared (Figure 2) the amplitude of the Chandler wobble with the number of large earthquakes and the associated energy release during this century. Kanamori realized, however, that such earthquake data fails to account for a significant long-period energy release by the greatest earthquakes of the century, such as the 1960 Chile and the 1964 Alaska events. He devised a magnitude scale appropriate for such large events, based on the earthquakes' moments. The corresponding energy release appears to correlate well with the large amplitudes of

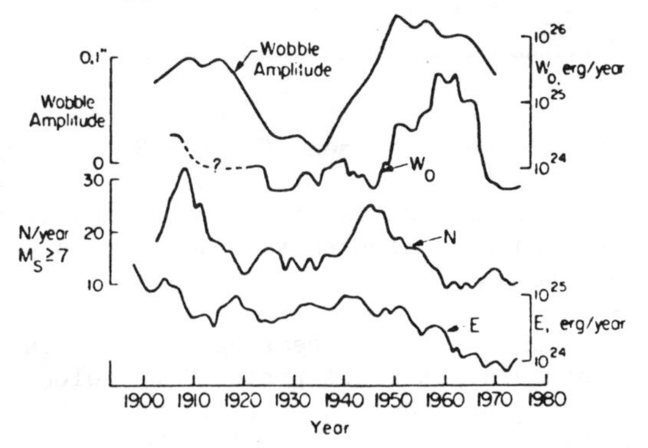

Figure 2. Comparison of earthquake and Chandler wobble time series by Kanamori (1977). N=seismicity, as number of events with surface-wave magnitude of 7 or greater; E=associated energy release, in erg/yr; W_o=energy release, erg/yr, based on earthquake moments.

the Chandler wobble, suggesting a connection exists between the wobble amplitude modulation and earthquake excitation. Unfortunately, the earthquake activity follows the peak wobble amplitude rather than leading it, implying the "wrong" cause-and-effect....

The fractional energy loss in a time series is often expressed inversely in terms of the quality factor Q; Q is directly proportional to the decay time. The uncertainties in wobble decay rate are equivalent to stating that the wobble Q is not well-determined. Corresponding to the amplitude modulation of the Chandler wobble in the time domain, the power spectrum of the wobble data (Figure 3) reveals two peaks at the Chandler frequency; determination of Q_w from the half-power width of the peak(s) is obviously problematic. Such a bifurcation in the spectrum could result from non-random excitation, as already mentioned, i.e. an excitation whose spectrum is not "white" in the vicinity of the Chandler frequency (cf. Brillinger, 1973). Alternatively, it could be indicating that the Earth possesses two natural wobble frequencies. Physically, this would correspond to two

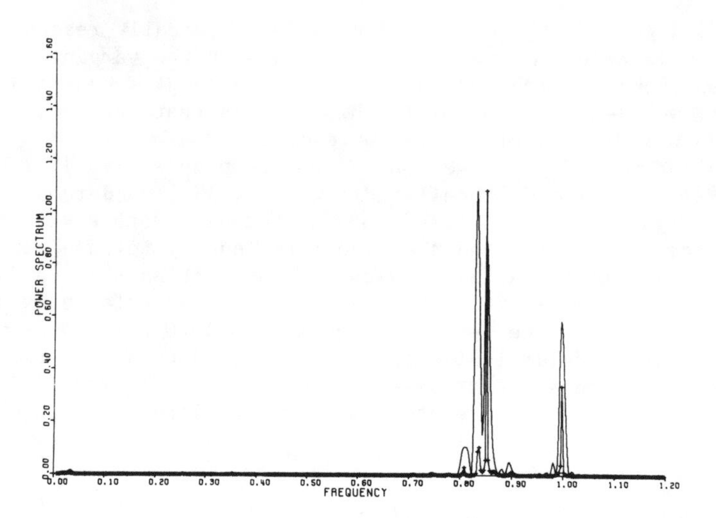

Figure 3. Unsmoothed prograde power spectrum of the ILS data following removal of the linear trend. The maximum entropy spectrum (hatched line) has been normalized to the largest value of the periodogram. Note the double peaks at the Chandler frequency, ∿0.84 cycles per year (cpy), revealed by both spectra.

partially coupled regions within the Earth "beating" against each other as they wobbled; such a hypothesis was first proposed by Colombo & Shapiro (1968), but without a theoretical analysis to support it, it remained only a vague possibility.

The first theoretical investigations of the hypothesis were, I believe, independent studies by myself and Zhang published in 1983. Zhang (1983) was in fact able to produce two Chandler frequencies in an elastic earth by decoupling the upper & lower mantle at a depth of ∿300 km. In contrast, I took the imperfectly coupled regions to be the oceans and solid earth (Dickman, 1983). With coupling 99.9% complete, I obtained wobble modes corresponding to a single-period Chandler wobble and the long-period Markowitz wobble; my Earth model did not yield two Chandler frequencies. You might be interested to hear about recent work of mine which is unpublished. By modelling the solid earth as having a lithosphere partially decoupled from the underlying mantle at the seismic low-velocity (or low-viscosity) zone, I can obtain a rotating system which possesses two wobble modes. If the low-velocity zone is treated as a boundary layer of uniform depth, two Chandler frequencies can be produced, while if the low-velocity zone is of variable depth--deeper beneath continents, shallower beneath the thin oceanic plates--a single Chandler frequency and a single Markowitz wobble frequency will result....

The difficulties in determining wobble Q reliably lead to a range of published Q values, as summarized in Table I. About the only generalization we can make is that determinations of Q_w since 1960 have tended to yield higher values than those (summarized in Munk &

MacDonald, 1960) prior to 1960. Estimates of Q_w or of the wobble damping time have been based on a variety of techniques, including maximum likelihood, least-squares fitting to the data, least-squares fits of the periodogram, maximum entropy, and combined autoregressive-moving average methods. Many of the determinations involved a short span of data, which artificially broadens the spectral peak and produces lower Q; only one of the analyses, by Ooe (1978), explicitly corrects for the finite duration of the data. A few of the analyses attempted to account for possible non-random excitation of wobble, usually by assuming the excitation spectrum was "red" rather than "white". Only one analysis (Zhang et al., 1983) actually found the Q of two Chandler peaks; most of the analyses instead smoothed the spectrum to artificially eliminate the twin peaks, or modelled the spectrum as containing a single peak. I believe the most reliable estimates to be those by Ooe (1978) and by Wilson and Vicente (1980), which is based on the better-quality ILS data, and which attempts to account for the effects of noise and non-random excitation; we may thus take the wobble Q to be $\sim$100-150, possibly much higher, corresponding to a decay time of $\sim$45 years or more.

There are a number of challenges confronting us if we are to resolve and understand the damping of the Chandler wobble. We have already discussed the problems underlying the determination of the actual amount of energy dissipation, also some of the associated theoretical problems. Another major challenge is to demonstrate convincingly whether or not tide data supports the hypothesis of a non-equilibrium oceanic response to wobble. Traditional data analysis techniques have great difficulty extracting reliable phase lags from noisy tide data--North and Baltic Seas data (Miller & Wunsch, 1973) being a major exception--yet the extent of damping by the oceans depends on how much the pole tide lags behind the wobble forcing (cf. Dickman, 1979, Table 2 or 5). Data analysis is more successful in determining the amplification of the tide (relative to an equilibrium response), but in order to infer how much the oceans lengthen the wobble period (such lengthening being the other major effect of the oceans on wobble), reliable amplitudes need to be obtained world-wide. The sparse distribution of data of long duration or high quality world-wide makes that need almost impossible to satisfy. The corresponding theoretical challenges are to develop rigorous fluid dynamical treatments of the pole tide in both shallow seas, where dissipative mechanisms must operate, and in the deep ocean, where the extent of dynamic behavior is unclear and the importance of dissipation

Table I (next page). Estimates of the Chandler wobble quality factor Q_w published since 1960. The various techniques employed to determine Q_w include maximum likelihood (MLM), least-squares fits to the data (LS) or the periodogram (LS/PER), maximum entropy (MEM) and autoregressive-moving average (AR.MA) methods. As discussed in the text, such factors as short data spans and smoothed spectra may interfere with determination of Q_w, while consideration of non-random excitation and noise effects may yield a more accurate estimate of Q_w.

TABLE I

ESTIMATES OF Q_W

author	*approach*	*preferred* Q_W	*range*	*comments*
MUNK & MACDONALD, 1960	(VARIETY)	30	6–60	short data spans
JEFFREYS, 1968	MLM	61 >80	37–186	short data span ; non-random excitation
SEKIGUCHI, 1972	LS	66	32–332	short data spans; red excitation
BRILLINGER, 1973	LS/PER	44	30–82	smoothed spectrum
YATSKIV, 1973	LS/PER	50	28–115	smoothed spectrum
SMYLIE ET AL., 1973	MEM	50		MEM; short data span
CURRIE, 1974, 1975	MEM	72	52–92	MEM; smoothed spectrum
WILSON & HAUBRICH, 1976	MLM	100	50–400	no noise bias
OOE, 1978	AR.MA	96	50–300	finite duration correction
WILSON, 1979	MLM/AR.MA	62	38–250	no noise bias; white or red excitation
WILSON & VICENTE, 1980	MLM/AR.MA	170	47–1000	BETTER DATA; no noise bias; white or red excitation
OKUBO, 1982	MEM	50	50–100	MEM; filter length assessed
- - - - - - - - - - - -	- - -	- - - - - - -	- - - - -	- -
ZHANG ET AL., 1983	MEM	302, 898		MEM; unsmoothed spectrum; TWO PEAKS

uncertain. All these points will be discussed more fully in a few minutes.

It is reasonable to assume that some Chandler wobble energy is dissipated through anelastic mechanisms within the mantle, since anelasticity also causes the attenuation of seismic waves. But it is likely that anelasticity, as represented by a mantle quality factor Q, is strongly frequency-dependent, and the frequencies of seismic waves are orders of magnitude higher than that of the Chandler wobble. Thus the relevance of seismic Q to Q_w is unclear. It is more reasonable to extrapolate the Chandler Q_w from the Q associated with earth tides; but this is difficult because earth tide observations must be corrected for the effects of ocean tide loading. The theoretical challenge corresponding to the frequency dependence of mantle Q is the need to extend the spectrum of anelastic mechanisms to frequencies well below the seismic band. The near- impossibility of doing so without the very information about Q(f) we would like to extrapolate <u>from</u> the spectrum can be illustrated with the aid of Figure 4.

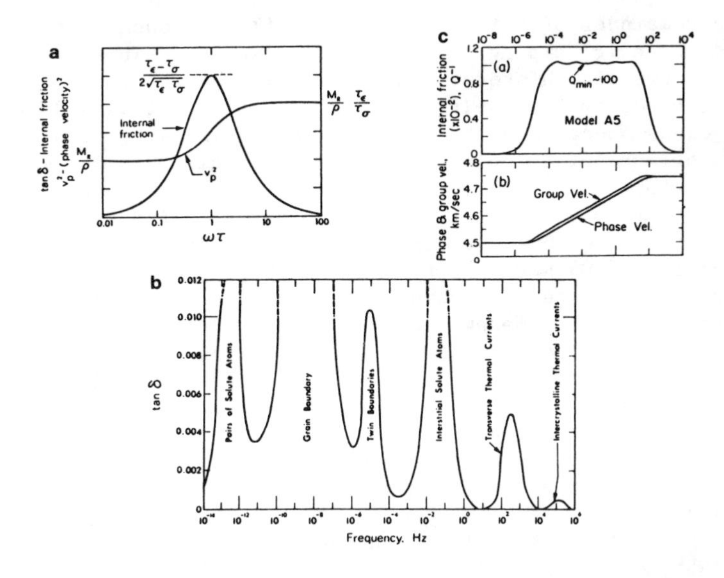

Figure 4. Illustrations by Liu et al. (1976) depicting some concepts involved in anelastic relaxation mechanisms. (a) The internal friction (proportional to Q^{-1}) versus frequency ω in a material possessing a single mechanism with relaxation time τ. (b) Schematic relaxation spectrum for metals, showing a superposition of different mechanisms. (c) Superposition model capable of producing a roughly frequency-independent Q within the seismic frequency band, as observed.

In an anelastic material the internal friction, which is inversely proportional to Q, is a highly non-linear function of frequency, for frequencies close to the characteristic "relaxation time" ($^{-1}$) of the material (Figure 4a). In most materials, a variety of anelastic mechanisms exist, each with its own characteristic time scale; the internal friction experienced by a material thus depends strongly on the frequency of the applied stress. For illustration, in the case of metals as shown in Figure 4b, grain boundary relaxation dominates other dissipative mechanisms if the applied frequency is ∿ 1 cycle per year. It was realized by Kanamori & Anderson (1977) that observations of a roughly constant Q in the seismic band are a consequence of a particular superposition of relaxation mechanisms (Figure 4c). The point is that the relaxation mechanism dominating at the Chandler frequency, and thus the mantle Q_w, need not have any connection with those at other frequencies. With such indeterminacy, constraining the relaxation mechanisms using a specified Q_w is likely to prove more successful than the reverse.

One of the goals of determining the amount of wobble dissipation by the oceans is in fact to obtain an independent estimate of the amount of internal friction in the mantle at the Chandler period, and thereby constrain models of low-frequency mantle anelasticity.

For the remainder of this talk I would like to concentrate on the hypothesis of oceanic damping of the Chandler wobble. This hypothesis has had a controversial history. In brief: It was the first damping mechanism proposed, by Lord Kelvin in 1876. Jeffreys (1915) discounted its effectiveness, because of the low molecular viscosity of water; but after the importance of turbulent viscosity in tidal friction was recognized (Taylor, 1919) the oceans appeared to be a viable candidate once again. Jeffreys (1920) then used a comparison with semi-diurnal tides to establish the ineffectiveness of oceanic damping. Eventually, Munk & MacDonald (1960) would point out that such a comparison was invalid; for example, the semidiurnal tide and the pole tide are of different harmonic type. Meanwhile, Bondi & Gold (1955) demonstrated that the oceanic body, because of its low moment of inertia, must be completely ineffective in damping wobble. Munk & MacDonald (1960), however, showed that their argument was not general; in fact, the oceanic effect need not depend on oceanic moments of inertia at all. They concluded that the question of oceanic damping was unresolved.

Since then, many researchers have considered it a "last resort": given the current state of knowledge, and level of understanding, it remained a legitimate possibility (Mansinha & Smylie, 1967; Lagus & Anderson, 1968; Rochester, 1970, 1973; Merriam, 1973; Miller, 1973; Miller & Wunsch, 1973; Stacey, 1977).

In 1974, Wunsch (1974) proposed a fluid dynamical theory of the pole tide in the North Sea which explained tide observations and predicted that shallow seas would dissipate a significant (if not dominant) portion of wobble energy. However, errors in that theory were revealed the following year (Wunsch, 1975), leaving the status of his theory indeterminate.

More recent efforts have still not resolved the question of oceanic damping. Three researchers (Dickman, 1979; Naito, 1979; Lambeck, 1980) showed nearly simultaneously on the basis of tide observations alone that such observations, which generally reveal non-equilibrium behavior in both shallow seas and the deep ocean, require the Chandler wobble to be largely damped by the pole tide. Such inferences are not unequivocal, however, given the noisy quality and sparse distribution of tide data. Smith & Dahlen (1981) did not completely rule out oceanic damping but argued against it, along lines similar to those of Jeffreys (1920). A rigorous theoretical investigation I have just published (Dickman, 1985) by no means settles the question either: I found the pole tide in global oceans essentially unable to damp the Chandler wobble; but the relevance of my results to the case of realistic oceans is unclear.

Based on data analyses such as those of Miller & Wunsch (1973), a distinction has arisen both in theory and observation between the pole tide in shallow seas and the pole tide in the open ocean. We will return to consider some of the most recent theoretical developments in the ocean damping controversy, concerning both the shallow seas and the deep ocean. In each case I will first discuss the observational evidence, and the potential consequences of such observations, then their theoretical basis.

2. THE SHALLOW-SEA POLE TIDE

2.1 Observations and Implications

Data analysis has repeatedly confirmed the existence of enhanced pole tides in the North and Baltic Seas. One of the most rigorous and comprehensive investigations is that by Miller and Wunsch (1973); I have depicted some of their results in Figure 5. They found the pole tide to undergo an "eastward intensification" along the southern coast of the North Sea, attaining amplitudes of 6 times equilibrium and more at the easternmost portion, and throughout the Baltic Sea. The tide in both Seas consistently lagged behind the predicted equilibrium response by $\sim 40^{\circ}$, i.e. about 1 1/2 months.

The pole tide in the North and Baltic Seas thus exhibits a distinctly non-equilibrium character, and may therefore be capable of dissipating a significant amount of Chandler wobble energy. Because the only fluid dynamic theory so far proposed to explain the North Sea pole tide contains unresolved difficulties, it may be worthwhile to first explore the implications of such tide characteristics without recourse to fluid dynamics.

The effects of the oceans on wobble are determined from the Liouville equations, which simply represent conservation of angular momentum for a deformable, wobbling earth. For an oceanless Earth the equations can be written

$$\dot{\underline{m}} - i\sigma_e \underline{m} = 0 \; ,$$

where $\underline{m} = m_x + i m_y$ is the dimensionless pole position in complex form. In this form we see that wobble is possible and would be characterized by a frequency σ_e. σ_e depends on the properties of the oceanless Earth (its dynamic flattening, elastic yielding, core fluidity, etc.;

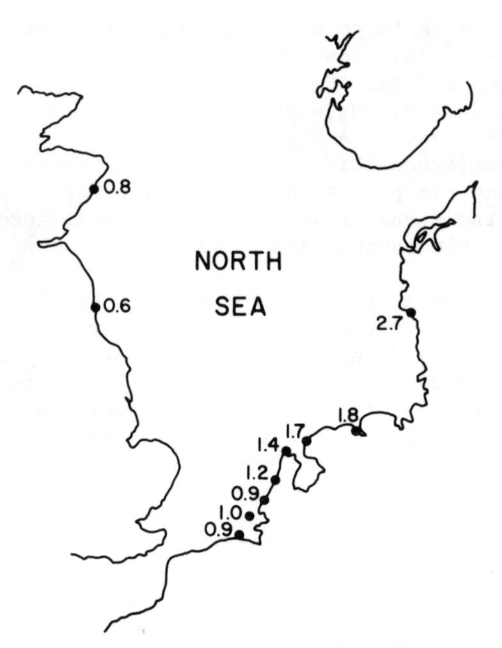

Figure 5. Height of the pole tide, in centimeters, along the coasts of the North Sea; from Miller & Wunsch (1973).

see Smith & Dahlen, 1981 or Dickman, 1982). When oceans are present the Liouville equations become

$$\dot{\underline{m}} - i\,\sigma_e \underline{m} = -i\,\frac{\Omega}{A_m + (1-\kappa)A_c}\,\underline{c}$$

where

$$\underline{c} = -\rho_w a^4 \int T \sin\theta \cos\theta e^{i\lambda}\, ds$$

(see Dickman, 1982 for details). These equations demonstrate that the oceans act to modify polar motion through the tidal products of inertia $\underline{c}$ (which are a function of the tide height T), and also through the relative angular momentum carried by the tide currents; the relative momentum is omitted here for simplicity. That portion of the products of inertia proportional to the wobble amplitude and in phase with it will feed back and alter the wobble frequency, while that portion out of phase with it will produce a decay in the wobble amplitude.

Thus, if we can model the products of inertia of the pole tide according to observation, the effects on wobble can be determined, with fluid dynamical considerations avoided.

I recently modelled the products of inertia accordingly for a number of shallow seas (Dickman, 1979, 1982). The equilibrium pole tide served as reference. It is possible to write the static tide in a complex separation-of-variables form, with the time dependence given by the rotation pole position ($\underline{m}(t)$) and the spatial dependence involving complex spherical harmonic functions. In general we would write

$$T = \underline{X}_{EQ}\underline{m} + \underline{X}^*_{EQ}\underline{m}^*,$$

where for a non-loading non-gravitating static tide $\underline{X}_{EQ}$ depends on the degree 2 order - 1 harmonic function (see Dickman, 1979 for details); all degrees are involved in the case of a loading, gravitating tide (see Dickman 1982, Dickman & Steinberg 1985). I assumed that the actual pole tide can, in the regions of interest, be modelled like the static tide, but with an amplification factor and phase lag corresponding to observation. If in the jth region the amplification factor is N_j and the phase lag is δ_j, then the tide height in that region is taken to be

$$T_j = N_j\underline{X}_{EQ}\underline{m}(t-\delta_j/\sigma) + N_j\underline{X}^*_{EQ}\underline{m}^*(t-\delta_j/\sigma)$$

(σ is the Chandler frequency). In the limiting case where a large number of regions are considered, such expressions would amount to nothing more than a discrete representation of the true pole tide. In general, wherever observations are not specified the tide must be assumed static, with a small correction added in the static regions to insure mass conservation--a necessity because of the enhancements elsewhere.

Some results are shown in Table II. The reference tide (case 0) lengthens the Chandler wobble period by 34 1/2 days [the reference tide here does not include self-gravitation and loading], and fails to damp the wobble at all. But, a tide enhanced in the North-Baltic Seas region with the observed average amplification and phase lag (case 1) has the potential, within the uncertainties, to completely damp the Chandler wobble!

When the tide there is modelled more realistically (case 2), with different average enhancements in the North and Baltic Seas, the results are not as dramatic; roughly, we might say that these Seas contribute on the order of 25% or more--depending on the Q_w assumed--to the attentuation of the Chandler wobble.

Pole tide observations are less definitive but still highly suggestive in other shallow seas. Marseille, whose tide data, smoothed periodogram and unsmoothed spectra are shown in Figure 6 for illustration, has been found by most analysts (e.g. Miller & Wunsch, 1973; ...) to support a pole tide enhanced by at least a factor of two. Although the data is obviously noisy, and contains much power at very long periods as well, such enhancements are based on amplitude estimates above noise level and also include statistical assessments. Data quality has a greater effect on phase estimates, which in most locations outside of the North, Baltic, and Mediterranean Seas are rendered quite uncertain by noise in the data. Table II presents a scenario (case "6") of shallow sea enhancements based on existing data analyses (see Dickman, 1982 for details). It is clear that, through their effects on the products of inertia, the shallow seas have the potential to provide $\sim$ 50% of Chandler wobble damping or more.

TABLE II

POTENTIAL EFFECTS OF SHALLOW-SEA TIDE ON WOBBLE

				Effects on Wobble		
Case #	*Enhanced Seas*	*Amplification Factors*	*Phase Lags*	*Period*	*Damping Time*	Q_w
0	none	N=1	δ=0	34.5 days	∞	∞
1	NORTH/BALTIC SEAS	N=5.5±2.5	δ=40°	35.4 d	75 yr (52,138)	200 (137,366)
2	NORTH SEA	N=3.5±2.5	δ=40°	34.8 d	163 yr (112,300)	433 (297,796)
	BALTIC SEA	N=7.5±2.5	δ=50°			
6	NORTH SEA	N=3.5±2.5	δ=40°	35.2 d	100 yr (68,189)	265 (180,503)
	BALTIC SEA	N=7.5±2.5	δ=50°			
	MEDITERRANEAN	N=2.0±1.0	δ=15°			
	ARABIAN SEA	N=2.0±1.0	δ=30°			
	"KETCHIKAN STRAITS"	N=2.0±1.0	δ=30°			
	BAY OF FUNDY	N=1.5±0.5	δ=5°			

Table II. The effects on wobble of a pole tide enhanced in selected shallow seas. The tidal products of inertia were calculated using the observed amplification factors and phase lags; the stated uncertainties in amplification yield the range of damping times and Q_w values shown in parentheses.

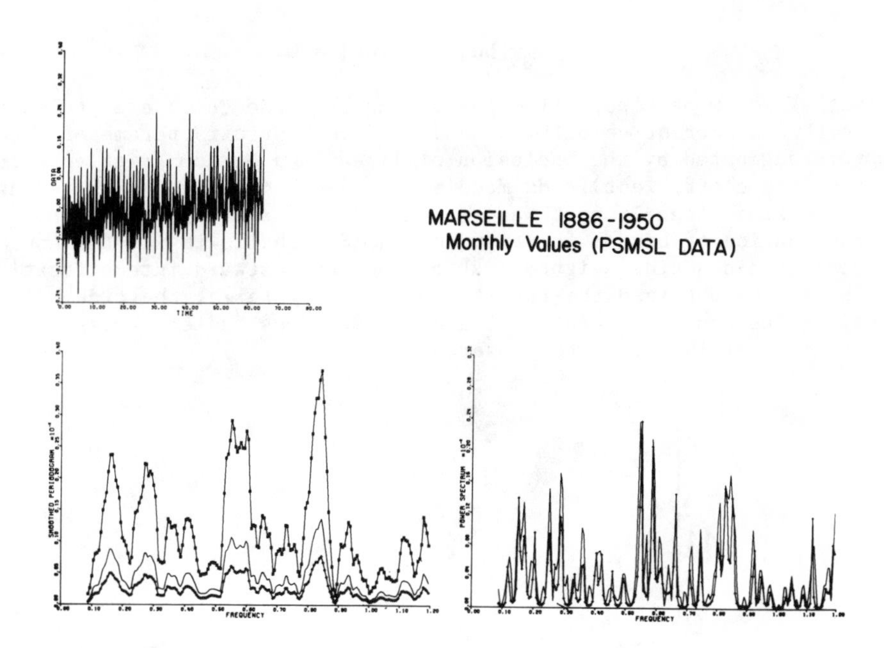

Figure 6. Illustration of the data and analysis techniques employed to determine shallow-sea pole tide characteristics. Monthly data (above left) from Marseille, 1886-1950, was not extended to include more recent values because of irregularities in the data after 1950. Power spectra (lower right), both Fourier and maximum entropy (hatched line), show much low-frequency noise. Smoothed periodogram (lower left), with 80% confidence limits above and below, reveals the pole tide peak at 0.84 cpy more clearly and allows statistical bounds to be placed on the amplitude estimates.

2.2 Theory

One major difficulty with the preceding conclusion is that, since it is based on data along the coasts of the seas only, it fails to include the effects of possible dynamic variability of the tide within the seas' interiors. A complete picture of the pole tide in a shallow sea can be obtained only through fluid dynamical theory. At present, however, our understanding of the fluid dynamic mechanisms responsible for shallow-sea enhancements is limited.

The only shallow-sea pole tide theory yet published is that by Wunsch (1974), proposed to explain the North and Baltic Seas observations. Wunsch began with a <u>steady</u> version of the Laplace Tidal Equations

$$-fv = -g\frac{\partial}{\partial x}[T-V/g]-Pu \qquad fu = -g\frac{\partial}{\partial y}[T-V/g] - Pv$$

$$\frac{\partial}{\partial x}[hu] + \frac{\partial}{\partial y}[hv] = 0$$

with V the wobble potential generating the tide (u, v are the eastward, northward current velocities, and f is the Coriolis parameter); the LTE were augmented by the inclusion of linearized bottom friction (with P the drag coefficient). He modeled the depth h of the North Sea as increasing linearly northward ($h = h_o + \ell' y$) and hoped that the combination of bottom friction and the Sea shallowing to the south would yield a tide height T exhibiting the eastward intensification. In fact he obtained the results shown in Figure 7a; the tide height along the southern coast of the North Sea does indeed increase in a manner reminiscent of the observations.

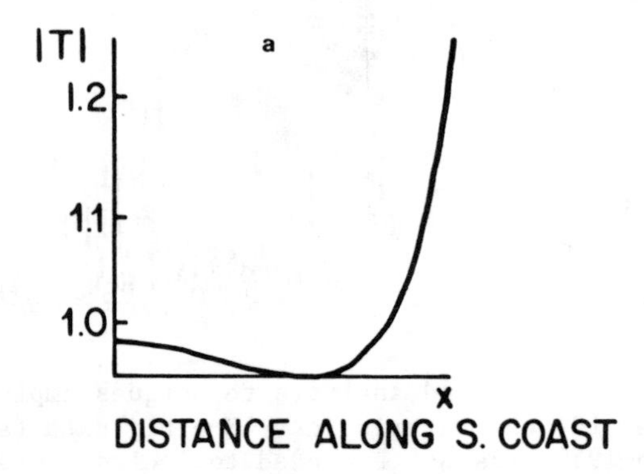

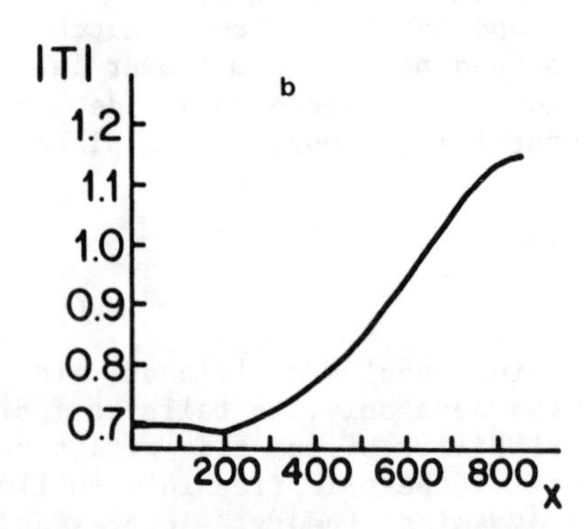

Figure 7. Theoretical predictions of the pole tide height (in centimeters) along the southern coast of the North Sea. (a) from Wunsch (1974), the original solution; (b) the amended solution by Wunsch (1975), as computed by Dickman & Preisig (1985). Note the greater range of tide height in (b) than in (a).

However, it transpired that his solution was erroneous. The following year Wunsch (1975) published an amended solution, without illustration and with only a brief statement indicating that the amended solution was physically inappropriate. In the hope of re-opening the question of North Sea pole tide dynamics, I and a student (J. R. Preisig) have recently reviewed and extended Wunsch's work. We looked first at the amended solution of Wunsch (1975), and discovered that it in fact possessed an eastward intensification more nearly in agreement with observation (Figure 7b). However, there are major difficulties with the solution, including its failure to satisfy one of the original tidal equations, and its failure to satisfy one of the imposed boundary conditions. The boundary conditions specified by Wunsch (1974) include kinematic, "no-flow" conditions at the coastlines, achieved by requiring the stream function to vanish there; and a condition on the tide height at the open northern boundary,

$$T - V/g = B_o \exp(ipx),$$

equivalent to forcing the North Sea tide by a non-equilibrium pole tide in the adjacent Atlantic.

Employing a slightly different depth model ($h = \ell y$) for the North Sea than Wunsch (1974) did and a bottom friction coefficient inversely proportional to the depth ($P = c/h$), we have obtained solutions for the stream function and tide height which satisfy <u>all</u> equations (see Dickman & Preisig, 1985); the tide height is given by

$$T - V/g = \frac{f}{g\ell} \exp\left(\frac{f\ell}{2c}x\right) \sum_n \bar{A}_n \left\{ \cos\left(\frac{n\pi}{b}x\right) \left[\frac{n\pi c}{f\ell b} \frac{\mathrm{SINH}(\lambda_n y)}{\lambda_n y}\right] + \right.$$

$$\left. + \sin\left(\frac{n\pi}{b}x\right)\left[\frac{\mathrm{SINH}(\lambda_n y)}{2\lambda_n y} - \mathrm{COSH}(\lambda_n y)\right]\right\}$$

(here x=b marks the easternmost boundary, and

$$\lambda_n^2 = \left(\frac{f\ell}{2c}\right)^2 + \left(\frac{n\pi}{b}\right)^2 \quad).$$

There are two points to note about our solutions. Firstly, the determination of the arbitrary coefficients $\bar{A}_n$, which is achieved through the "open-water" boundary condition on T, is non-unique; this is a consequence in part of the solution itself, i.e. its x-dependence, and it can be shown that the non-uniqueness remains for any physically reasonable alternate version of the boundary condition. The non-uniqueness is illustrated in Figure 8, which shows the solutions (i, ii, iii) when $\bar{A}_n$ are obtained in different ways from (essentially) the same boundary condition. All solutions show a strong eastward intensification. By imposing a smaller amount of forcing (B_o) at the open boundary than Wunsch (1974) had specified, the amplification can be scaled down--see solutions iv and v, which match the North Sea observations well; but in any event the non-uniqueness is a property of the solution.

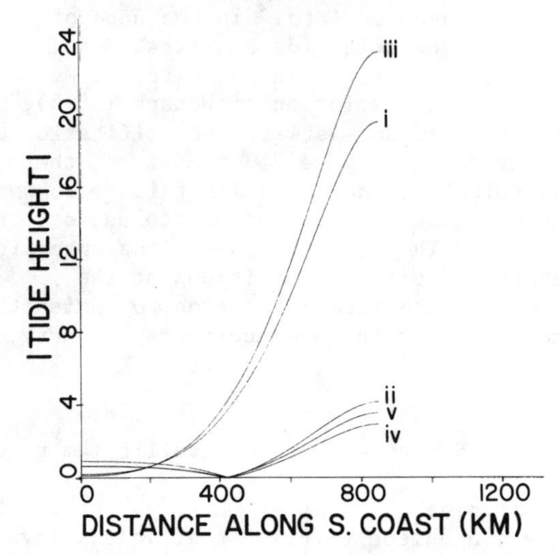

Figure 8. Theoretical non-uniqueness of the pole tide height (in centimeters) along the southern coast of the North Sea. The five different solutions depicted here all satisfy the original tide equations but apply to a slightly different model of the Sea than in Figure 7. (i) - (iii) reflect different determinations of the arbitrary coefficients $\bar{A}_n$ from essentially the <u>same</u> boundary condition at the northern edge of the Sea; (iv) - (v) are scaled-down versions of (iii) which are more or less consistent with observations.

The second point to note is that these solutions imply quite a lot of dissipation of energy (Table III); in fact, any one of these dissipates orders of magnitude more energy than the Chandler wobble does (Chandler dissipation does not exceed 10^8 Watts). This excessive energy dissipation and the non-uniqueness of the solutions both stem from the same source: the failure of the solutions to be constrained, and restrained, by the open-water boundary condition at the northern edge of the Sea. There is enormous dissipation in the northern reaches because the tide height has grown "uncontrollably" and the associated tidal currents have become extremely strong.

At the recent AGU meeting in Baltimore, I described a number of analogies these difficulties share with Stokes' Paradox, a fluid dynamics situation involving creeping flow past a cylinder. The main conclusion to be drawn from the analogy is that the non-uniqueness can probably be removed by including inertia forces in the original tidal equations. When such equations are solved and the open-water boundary condition is at last satisfied, it is likely that--in addition to the solution exhibiting an eastward intensification--the tidal dissipation of wobble energy will be reduced to less than overwhelming amounts. In brief, though, the best that theory can do at this time is only to suggest, in agreement with our earlier discussion, that shallow-sea dissipation of Chandler energy should be significant.

TABLE III

solution	pole tide dissipation
(i)	2.8×10^{12} Watts
(ii)	1.3×10^{12}
(iii)	7.0×10^{12}
(iv)	1.7×10^{11}
(v)	3.4×10^{11}

Table III. Rate of energy dissipation of the North Sea pole tide computed according to the fluid dynamic analysis of Dickman & Preisig (1985). The actual rate of dissipation of Chandler energy is less than 10^8 Watts. The analysis was based more or less on the model of Wunsch (1974), in which the tide behavior results from a combination of bottom friction and the Sea shallowing to the south. As mentioned in the text, the 5 solutions represent different approaches to implementing the northern boundary condition.

3. THE POLE TIDE IN THE DEEP OCEAN

3.1 Observational Indications

Questions concerning the open-ocean pole tide are, in several respects, of a completely different sort than those relating to the shallow-sea tide. First of all, the potential of the open-ocean pole tide to affect wobble is much greater. This is illustrated in Table IV, which gives the effects on wobble period and decay time of a pole tide enhanced uniformly by a factor N over equilibrium and with a uniform phase lag δ_o world-wide [here, core fluidity was not accounted for; roughly, it would increase all effects listed in Table IV by $\sim$ 12.8%]. Enhancements by even a factor of 2, on the average, would require major revision of current models of the Earth's interior because of the effect such a tide would have on the Chandler period. Similarly, small phase lags of the pole tide, world-wide, are sufficient to completely damp the Chandler wobble, even if the actual Q_w is quite low. Such conclusions would have to be qualified if, as is likely, the oceanic pole tide exhibits regional variations of its characteristics (see Dickman, 1979, 1982); nevertheless, it is improbable that the large enhancements at mid-ocean sites (e.g. N $\sim$ 5 at Honolulu) occasionally reported in the literature (see Dickman, 1982 for a review) are accurate. Conversely, of course, the true enhancements need not be large for their consequences to be great.

The problem evidently reduces to the question of what the actual mid-ocean pole tide characteristics really are. Here it is important that the tide observations be compared with as realistic an equilibrium tide as possible, i.e. one calculated including the effects of self-gravitation and loading (self-gravitation and loading are unlikely to much affect the shallow-sea tide). Although the theory of such a "realistic" static tide was developed as early as 1973 (Merriam, 1973; see also Dahlen, 1976), proper comparisons have only recently been attempted. Following a suggestion by Dahlen (1976) which was not pursued by him, Daniel Steinberg and I have developed an algorithm for calculating the self-gravitating and loading pole tide in non-global oceans overlying an elastic earth (Dickman & Steinberg, 1985). Figure 9a depicts the r.m.s. values (over a cycle of wobble) of this tide in response to a wobble of typical amplitude (0.16"); continental outlines are crude because they are based on an ocean function expanded in spherical harmonics only out to degree & order 24. What is significant about our algorithm is that it greatly reduces the artificial distortions in the tide height produced by such a truncated ocean function; Figure 9b, in contrast, shows the same tide but calculated by us from the old algorithm of Dahlen (1976).

TABLE IV

GLOBALLY ENHANCED POLE TIDE

δ_o	0°	15°	30°	45°	60°
	Contribution to Chandler Period				
N=1	31 days	30	27	23	17
N=2	57	56	53	46	36
N=4	102	100	97	91	82
N=6	136	136	136	138	NR
N=8	165	166	170	198	NR
	Damping of Chandler Wobble				
N=1	∞	9.4 yr	4.8	3.3	2.6
N=2	∞	4.7	2.4	1.6	1.2
N=4	∞	2.3	1.1	0.7	0.5
N=6	∞	1.5	0.7	0.4	NR
N=8	∞	1.1	0.5	0.2	NR

Table IV. Hypothetical effects on wobble of a pole tide enhanced uniformly world-wide. Tidal products of inertia were calculated using the specified amplification factors (N) and phase lags (δ_o).

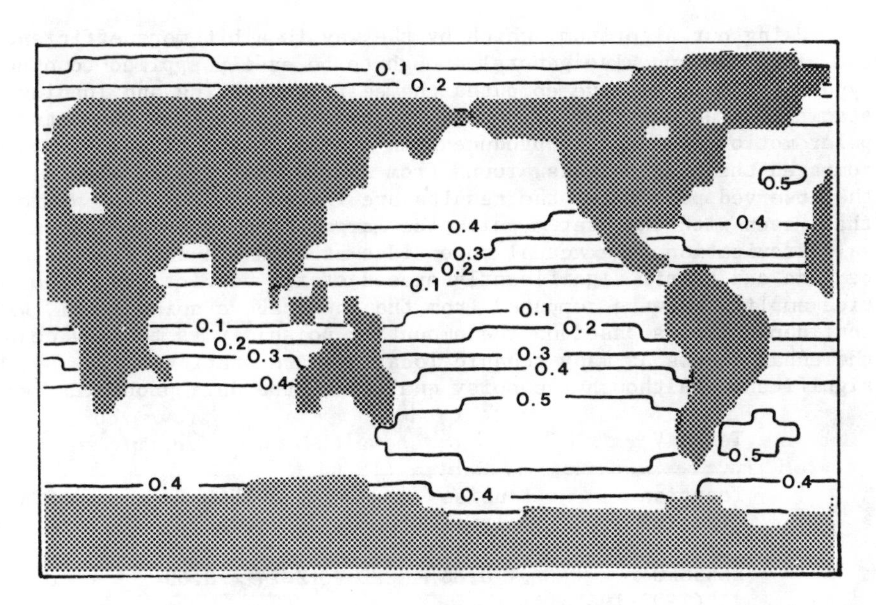

Figure 9a. The rms static pole tide, averaged over a cycle of wobble of amplitude 0.16", including the effects of self-gravitation and loading. Tide heights are contoured in centimeters.

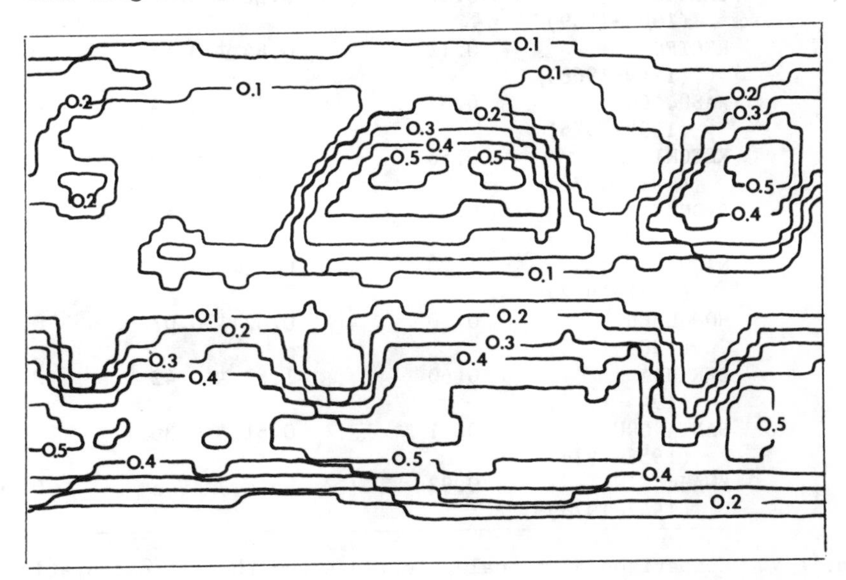

Figure 9b. Same as Figure 9a but computed using an older algorithm which is strongly dependent on the extent of ocean-function truncation. The tide height shown here has been distorted by a Gibbs phenomenon and fails to vanish completely on land.

Using our algorithm, which by the way is a bit more efficient than the older one and also general enough to be easily applied to other types of tides, we have computed the self-gravitating and loading static tide at a number of ports; in these computations the actual polar motion was used to produce time series of the tides. We then compared the tide heights, found from Fourier spectral analysis, with the observed pole tide; the results are listed in Table V. We found that, even with the static pole tide larger than otherwise due to self-gravitation, the actual pole tides at such mid-ocean sites as Bermuda and Hawaii significantly exceed static predictions. The actual tide amplitudes were computed from the power above noise level, with confidence limits based on the amount of noise; it is thus evident that the enhancements at many oceanic locations are statistically significant. Although the noisy quality of the data prohibits reliable

TABLE V

location (years)	noise level	ACTUAL tide	STATIC tide
ESBJERG (1900-1965)	0.86	2.74 ± 0.60	0.41 cm
ABERDEEN (1916-1965)	0.53	0.90 ± 0.37	0.39
BALBOA (1908-1969)	0.86	0.26 ± 0.61	0.12
RECIFE (1949-1968)	0.19	0.84 ± 0.14	0.13
WASH. DC (1932-1975)	0.34	1.03 ± 0.24	0.42
KETCHIKAN (1919-1972)	0.56	1.11 ± 0.40	0.36
SYDNEY (1900-1977)	0.17	0.65 ± 0.12	0.41
MIDWAY (1947-1972)	1.20	below noise	0.44
HONOLULU (1905-1975)	0.10	0.52 ± 0.07	0.30
BERMUDA (1944-1972)	0.60	1.67 ± 0.42	0.49
SANTA CRUZ (1958-1974)	0.51	0.61 ± 0.36	0.53
KWAJALEIN (1947-1975)	0.93	below noise	0.18

Table V. Comparison of actual pole tides and theoretical static tides at various ports. Monthly time series of the self-gravitating, loading static pole tide were constructed for the same time spans as the actual data. Amplitudes were determined through Fourier spectral analysis; uncertainties were based on noise levels near the Chandler frequency.

determination of phase lags, it makes sense that these enhancements--a reflection of the non-equilibrium nature of the actual tide--must be accompanied by non-zero phase lags. Therefore, some amount of dissipation of wobble energy by the open-ocean pole tide is indicated.

Because of the small effect of shallow-sea enhancements on wobble period and ellipticity, any departures of the observed period and ellipticity from that calculated assuming a static open-ocean pole tide could be used to infer the average world-wide extent of enhancement (see Smith & Dahlen, 1981 and Zschau, this issue, for alternative approaches). Due to lack of time we will not explore this aspect here. In summary, however, from the works of Dickman & Steinberg (1985) and perhaps Daillet (1979) we might infer a globally averaged $N \cong 1.2$ and $\delta_o \sim 10^o$; these would suggest a wobble decay time by the open-ocean tide of $\sim$ 15 years ($Q_w \sim 40$).

3.2 Theory

As in the case of the shallow-sea tide, we must turn to fluid dynamic theory in order to obtain a spatially complete picture of the pole tide, as well as to explain those observations of enhancement which are statistically significant. Few theoretical treatments of the open-ocean pole tide have been published; due to their approximate natures they must all be taken as inconclusive (see Dickman, 1985 for a summary). I have recently published a theory of the tide (Dickman, 1985); it begins with the Laplace Tidal Equations

$$\dot{u}_\theta - fu_\lambda = -g\frac{\partial}{a\partial\theta}[T - V/g] - Pu_\theta$$

$$\dot{u}_\lambda + fu_\theta = -g\frac{\partial}{a\sin\theta\,\partial\lambda}[T - V/g] - Pu_\lambda$$

$$\dot{T} = -\frac{1}{a\sin\theta}\left\{\frac{\partial}{\partial\theta}[hu_\theta\sin\theta] + \frac{\partial}{\partial\lambda}[hu_\lambda]\right\}$$

including bottom friction as before, and now in spherical coordinates. I make no approximations to these equations, but my theory is limited to the case of global oceans of uniform depth. Using a spherical harmonic approach, I can write the LTE as a matrix equation whose inversion yields the tide height coefficients directly. Because the tide magnitude depends on the imposed frequency, but the wobble frequency is affected by the tide, this equation must be solved jointly with the Liouville equation, which determines how the tide affects wobble. Now our wobble equation must include the relative angular momentum $\underline{\ell}$ associated with tidal currents;

$$\dot{\underline{m}} - i\sigma_e \underline{m} = -i \frac{\Omega}{A_m + (1-\kappa)A_c} \{\underline{c} + \underline{\ell}/\Omega\}$$

where

$$\underline{\ell} = -2\rho_w a^3 h \int e^{i\lambda} \cos\theta [u_\lambda - iu_\theta \cos\theta] ds.$$

As expressed here it accounts for core fluidity, which largely decouples the core from mantle wobble ($\kappa \approx 1$; see Dickman, 1982 or Smith & Dahlen, 1981) and therefore boosts the oceanic effect.

TABLE VI

SELF-CONSISTENT POLE TIDE
IN
GLOBAL OCEANS

(M_P = 0.16", $2\pi/\sigma_e$ = 364.1 days)

P	$\underline{\sigma}$		T_{MAX}
1.5x10-3sec-1	433.40 days	25 years	0.8532 cm
$1.5\text{x}10^{-4}$	433.44	166	0.8539
$1.5\text{x}10^{-5}$	432.93	182	0.8496
$1.5\text{x}10^{-6}$	432.50	847	0.8444
→ $1.5\text{x}10^{-7}$	432.49	8345	0.8442
→ $1.5\text{x}10^{-8}$	432.49	82485	0.8442
$1.5\text{x}10^{-9}$	432.49	734859	0.8442

Table VI. Characteristics and effects on wobble of the dynamic pole tide in global uniform oceans calculated as the joint solution to the Laplace tidal equations and the Liouville equations. Results are shown for a range of bottom friction strengths, the most likely values of which are indicated with arrows, and for a wobble amplitude of 0.16 arcsec. With the earth model employed, a static pole tide would be of maximum amplitude 0.8537 cm, and would produce a Chandler period of 433.3 days and an infinite decay time.

The results of the joint solution are summarized in Table VI. We see that the pole tide in global uniform oceans would, for realistic values of bottom friction, be very nearly equilibrium; it affects the wobble period by only $\sim$ 1 day differently than a static tide would, and it provides only marginal dissipation of wobble energy. As stated earlier, of course, the relevance of these results to the actual pole tide in realistic oceans is yet unclear. I have recently begun to investigate the case of oceans with continents present; I have managed to obtain a solution to the LTE in this case, by matrix inversion as before, but I still need to implement the boundary conditions....

4. FINAL REMARKS

In conclusion, the combination of theoretical and observational analyses strongly suggests that the oceans are responsible for a significant fraction of wobble energy dissipation. Whether or not they are the primary "sink" of wobble energy is so far unresolved. Depending on the true Q_w, it is possible that the shallow seas alone dissipate the bulk of wobble energy.

We obviously need to obtain a longer duration of good-quality polar motion data in order to accurately determine Q_w. Accurate estimation of the solid-earth tidal Q may help to constrain Q_w as well. And, more good-quality deep-ocean tide data need to be collected in order to establish the role of the world-wide pole tide. Finally, it is possible that the next few years may see rigorous elucidation of the fluid dynamic basis for the pole tide, in both the shallow seas and the deep ocean.

Thank you, and I apologize for going over my time limit!

5. REFERENCES

Anderson, D. L., 1974. 'Earthquakes and the rotation of the Earth', Science, **186**, 49-50.

Bondi, H. and T. Gold, 1955. 'On the damping of the free nutation of the earth', Mon. Not. Roy. Astr. Soc., **115**, 41-46.

Braginskiy, S. I. and V. M. Fishman, 1976. 'Electromagnetic interaction of the core and mantle when electrical conductivity is concentrated near the core boundary', Geomag. Aeron., **16**, 443-446.

Braginskiy, S. I. and V. V. Nikolaychik, 1973. 'Estimation of the electrical conductivity of the Earth's lower mantle from the lag of an electromagnetic signal', Izvestiya Akad. Nauk SSSR (Fizika zemli), **9**, 77-79.

Brillinger, D. R., 1973. 'An empirical investigation of the Chandler Wobble and two proposed excitation processes', Bull. Int. Statist. Inst., **45** (3), 413-434.

Bukowinski, M. and L. Knopoff, 1976. 'Electronic transition in iron and the properties of the core', in The Physics and Chemistry of Minerals and Rocks, ed. R. G. J. Strens, 491-508, John Wiley & Sons, London.

Colombo, G. and I. I. Shapiro, 1968. 'Theoretical model for the Chandler wobble', Nature, **217**, 156-157.

Currie, R. G., 1974. 'Period and Q_w of the Chandler Wobble', Geophys. J. Roy. Astron. Soc., **38**, 179-185.

Currie, R. G., 1975. 'Period Q_p and amplitude of the pole tide', Geophys. J. Roy. Astron. Soc., **43**, 73-86.

Dahlen, F. A., 1976. 'The passive influence of the oceans upon the rotation of the earth', Geophys. J. Roy. Astron. Soc., **46**, 363-406.

Daillet, S., 1981. 'Secular variation of the pole tide: correlation with Chandler wobble ellipticity', Geophys. J. Roy. Astr. Soc., **65**,

Dickman, S. R., 1979. 'Consequences of an enhanced pole tide', J. Geophys. Res., **84**, 5447-5456.

Dickman, S. R., 1982. 'The pole tide and its geophysical consequences' (review article), Geophys. J. R. astr. Soc., submitted.

Dickman, S. R., 1983. 'The rotation of the ocean - solid earth system', J. geophys. Res., **88**, 6373-6394.

Dickman, S. R., 1985. 'The self-consistent dynamic pole tide in global oceans', Geophys. J. R. astr. Soc., **81**, 157-174.

Dickman, S. R. & J. R. Preisig, 1985. 'Another Look at North-Sea Pole Tide Dynamics: Research Note', Geophys. J. R. astr. Soc., submitted.

Dickman, S. R. & D. J. Steinberg, 1985. 'New aspects of the equilibrium pole tide', Geophys. J. R. astr. Soc., to be submitted.

Gans, R. F., 1972. 'Viscosity of the Earth's Core', J. Geophys. Res., **77**, 360-366.

Herring, T. A., C. R. Gwinn, and I. I. Shapiro, 1985. 'Geophysics by radio interferometry: Power-spectral-density function of the earth's nutations', EOS, **66**, 245.

Jeffreys, H., 1915. 'The viscosity of the Earth', Mon. Not. R. astr. Soc., **75**, 648-658.

Jeffreys, H., 1920. 'The chief cause of the lunar secular acceleration', Mon. Not. R. astr. Soc., **80**, 309-317.

Jeffreys, H., 1968. 'The Variation of Latitude', Mon. Not. Roy. Astr. Soc., **141**, 255-268.

Kanamori, H., 1977. 'The energy release in great earthquakes', J. Geophys. Res., **82**, 2981-2988.

Kanamori, H. and D. L. Anderson, 1977. 'Importance of physical dispersion in surface-wave and free-oscillation problems: Review', Rev. Geophys. Space Phys., **15**, 105-112.

Kelvin, Lord (Sir William Thomson), 1876. 'Effects of elastic yielding on precession and nutation' (Presidential Address, British Association), reprinted in Mathematical Physical Papers, **3**, 320-335, Cambridge University Press.

Lagus, P. L. & Anderson, D. L., 1968. 'Tidal dissipation in the Earth and planets', Phys. Earth planet. Int., **1**, 505-510.

Lambeck, K., 1980. The Earth's Variable Rotation: Geophysical Causes and Consequences, Cambridge University Press.

Liu, H.-P., D. L. Anderson, and H. Kanamori, 1976. 'Velocity Dispersion due to Anelasticity; Implications for Seismology and Mantle Composition', Geophys. J. Roy. Astr. Soc., **47**, 41-58.

Mansinha, L. & Smylie, D. E., 1967. 'Effect of earthquakes on the Chandler wobble and the secular polar shift', J. geophys. Res., **72**, 4731-4743.

Merriam, J. B., 1973. Equilibrium tidal response of a non-global self-gravitating ocean on a yielding earth, M.S. thesis, Memorial University of Newfoundland, St. John's.

Miller, S. P., 1973. Observations and interpretation of the pole tide, M.Sc. thesis, Massachusetts Institute of Technology.

Miller, S. P. & Wunsch, C., 1973. 'The pole tide', Nature Phys. Sci., **246**, 98-102.

Munk, W. H. & MacDonald, G. J. F., 1960. The Rotation of the Earth, Cambridge University Press (reprinted with corrections, 1975).

Naito, I., 1979. 'Effects of the pole tide on the Chandler wobble', J. Phys. Earth, **27**, 7-20.

O'Connell, R. J. and A. M. Dziewonski, 1976. 'Excitation of the Chandler wobble by large earthquakes', Nature, **262**, 259-262.

Okubo, S., 1982. 'Theoretical and observed Q of the Chandler wobble - Love number approach', Geophys. J. R. astr. Soc., **71**, 647-657.

Ooe, M., 1978. 'An optimal complex AR.MA model of the Chandler Wobble', Geophys. J. Roy. Astron. Soc., **53**, 445-457.

Rochester, M. G., 1970. 'Polar wobble and drift: A brief history', in Earthquake Displacement Fields and the Rotation of the Earth, edited by L. Mansinha, D. E. Smylie, and A. E. Beck, p. 3-13, D. Reidel, Hingham, Mass.

Rochester, M. G., 1973. 'The Earth's rotation', EOS, **54**, 769-781.

Rochester, M. G. and D. E. Smylie, 1965. 'Geomagnetic core-mantle coupling and the Chandler wobble', Geophys. J. Roy. Astr. Soc., **10**, 289-315.

Sekiguchi, N., 1972. 'On some properties of the excitation and damping of the polar motion', Publ. Astron. Soc. Japan, **24**, 99-108.

Smith, M. L. & Dahlen, F. A., 1981. 'The period and Q of the Chandler wobble', Geophys. J. R. astr. Soc., **64**, 223-281.

Smylie, D. E., G. K. C. Clarke and T. J. Ulrych, 1973. 'Analysis of

Irregularities in the Earth's Rotation', in Methods in Computational Physics, **13**, 391-430.
Stacey, F. D., 1977. Physics of the Earth, 2nd edn., Wiley, New York.
Taylor, G. I., 1919. 'Tidal Friction in the Irish Sea', Phil. Trans. Roy. Soc. Lond., A, **220**, 1-33.
Toomre, A., 1974. 'On the "Nearly Diurnal Wobble" of the Earth', Geophys. J. Roy. Astr. Soc., **38**, 335-348.
Wilson, C. R., 1979. 'Estimation of the parameters of the Earth's polar motion', in Time and the Earth's Rotation (**I.A.U. Symp. No. 82**), ed. D. D. McCarthy & J. D. H. Pilkington, 307-312, D. Reidel, Dordrecht.
Wilson, C. R. and R. Haubrich, 1976. 'Meteorological excitation of the Earth's wobble', Geophys. J. Roy. Astr. Soc., **46**, 707-743.
Wilson, C. R. and R. O. Vicente, 1980. 'An analysis of the homogeneous ILS polar motion series', Geophys. J. R. Astr. Soc., **62**, 605-616.
Wunsch, C., 1974. 'Dynamics of the pole tide and the damping of the Chandler wobble', Geophys. J. Roy. Astron. Soc., **39**, 539-550.
Wunsch, C., 1975. 'Errata', Geophys. J. Roy. Astron. Soc., **40**, 311.
Yatskiv, Ya.S., 1973. 'Chandler Wobble and Free Diurnal Nutation Derived from Latitude Observations', Proc. 2nd Int. Symp. Geodesy & Physics of the Earth, Potsdam, 143-151.
Zhang, H., 1983. 'On the Chandler Wobble'. I. Theory, Chin. Astr. Astrophys., **5**, 460-468 (1981).
Zhang, H., Y. Han, and D. Zheng, 1983. 'On Chandler Wobble--a result on analysis of observations', Scientia Sinica, A, **26**, 181-192.

THE INFLUENCE OF EARTHQUAKES ON THE POLAR MOTION

Annie Souriau
Centre National d'Etudes Spatiales
Groupe de Recherche de Géodésie Spatiale
18, Avenue Edouard Belin
31055 TOULOUSE CEDEX - FRANCE

ABSTRACT

The high accuracy of the recent polar data obtained by space technics, and the availability of the seismic moment tensors of the strong and moderate earthquakes, make it possible to investigate with an increased confidence the influence of earthquakes on the excitation of the Chandler wobble. The comparison of the observed Chandler wobble for 1977-1983, with the synthetic one due to the cumulative effect of the worldwide seismicity during that period, leads to discard the direct effect of earthquakes as a major source of the maintain of the Chandler wobble.

The lack of very strong earthquakes during the last two decades has made impossible the observation of the effect of a single earthquake on the pole path ; however, with the present polar data accuracy, the polar shift of about 10 m.a.s. due to an earthquake of seismic moment $\sim 10^{30}$ dyne.cm would certainly be detected, as well as possible pre or post-seismic deformations.

Various physical mechanisms likely able to induce the observed Chandler wobble amplitude variations can be modelled as step perturbations of the Earth inertia tensor. If such is the case, a secular drift will result from the Chandler wobble excitation. The modelization of this drift for an excitation by random steps shows that the polar drift varies roughly as $\sqrt{t}$ (t = time), whereas the observed one is nearly linear in time. This indicates that Chandler wobble and secular drift may each have specific excitation mechanisms. However, because of their interdependence through the excitation functions, they have to be modelled simultaneously.

INTRODUCTION

The mechanism of excitation of the Chandler wobble, the 14-month period free nutation of the Earth, has been for half a century the

A. Cazenave (ed.), Earth Rotation: Solved and Unsolved Problems, 229–240.

subject of many investigations. Seismic activity as a possible source of excitation relies upon the estimation that the energy required to maintain the wobble is several orders less than the energy released by the earthquakes. The numerous controversies about the possible influence of earthquakes on the Chandler wobble during the last two decades (see Munk and MacDonald, 1960 and Lambeck, 1980 for a review) are due to theoretical and experimental reasons. First, the accuracy of the polar data deduced from astronomical observations was rather poor, around 0.01 arc sec, so that the identification of possible breaks in the pole path related to earthquakes was not an easy task (Smylie and Manshina, 1968, 1970 ; Haubrich, 1970 ; Mansinha et al., 1979). Secondly, it was necessary to develop models of seismic source convenient to represent the earthquake displacement field at teleseismic distances. Successive refined models are given by Mansinha and Smylie, 1967 ; Ben Benahem and Israel, 1970 ; Smylie and Mansinha, 1971 ; Dahlen, 1971, 1973 ; Israel et al., 1973. Another difficulty is the quantification of the energy realeased in the earthquakes. It is now well established that most of the magnitude scales are inadequate to describe the very strong earthquakes. The use of empirical moment-magnitude relationships has lead to overestimations of the seismic moment tensors for earthquakes in the past (Kanamori, 1976, 1977) ; this incites to be suspicious about the modelizations and correlations based on the historic seismicity.

The Earth rotation parameters are now determined with a high accuracy thanks to space technics, in particular Laser and VLBI ($\sigma < 0.001$ arc sec for the pole position ; $\sigma \sim 50$ µs for the length of day). On the other hand, earthquakes are now well quantified from their seismic moment tensors and focus parameters. Then time is up to consider once more the influence of earthquakes on the polar motion, on the basis of these new data.

THE DIRECT EFFECT OF EARTHQUAKES

Effect of a single earthquake

The mass redistribution induced by an earthquake is usually modelled as a step perturbation of the Earth inertia tensor occurring instantaneously. Such a representation takes into account only the static displacement field due to the rupture, but ignores the pre or post-seismic displacements. This "direct effect" of earthquakes can theoretically be observed through three effects :

1) it perturbs the amplitude and phase of the Chandler wobble,
2) it induces a shift of the mean pole position and then contributes to the polar secular drift,
3) it induces a change in the length of day.

The change in the length of day is controlled by the term I_{33} of the Earth inertia tensor (axis 3 is the mean rotation axis ; axes 1 and 2 are in the equatorial plane). It is of the order of 10 µs for an earthquake of seismic moment tensor 10^{30} dyne.cm, i.e. for a very strong earthquake such as the 1960 Chilean or the 1964 Alaskan earthquakes (O'Connell and Dziewonski, 1976). This lies within the noise of the data ; moreover, as short term atmospheric effects have a much more important contribution and are not modelled with such an accuracy, it is hopeless to detect any seismic effect in the length of day at the present time.

The Chandler wobble and the secular drift excitation are controlled by the terms I_{13} and I_{23} of the Earth inertia tensor. The perturbations I_{13} and I_{23} induced by an earthquake are (Dahlen, 1973) :

$$\Delta I_{13} = \sum_{k=1}^{3} \Gamma_k (h)\, f_{13}^k (S)$$
$$\Delta I_{23} = \sum_{k=1}^{3} \Gamma_k (h)\, f_{23}^k (S) \qquad (1)$$

where f_{13} and f_{23} are functions of the source coordinates and source parameters, and the Γ_k are functions of the depth h of the source and of the Earth model. The components (ψ_1, ψ_2) of the relevant excitation function $\underline{\psi}$, which give the shift of the mean pole position, are proportional to these perturbations.

Seismic moment tensors are systematically determined for moderate and major earthquakes since 1977 with the centroid moment tensor method (Dziewonski et al., 1981). This method derives, mainly from mantle waves, the best point source equivalent to a seismic source of finite size. Other moment tensor and fault plane solutions are also published for major earthquakes (see for example the "monthly listings of preliminary determination of epicenters", U.S.G.S. bulletins). Table 1 gives the main polar shifts obtained for single earthquakes from 1977 to 1985. The main shift, for the 1977 Sumbawa earthquake, does not exceed 0.19 m.a.s. (0.59 cm) at a time where the polar data accuracy was $\sigma \sim 6$ m.a.s. at best (Lageos series from Austin). For the 1985 Chilean earthquake, the shift of 0.17 m.a.s. also lies within the accuracy of the present data ($\sigma \gtrsim 3$ m.a.s.). Note that the geometry and location of the rupture are very important : for example, the 1977 Rumanian and the 1979 Ecuadorian earthquakes give the same polar shift, with a seismic moment for the second one 10 times higher than for the first one.

The unability to detect any break in the pole path related to earthquakes, in spite of the improvement of the polar data accuracy, is due to the lack of very strong event in the seismic activity during the last two decades. The highest seismic moments have been around 10^{28} dyne.cm. On the contrary, earthquakes of strong magnitudes corresponding to seismic moments 10^{30}-10^{31} dyne.cm have been frequently

Year	Month	Day	Origin	Seismic moment * 10^{27} (dyne-cm)	Shift amplitude (mas)	Shift direction (° E)
1977	8	19	Sumbawa	35.9	0.19	157
1985	3	3	Chile	10.0	0.17	110(*)
1977	6	22	Tonga	14.3	0.12	197
1983	5	26	Honshu	4.55	0.093	135
1985	9	19	Mexico	16.3	0.076	314(*)
1980	7	17	Sta Cruz	4.84	0.044	329
1977	11	23	Argentina	1.86	0.041	114
1983	10	4	Chile	3.38	0.040	110
1983	3	18	New Ireland	4.63	0.034	279
1978	6	12	Honshu	2.04	0.033	132
1978	12	6	Kuril	6.40	0.033	125
1979	12	12	Ecuador	16.9	0.033	220
1978	3	24	Kuril	2.28	0.031	136
1977	3	4	Rumania	1.61	0.030	20

Table 1. Main polar shifts induced by earthquakes between January 1977 and November 1985. Centroid moment tensor solutions ; (*) : preliminary determinations. SNREI Earth model (Dahlen, 1968)

reported in the past. Some of them have certainly been overestimated (see Kanamori, 1976, 1977) ; however, the most recent ones are rather well documented. For example, the 1960 Chilean earthquake, with a seismic moment around 0.8 10^{30} dyne.cm, induces a polar shift $\sim$ 20 m.a.s. (see Table 1 in Mansinha et al., 1979), which would be readily observed on the present polar data. Even more, it would probably be possible, for such very strong earthquakes, to choose between several focal solutions or between different rheological parameters of upper mantle models.

Cumulative effect of earthquakes

Even if a single earthquake does not have a significant contribution to the excitation of the Chandler wobble, one can question whether the cumulative effect of the whole world seismicity would be able to maintain the Chandler wobble. The pole position $\underline{m}$ (t) resulting from a series of earthquakes occurring at times t_i is given by (Mansinha and Smylie, 1967) :

$$\underline{m}(t) = \underline{m}_0 \; e^{j2\pi\sigma_0 t} + \sum_{0<t_i<t} \underline{\psi}_i - \sum_{0<t_i<t} c \; \underline{\psi}_i \; e^{j2\pi\sigma_0 (t-t_i)} \qquad (2)$$

where σ_0 is the complex Chandler frequency which takes into account the quality factor, $\underline{m}_0$ is the pole position at time t = 0 (to be defined), and c is a constant. In the right hand side of this equation, the first term represents the damped free oscillation of the pole in the absence of excitation, the last term represents the amplitude and phase perturbations of the Chandler term, and the second term is a secular drift.

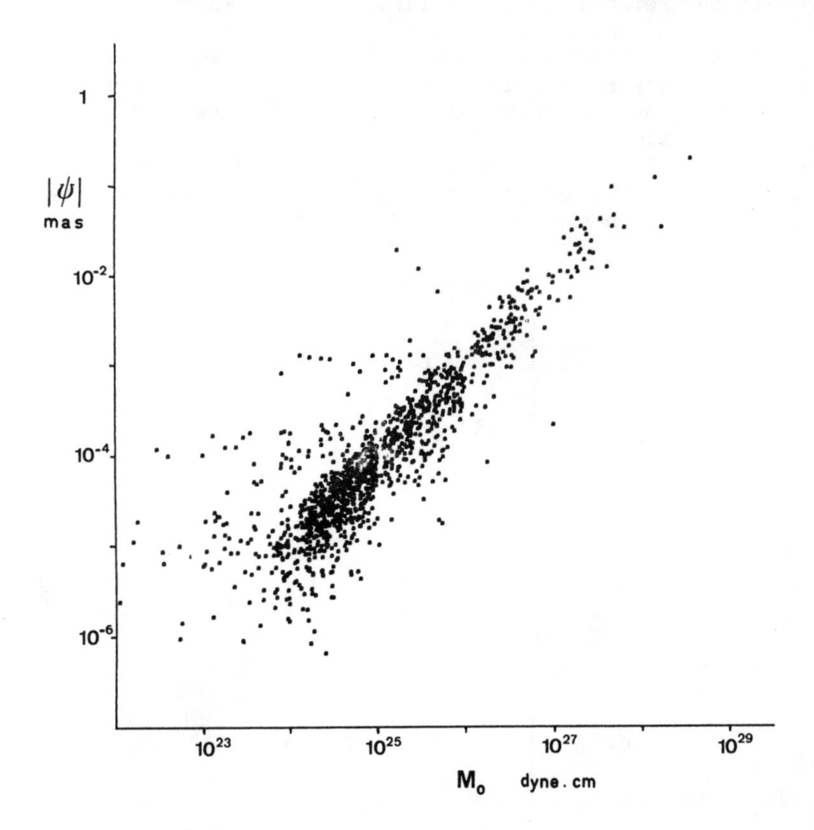

Figure 1. Relationship between the amplitude of the polar shift and the seismic moment for major and moderate earthquakes occurring between 1977 and 1983 (1287 events). A slope of 1 is expected for random geometries of the focal solutions.

An interesting modelization of the Chandler wobble for the period 1901-1970 has been carried out by O'Connell and Dziewonski (1976), on the basis of large earthquake historic seismicity, with constraints on the fault geometry given by plate tectonics. However, the success of this modelization is questionable because of the overestimate by a factor $\sim$ 100 of most of the seismic moments used by the authors (Kanamori, 1976) and because of the incompatibility of the relevant theoretical secular drift with the observed one (Souriau and Cazenave, 1985).

For the period 1977-1983, the seismic parameters computed with the centroid moment tensor method are available for all the major and moderate earthquakes, i.e. for 1287 events with seismic moments $\geq 10^{23}$ dyne.cm (Harvard University data). Figure 1 gives the relevant polar shifts $|\psi|$ as a function of the seismic moments M_0. It is consistent with the relation $\ln |\psi| = \ln(M_0) + cst$, expected for source solutions with random geometries. However, the theoretical secular drift due to these earthquakes shows that the fault plane solutions are not random for the very strong earthquakes (Figure 2) : the drift is mainly controlled by a very small number of major events ; it is strongly polarized, it does not exhibit the features of a two-dimensional random walk as would be expected for random shifts. Note that this polar drift is completely negligible compared to the observed one, roughly 3.3 m.a.s. per year.

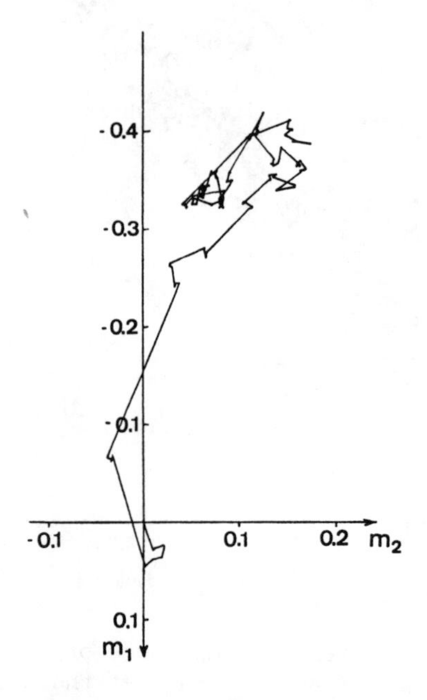

Figure 2. Cumulative polar shift induced by the major and moderate earthquakes (1287 events) for the period 1977-1983. m_1 to the origin meridian ; m_2 to the East ; values in m.a.s..

The synthetic Chandler wobble due to the cumulative effect of the 1287 earthquakes has been compared to the Chandler term for the same period (1977-1983), obtained by filtering the BIH data for the annual term and the secular drift (Souriau and Cazenave, 1985). The synthetic curve does not reproduce the observed one ; in particular, it is unable to account for the rapid amplitude increase of the Chandler wobble between 1980 and 1983 (figure 3). A study by Gross (1985) leads to a similar conclusion. These results definitely dismiss the direct effect of earthquakes as a major source of excitation of the Chandler wobble.

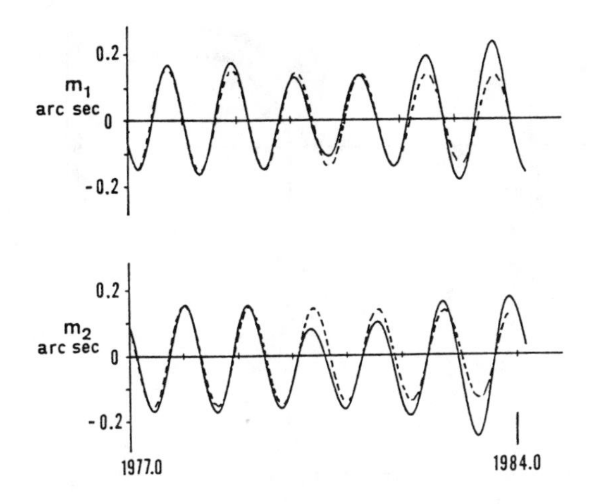

Figure 3. Comparison of the observed Chandler wobble deduced from BIH data (circular D) with the synthetic wobble due to the seismic activity between 1977 and 1983. m_0, in equation (2), is the chosen equal to the observed value at the beginning of the serie, January 1977, and the quality factor Q = 100.

Pre and post-seismic deformations

Theoretical studies have considered the effect of the mantle rheology on the post-seismic deformation and wobble excitation (Sabadini et al., 1984 ; Slade et al., 1984). The viscosity of the upper mantle induces both an amplification and a time lengthening of the perturbations of the Earth inertia tensor. These effects are by far not sufficient to allow for an explanation of the wobble excitation by earthquakes ; nevertheless, it is hopeful to detect them on the polar data for very strong earthquakes.

Slow aseismic deformations preceding or following strong earthquakes have often been reported. If these deformations displace a significant amount of material out of isostatic equilibrium and if they have a short time constant compared to the wobble period, they may be observed on the pole path. Figure 4 gives the pole path from January 1983 to October 1985 (Lageos series, Austin University data). No obvious perturbation can be observed in relation to a major earthquake, except perhaps for the 1985 Chilean earthquake. Besides, some breaks can be observed inside aseismic periods (July-August 1983, for example).

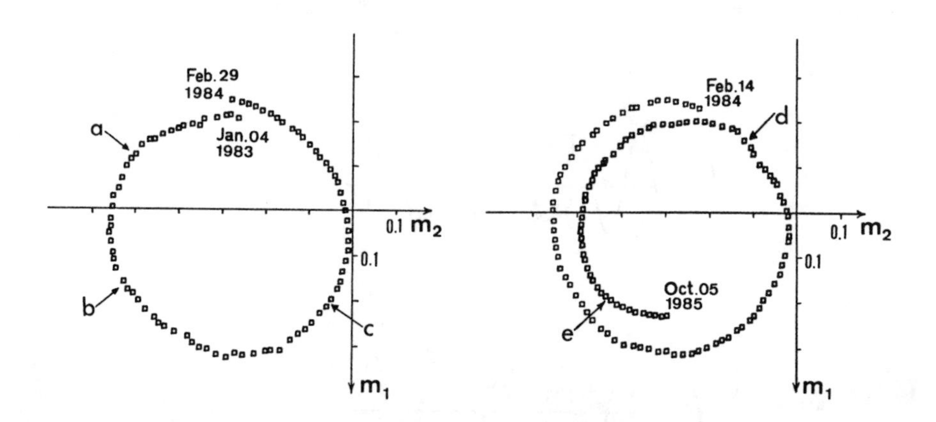

Figure 4. Lageos serie from Austin Univ. (ERP (CSR) 84L02) from January 1983 to October 1985, with approximately one point each 5 days. m_1 to the origin meridian, m_2 to the East, values in arc sec. Arrows indicate the main earthquakes (see table 1) : (a) New Ireland, March 18, 1983 ; (b) Honshu, May 26, 1983 ; (c) Chile, October 4, 1983 ; (d) Chile, March 03, 1985 ; (e) Mexico, September 19, 1985.

At a secular scale, correlations have been observed between the seismic activity and the amplitude of the Chandler wobble or the length of day (Myerson, 1970 ; Anderson, 1974 ; Press and Briggs, 1975 ; Kanamori, 1977). These correlations have suggested an influence of the pre and post-seismic deformations on the Chandler wobble excitation. However, as energy is necessary mainly during the periods of rapid increase or decrease of the wobble amplitude, a correlation of the seismicity with the time derivative of the energy in the Chandler oscillation would be expected in that case. An attempt for such a correlation (Myerson, 1970) does not lead to a clear conclusion. A different interpretation is proposed by Kanamori (1977), who noticed that the wobble amplitude variations slightly preced the variations of the seismic released energy. He then speculated that earthquakes may be a consequence -and not the cause- of the wobble amplitude variations, through a mechanisms which involves a decoupling between tectonic plates.

SECULAR DRIFT RELATED TO THE EXCITATION OF THE CHANDLER WOBBLE

The source of excitation of the Chandler wobble remains an open question. None of the various proposed mechanisms, as for example atmospheric displacements (Wahr, 1983), oceanic effects (Chao, 1984) or convection in the core (Le Mouël et al., 1985) has led up to now to a satisfying explanation of the wobble amplitude variations. Number of various possible excitation mechanisms can be modelled as step perturbations of the Earth inertia tensor. It is then of interest to consider the effect of this kind of excitation on the mean pole position.

As shown by equation (2), the excitation of the Chandler wobble by step perturbations of the Earth inertia tensor also induces a secular drift. If the excitation vectors $\underline{\psi}$ are random in direction, the mean pole position will describe a two-dimensional random walk. Mathematical formulations give the distance to origin reached with the maximum probability at the end of the walk when the step lengths (i.e. the $\underline{\psi}$-amplitudes) follow simple distributions (Chandrasekhar, 1943 ; Feller, 1968). For step lengths following a gaussian distribution, the most probable distance to origin reached by the two-dimensional random walk after N steps is proportional to $\sqrt{N}$. This means that, if the Chandler wobble is maintained by such random excitations occurring nearly uniformly in time, the amplitude of the polar secular drift will be proportional to $\sqrt{t}$.

Statistical modelizations of the polar motion data since 1901 (Mansinha and Smylie, 1967 ; Souriau, 1986) show that the secular drift resulting from the excitation of the Chandler wobble by random steps is of the same order of magnitude as the observed one. However, the computed secular drift is roughly in $\sqrt{t}$, whereas the observed one is roughly proportional to time. There is then an incompatibility between the observed secular drift and a polar excitation by random steps only, suggesting that some independent excitation mechanisms have to be invoked for Chandler wobble and for secular drift.

Lithospheric rebound related to deglaciation has been proposed to explain the polar secular drift (Dickman, 1979 ; Nakiboglu and Lambeck, 1980 ; Sabadini and Peltier, 1981 ; Sabadini et al., 1982 ; Peltier and Wu, 1983). These modelizations assume implicitely that Chandler wobble and secular drift are completely independent. However, it is likely that part of the secular drift is a consequence of the excitation of the Chandler wobble. It is then important, in the future, to model simultaneously Chandler wobble and secular drift.

CONCLUSION

Thanks to high accuracy polar data and seismic moment tensor solutions, it has been possible to discard the direct effect of earthquakes as a significant source of Chandler wobble excitation. However, with the improvement of space technics, it is hopeful to observe on the pole path the effect of a strong earthquake and perhaps to get constraints on mantle rheology and seismic source parameters.

The possible contribution of slow aseismic deformations is not yet quantified ; new data on global positioning and tectonic deformations at the earth surface will probably be an important element to answer this question.

The source of the Chandler wobble excitation is still enigmatic. However, as Chandler wobble excitation and secular drift are not independent of each other, these two phenomena will have to be modelled simultaneously. In particular, any excitation mechanism proposed to maintain the Chandler wobble will have to include an evaluation of the secular drift which can result from this excitation.

REFERENCES

Anderson, D.L., 1974. Earthquakes and the rotation of the Earth, Science, **186**, 49-50.

Ben Menahem, A. and Israel, M., 1970. Effect of major seismic events on the rotation of the Earth, Geophys. J. R. Astron. Soc., **19**, 367-393.

Chao, B.F., 1984. The excitation of the Chandler wobble by the Southern oscillation, based on ILS data, Eos, Trans. Am. Geophys. Un., **65**, 859.

Chandrasekhar, S., 1943. Stochastic problems in Physics and Astronomy, Rev. of Modern Physics, **15**, 2-91.

Dahlen, F.A., 1968. The normal modes of a rotating, elliptical, Earth Geophys. J. R. Astron. Soc., **16**, 329-367.

Dahlen, F.A., 1971. The excitation of the Chandler wobble by earthquakes, Geophys. J. R. Astron. Soc., **25**, 157-206.

Dahlen, F.A., 1973. A correction to the excitation of the Chandler wobble by earthquakes, Geophys. J. R. Astron. Soc., **32**, 203-217.

Dickman, S.R., 1979. Continental drift and true polar wandering, Geophys. J. R. Astron. Soc., **57**, 41-50.

Dziewonski, A.M., Chou, T.A. and Woodhouse, J.H., 1981. Determination of earthquake source parameters from waveform data for studies of global and regional seismicity, J. Geophys. Res., **86**, 2825-2852.

Feller, W., 1968. An introduction to probability theory, J. Wiley, New-York, 511 pp.

Gross, R.S., 1985. The influence of earthquakes in the Chandler wobble during 1977-1983, Submitted to the Geophys. J. R. Astron. Soc.

Haubrich, R.A., 1970. An examination of the data relating pole motion to earthquakes, in : Earthquake Displacement Fields and the Rotation of the Earth, L. Mansinha et al., eds., pp. 149-157, Reidel, Dordrecht.

Israel, M.A., Ben Menahem, A. and Singh, S.J., 1973. Residual deformation of real Earth models with application to the Chandler wobble, Geophys. J. R. Astron. Soc., **32**, 219-247.

Kanamori, H., 1976. Are earthquakes a major cause of the Chandler wobble?, Nature, **262**, 254-255.

Kanamori, H., 1977. The energy release in great earthquakes, J. Geophys. Res., **82**, 2981-2987.

Lambeck, K., 1980. The Earth's variable rotation: geophysical causes and consequences, Cambridge University Press, 449 pp.

LeMouël, J.L., Gire, C. and Hinderer, J., 1985. The flow of the core fluid at the core mantle boundary as a possible mechanism to excite the Chandler wobble (in French), C.R. Acad. Sc. Paris, **301**, Serie 2, 27-30.

Mansinha, L. and Smylie, D.E., 1967. Effect of the earthquakes on the Chandler wobble and the secular polar shift, J. Geophys. Res., **72**, 4731-4743.

Mansinha, L. and Smylie, D.E., 1970. Seismic excitation of the Chandler wobble, in : Earthquake Displacement Fields and the Rotation of the Earth, L. Mansinha et al., eds., pp. 122-135, Reidel, Dordrecht.

Mansinha, L., Smylie, D.E. and Chapman, C.H., 1979. Seismic excitation of the Chandler wobble revisited, Geophys. J. R. Astron. Soc., **59**, 1-17.

Munk, W.H. and M. Donald, G.J.F., 1960. The Rotation of the Earth, 323 pp., Cambridge University Press.

Myerson, R.J., 1970. Evidence for association of earthquakes with the Chandler wobble, using long term polar data of the ILS-IPMS. In: Earthquake Displacement Fields and the Rotation of the Earth, L. Mansinha and al., eds., Reidel, Dordrecht, Holland, 159-168.

Nakiboglu, S.M. and Lambeck, K., 1980. Deglaciation effects on the rotation of the Earth, Geophys. J. R. Astron. Soc., **62**, 49-58.

O'Connell, R.J. and Dziewonski, A.M., 1976. Excitation of the Chandler wobble by large earthquakes, Nature, **262**, 259-262.

Peltier, W.R. and Wu, P., 1983. Continental lithospheric thickness and deglaciation induced true polar wander, Geophys. Res. Letters, **10**, 181-184.

Press, F. and Briggs, P., 1975. Chandler wobble, earthquakes, rotation and geomagnetic changes, Nature, **256**, 270-273.

Sabadini, R. and Peltier, W.R., 1981. Pleistocene deglaciation and the Earth's rotation: implications for mantle viscosity, Geophys. J. R. Astr. Soc., **66**, 553-578.

Sabadini, R., Yuen, D.A. and Boschi, E., 1982. Polar wandering and the forced responses of a rotating, multilayered viscoselastic planet, J. Geophys. Res., **87**, 2885-2903.

Sabadini, R., Yuen, D.A. and Boschi, E., 1984. The effects of post-seismic motions on the moment of inertia of a stratified visco-elastic Earth with an asthenosphere, Geophys. J. R. Astron. Soc., **79**, 727-745.

Slade, M.A., Lyzenga, G.A. and Raefsky, A., 1984. Constraints on mantle rheology from post-seismic Chandler wobble excitation, Eos, Trans. Am. Geophys. Un., **65**, 1003.

Smylie, D.E. and Mansinha, L., 1968. Earthquakes and the observed motion of the rotation pole, J. Geophys. Res., **73**, 7661-7673.

Smylie, D.E. and Mansinha, L., 1971. The elasticity theory of dislocations in real Earth models and changes in the rotation of the Earth, Geophys. J. R. Astron. Soc., **23**, 329-354.

Souriau, A., 1986. Random walk of the Earth pole related to the Chandler wobble excitation, Geophys. J. R. Astron. Soc., to be published.

Souriau, A. and Cazenave, A., 1985. Reevaluation of the seismic excitation of the Chandler wobble from recent data, Earth Planet. Sci. Lett., **75**, 410-416.

Wahr, J.M., 1983. The effects of the atmosphere and oceans on the Earth's wobble and on the seasonal variations in the length of day; II-Results, Geophys. J. R. Astron. Soc., **74**, 451-487.

Flow in the fluid core and Earth's rotation

C. Gire and J.L. Le Mouël
Institut de Physique du Globe de Paris
4, Place Jussieu
75230 PARIS
France

ABSTRACT. We address in this paper the possible effects of motions in the fluid core on the Earth's rotation, both length of day and pole motion. The geomagnetic field is thought to be generated by dynamo action in the conducting core ; information on the fluid motion is then obtained through the study of the secular variation of this field.

A correlation between the time variations of some parameters of the geomagnetic field and the time variations of the rotation rate of the Earth which has been proposed some years ago (Le Mouël et al.,1981) is extended using recent data which do seem to buttress it. Magnetic variations lead rotation variations by some ten or fifteen years. Two mechanisms of core mantle coupling which would allow exchange of angular momentum between the core and the mantle and account for the observed correlation are briefly discussed.

The core motions can also have an effect on the pole motion : the pressure field associated with the time varying motions at the core-mantle boundary, which generate the secular variation, can act on the elastic mantle and change its inertia products. The excitation function corresponding to this mechanism is shown to have such an order of magnitude that it could contribute to maintain the Chandler wobble again dissipation. Its efficiency depends on the time constant of the secular variation.

INTRODUCTION

This paper presents possible relationships between core motions and Earth's rotation. Information on the fluid motions at the top of the core is provided by the variations of the geomagnetic field over the decade time scale (the secular variation). For such a time scale, indeed, the core fluid can be considered as a perfect conductor (e.g. Voorhies and Benton, 1982) ; in this approximation, the so-called frozen flux approximation first introduced by Roberts and Scott (1965), the secular variation field only depends on the motion of the fluid at the top of the core.

As for the observed decade variations in length of day (ℓ.o.d.), they are generally attributed to an exchange of angular momentum

A. Cazenave (ed.), Earth Rotation: Solved and Unsolved Problems, 241–258.

between the solid mantle and the fluid core as there are no other reasonable candidates (Hide, 1977 ; Runcorn, 1982). Therefore these decade variations are expected to be in relation with the geomagnetic field generating process. Different correlations between the geomagnetic secular variation and the decade variations in ℓ.o.d. have been proposed (e.g. Vestine, 1952 ; Runcorn, 1982 ; Backus, 1983 ; Le Mouël et al., 1981 ; Le Mouël and Courtillot, 1981, 1982). In Section 1 of this paper we resume and extend the correlation advocated by Le Mouël et al., which seems to be buttressed by recent data, and we discuss the possible coupling mechanism between the core and the mantle which could account for this correlation.

In Section 2 we deal with the excitation of the Chandler wobble. If the flow of the fluid at the core mantle boundary (CMB) is regarded as geostrophic, the corresponding overpressure field can be computed. This overpressure field, which varies with time, acts on the inner boundary of the elastic mantle, deforms it and alters its products of inertia. We examine whether the order of magnitude of the excitation function due to this redistribution of matter is large enough to contribute significantly to the excitation of the Chandler wobble.

1. Decade fluctuations of the Earth's rotation and core-mantle coupling

1.1. Observations

It has been well known that the periodic variations in the Earth's rotation rate with a period up to two years and some more irregular variations as the ones related to El Nino events are accounted for by atmospheric circulation (Cazenave, 1975 ; Hide et al., 1980 ; Eubanks, 1985 ; Hide, 1984). On the contrary, the so-called decade variations are attributed to an exchange of angular momentum between the mantle and the liquid core, these longer period fluctuations being too large to be ascribed to any other source (Eubanks, 1985 ; Runcorn, 1982 ; Hide, 1977, 1984). In some former papers (Le Mouël et al., 1981 ; Le Mouël and Courtillot, 1982), a correlation between a certain parameter of the geomagnetic field and the decade variations in the Earth's rotation rate was proposed. The end of the considered time span was 1978 ; let us examine again this correlation when adding data from the 1978-1984 period.

The chosen magnetic indicator is the first time derivative $\delta D/\delta t$ of the declination D of the geomagnetic field $\underline{B}$ as recorded in European observatories. The rotation data are annual mean values taken from Morrisson (1979) for the 1900-1978 period, from B.I.H. since 1979. Figure 1 displays the correlation between $\delta D/\delta t$ and $\Delta\Omega/\Omega$, the relative change in the Earth's rotation rate : the magnetic variations appear to lead the rotation variations, a period of increasing (respectively decreasing) $\delta D/\delta t$ is followed by a period of increasing (respectively decreasing) rotational velocity of the Earth, the time lag being something like 10-15 years. From the examination of the part of Figure 1 corresponding to the 1900-1978 period, Le Mouël et al. (1981) predicted that an acceleration of the Earth's rotation was to be expected around 1980 ; the recent data (up to 1984) do show that such an acceleration occurred (Figure 1).

It can be argued that comparing a local parameter such as the declination in Western Europe to a global parameter such as the Earth's rotation rate is of little meaning. But, in fact, D variations in Europe are expected to be quite sensitive to changes in the zonal motions of the fluid at the core-mantle boundary, due to the general North-South orientation of the D (or the closely associated East component Y) contours in the European area.

Indeed, let us write the temporal variation of D, $\delta D/\delta t$, in the classical form (e.g. Nagata, 1965 ; Yukutake and Tachinaka, 1969 ; Le Mouël et al., 1981) :

$$\delta D/\delta t = \partial D/\partial t + u_\phi \, \partial D/\partial \phi \qquad (1)$$

where $\partial D/\partial \phi$ is the azimutal derivative of D (ϕ is longitude), and $\partial D/\partial t$ is sometimes called the "non drifting" part of D. u_ϕ is the westward drift rate of the geomagnetic field as seen at the Earth's surface when considering the D component. u_ϕ is estimated by maximizing the contribution of the drifting term $u_\phi \, \partial D/\partial \phi$ to the secular variation $\delta D/\delta t$ in (1). It can be checked (Gire et al., 1984) that this drifting term represents about 70% of the total time derivative $\delta D/\delta t$. This observation is simply illustrated in Figure 2 which represents sketchy planetary charts of $\delta Y/\delta t$ and $\partial Y/\partial \phi$. The similarity in the drawings of the zero lines is indeed striking.

Therefore, if the westward drift of Y (or D), as observed at the Earth's surface, is related to a drift of the external layers of the core, as generally assumed, the secular variation $\delta D/\delta t$ of the declination in Europe will be a good indicator of this drift and of the exchange of angular momentum between the rigid mantle and the liquid core.

This classical presentation is certainly over-simplified : the problem of determining u_ϕ from $\underline{B}$ and $\partial \underline{B}/\partial \phi$ is an ill-posed problem, and this quantity u_ϕ is not a uniquely determined quantity (Gire et al., 1984). Nevertheless, it remains that the time variations of $\delta D/\delta t$, as recorded in Europe, can be expected to provide us with an estimation of the variations of the intensity of the zonal flow of the core fluid at the CMB. If we compute the velocity field of the fluid at the core mantle boundary in the geostrophic approximation (Le Mouël et al.,1985; see Section 2), and if we change the intensity of this flow, we indeed observe a variation of D in Europe, $\delta D/\delta t$, whose geographic distribution is quite similar to the $u_\phi \partial D/\partial \phi$ one.

Remark 1. The maps of Figure 2 are derived from planetary models (spherical harmonics expansions) and the conclusions can depend to some extent on the degree of the expansion (G. Backus, personal communication). Figure 3 shows how good the similarity between the geometries of $\delta D/\delta t$ and $\partial D/\partial \phi$ is in Western Europe (the contours being directly derived from observatory data).

Remark 2. A change of slope of the ($\delta D/\delta t$) curve similar to the one observed in 1969 (the 1969 jerk), occurred in 1979 (Figure 1). This event seems to be of planetary extent (Gavoret et al., 1985 ;B. Langel,

D. Barraclough, personnal communications), as the 1969 one. If true, and if the correlation we propose between magnetic and rotation variations is genuine, a deceleration of the Earth rotation rate should follow the acceleration which has been observed since 1980-1983, the change of trend occurring in the 1988-1993 interval.

1.2. Models of core mantle coupling

Le Mouël and Courtillot (1982) tried to build a model accounting for the proposed correlation between the geomagnetic parameter $\delta D/\delta t$ (i.e. u_ϕ in the hypothesis discussed above) and the $\Delta\Omega/\Omega$ curves. It is a simple three layers Bullard model (the lower core ($r \leq a$), the top layers of the core ($a \leq r \leq b$) and the mantle, the three layers being electromagnetically coupled. The main observation underlying the model is the following (Figure 1) : an acceleration of the Earth's rotation rate lags an acceleration of the westward drift u_ϕ ($\delta D/\delta t$) by 10-15 years, after a flat plateau where the rotation of the mantle is nearly constant. Hence, during this time interval the westward (eastward) acceleration of the velocity of the top layers of the core must be compensated by an eastward (westward) acceleration of the lower core in order to keep the total angular momentum constant. Then let us suppose that an angular momentum amount $-P(t)\ H(t)$ is transferred from the inner core to the outer shell of the core ($H(t)$ is the Heaviside function and the origin of times is for example 1969). A differential rotation accross the boundary ($r = a$) between the inner core and the outer shell results which generates a toroidal magnetic field the sources of which are distributed on the sphere ($r = a$). This toroidal field diffuses through the outer shell (conductivity σ_c), then through the mantle (conductivity σ_m). At the same time, an other toroidal field is generated by the differential rotation at the boundary ($r = b$) between the outer shell and the mantle (CMB), which diffuses through the mantle. Electromagnetic torques result from the interaction between the electric current sustaining these toroidal fields and the poloidal main field, and thus couple the three layers.

The solution of the differential system governing the rotation of the three layers depends on two time constats ; the longer one, τ_1, is indeed of the order of 15 years. The main result is that a westward acceleration of the outer shell is followed by an increase (eastward acceleration) of the rotation of the mantle, the time lag being of the order of τ_1.

Although this model accounts for the general features of the correlation displayed in Figure 1, it suffers from a major limitation : the toroidal fields generated by the differential rotations on the two boundaries (mantle-outer shell and outer shell-lower core) are computed in a steady state case (i.e. diffusion in the conducting core and the mantle balances the creation of magnetic field on the two spherical boundaries). So the time constant τ_1 is only representative of the mechanical time constant of the system, depending on the moment of inertia of the three layers. And, as a consequence, even though the time lag is of the right order of magnitude, the temporal response of the mantle rotation in the model, proportionnal to $(1 - \exp(-t/\tau_1))$, does not fit the way the mantle actually accelerates after the outer shell of the core has accelerated westwards. In fact, the transport of the toroidal magnetic fields has been neglected. In the lower mantle (adopting $\sigma_m \simeq 300\ \Omega^{-1}\ m^{-1}$ for the mean conductivity of the 700 lowest km in the mantle (Achache et al., 1981)), the layer filled by the toroidal magnetic field, through diffusion, at time t, will be of the order of :

$$d_m \simeq (t/\mu\sigma_m)^{1/2} \simeq 300\ t^{1/2}\ \text{km} \qquad (2)$$

(t being in years in the last term, μ being the magnetic permability). The time constant of the diffusion in the lower mantle can thus be considered as small with regard to the time constant involved in the present problem (15 years) and the steady state approximation is reasonnable as far as the mantle is concerned.

In the core (with $\sigma_c \simeq 3\ 10^5\ \Omega^{-1}\ m^{-1}$) the thickness of the corresponding layer is :

$$d_c \simeq (t/\mu\sigma_c)^{1/2} \simeq 10\ t^{1/2}\ \text{km} \qquad (3)$$

(t being in years in the last term) ; d_c is then of the order of 40 km for $t \simeq 15$ years. But the order of magnitude of the thickness of the outer shell - although there is neither any direct way to evaluate it, nor any definite evidence of the existence of this layer - is generally assumed to be larger than 100 km (e.g. Vestine et al., 1968 ; Backus, 1983 ; Le Mouël and Courtillot , 1981 ; Runcorn, 1982). In fact, the estimate of this thickness h relies on a simple balance equation between the angular momentum variations of the different layers involved (mantle, outer core, lower core) :

$$I_m\, \Delta\Omega_m + I_o\, \Delta\Omega_o + I_i\, \Delta\Omega_i = 0 \qquad (4)$$

(I_m, I_o, I_i are respectively the moments of inertia of the mantle, the outer shell and the inner core ; Ω_m, Ω_o, Ω_i the respective angular velocities).

The moment of inertia I_o of the outer shell is :

$$I_o \simeq (8/3)\pi\rho b^3 h \quad (\rho \simeq 10^4\ \text{kgm}^{-3}\text{, the outer core density})$$

From the rotation data (Figure 1) we obtain $\Delta\Omega_m \simeq 3\ 10^{-12}$ rad.s^{-1} ; from the secular variation data, we get $\Delta\Omega_o \sim 1.5\ 10^{-10}$ rad.s^{-1}. Balancing the variations of the angular momentums of the mantle and of the outer shell, we get indeed $h \simeq 120$ km ($I_m \simeq 7\ 10^{35}$ kgm^2). Although the order of magnitude of d_c and h are comparable, the transport of the toroidal magnetic field by diffusion from the Bullard discontinuity of the model (r = a) to the CMB (r = b) is probably too slow to account for the observed time lag between the westward (resp. eastward) acceleration of the outer core and the eastward (resp. westward) acceleration of the mantle. If such a model is to be maintained, with an electromagnetic coupling between its three layers - and particularly with a torque exerted on the mantle by the lower core, which finally accelerates the mantle together with the lower core -, an other transport mechanism of the magnetic field should be invoked. One could think of different kinds of waves influenced by either Coriolis force or Lorentz force or both.

We will propose now quite a different mechanism which could be capable of coupling the mantle with the lower core. This second model is certainly quite speculative at the moment, and difficult to handle quantitatively. If the flow at the top of the core is supposed to be geostrophic (Le Mouël, 1984 ; Le Mouël et al., 1985), it can be derived from secular variation data in a reasonably unique way (Backus and Le Mouël, 1985). The poloidal part of the flow, which allows an exchange of fluid with the deeper core, is illustrated by Figure 4. It is essentially characterized by one upwelling and one downwelling, both equatorial and antipodal one of the other. The toroidal ingredient contains the westward drift term we considered earlier.

The pattern of the poloidal flow appears to be constant, although its intensity varies, and fixed with respect to the mantle, all over the period for which it can be derived from secular variation data -and of course within the accuracy provided by the inversion-. Some considerations on the figure of the geoid suggest that the two jets of Figure 4 (upwelling and downwelling) could have remained fixed with respect to the mantle for hundreds of thousands years (Allegre et al., 1985).

These jets are linked to the lower core, since they do not drift. Then, they tend to maintain the lower core and the mantle tightly attached one to the other. The detailed coupling mechanism could be a variant of the topographic coupling proposed by Hide and Horai (1968) and Malin and Hide (1982) ; but here, instead of a net torque, we have a restoring couple which prevents the mantle and the lower core to have any important differential rotation one with respect to the other. The principle of the coupling mechanism is illustrated by Figure 5, which has no claim to be realistic. Now, coming back to the observations (Figure 1), the westward acceleration of the outer shell of the core must be first balanced by an eastward acceleration of the lower core, as discussed in the frame of the former mechanism. Then the lower core drags the mantle with it, due to the restoring couple illustrated by Figure 5, with a time lag of some 10-15 years.

This supposes of course that the relative displacement between the lower core and the mantle at the CMB allowed by the restoring couple is quite small : this displacement is only a few kilometers in 15 years. But if the hypothesis advanced by Allegre et al. is correct, this relative displacement was less than 1000 km in some 10^6 years !

2. The flow at the core-mantle boundary as a possible source of excitation of the Chandler wobble

The question of the excitation of the Chandler wobble, although frequently adressed for more than 80 years, has not yet received any definitive answer. The two main mechanisms capable of maintaining the wobble against dissipation which have been proposed up to now are
1/ the motion of the atmosphere (Jeffreys, 1940 ; Rudnik, 1956 ; Wilson, 1975), 2/ the earthquakes (Cecchini, 1928 ; Stoyko and Stoyko,

1966 ; Smylie and Manshina, 1968 ; Haubrich, 1970 ; Anderson, 1974 ; Kanamori, 1977 ; O'Connel and Dziewonsky, 1976). Both have been questionned : Munk and Hassan (1961) find that the meteorological mechanism can only play a minor role in maintaining the wobble. But recently R. Hide (1984), examining meteorological data over a time interval corresponding to nearly three Chandlerian periods (1980-March 1984), concluded that the atmospheric excitation could make a major contribution to the observed polar motion over that interval. But Chao (1985) has questionned Hide's analysis and its conclusions. Souriau and Cazenave (1985) have shown that, at least for the 1977-1982 time span (for which good rotation data exist and a set of hypocentral coordinates and moment tensors for 1287 strong and moderate earthquakes is avalaible), the seismic mechanism is by two orders of magnitude too weak to excite the wobble. Thus, it seems that the question is still open.

A third mechanism has been proposed by Runcorn (1970, 1982) who calls for impulsive electromagneitc torques resulting from motions in the fluid core. We will show in the present paper that an other mechanism can contribute to maintaining the wobble : the direct effect of the pressure variations associated with the time varying motions of the fluid core at the CMB.

2.1. Motions at the core-mantle boundary

In some former papers (Le Mouël, 1984 ; Le Mouël et al., 1985) it was argued that the motion of the fluid at the CMB is geostrophic (except in a thin strip around the geographic equator). Where this approximation holds, the (horizontal) velocity field $\underline{u}$ can be written :

$$\underline{u} = (1/2\rho\Omega\cos\theta)\ \underline{n} \times \underline{\nabla}\ p \qquad (5)$$

ρ is the core fluid density, Ω the Earth's rotation ($\Omega \simeq 7\ 10^{-5}$ rad.s^{-1}), θ the colatitude, $\underline{n}$ the unit outward radial vector, p the pressure, $\underline{\nabla}$ the gradient. The geostrophic constraint (5) strongly reduces the ambiguity one is faced to when one tries to compute $\underline{u}$ from geomagnetic secular variation (SV) data, assuming that the SV field results from the interaction of the fluid velocity field $\underline{u}$ with the main field $\underline{B}$ at the CMB (Roberts and Scott, 1965 ; Backus, 1968 ; Backus and Le Mouël, 1985). It is then possible to get an estimate of the motion $\underline{u}$ at the CMB. Let us take for a typical magnitude U and a typical length scale L of $\underline{u}$:

$$U \simeq 10^{-4}\ \mathrm{ms}^{-1} \qquad L \simeq 3\ 10^{6}\ \mathrm{m}$$

(Le Mouël et al., 1985).

Hence, from (5), we get an estimate of the magnitude P of the overpressure field p related to the velocity field $\underline{u}$:

$$P \simeq 500\ \mathrm{Pa}$$

Furthermore, it appears that $\underline{u}$ varies with time, particularly its magnitude which approximately doubled from 1970 to 1980 (in the retained approximation). Thus temporal variations in p of the order of P can be expected at the CMB.

2.2. The excitation of the Chandler wobble

The varying pressure p, acting on the inner boundary of the elastic mantle, can deform it and, depending on its geometry, can alter the products of inertia I_{13} and I_{23} of the mantle (as usual the index 1 is related to the equatorial axis in the Greenwich meridian, 2 to the equatorial axis in the 90° East meridian, 3 to the principal axis).

Let us look in more detail at the motion $\underline{u}$ derived from SV data in the geostrophic approximation (Le Mouël et al., 1985 ; Figure 4) ; it appears that a dominant part of the motion is symmetrical with respect to the equator ; the corresponding pressure field is unable to excite the wobble. But some components which are antisymmetrical with respect to the equator also enters into $\underline{u}$; particularly some components which correspond in (5) to the elementary pressure fields :

$$\Pi_{2c}(t)\ P_{21}(\cos\theta)\cos\phi$$

$$\Pi_{2s}(t)\ P_{21}(\cos\theta)\cos\phi$$

(P_{21} being the Legendre function of degree 2 and order 1 ; see Le Mouël and al., 1985 for the elementary motions associated with these pressure fields). The variations $\delta\Pi_{2c}$ and $\delta\Pi_{2s}$ in the coefficients Π_{2c} and Π_{2s} generate variations δI_{13} and δI_{23} in the inertia products I_{13} and I_{23}. Using a pressure Love number adapted to the problem of the response to an inner pressure (e.g. Sasao et al., 1977, 1980 ; Le Gros et al., 1985), it comes :

$$\delta I_{13} \simeq 6\ 10^{25}\ \delta\Pi_{2c}$$

$$\delta I_{23} \simeq 6\ 10^{25}\ \delta\Pi_{2s}$$

(Le Mouël et al., 1985).

Using the usual complex notation ($\delta\underline{I} = \delta I_{13}+i\ \delta I_{23}$, $\underline{\Psi}_D = \Psi_1+i\Psi_2$) the corresponding excitation function is :

$$\underline{\Psi}_D = K(K-k)^{-1}(I_{33}-I_{11})^{-1}(\delta\underline{I}-(i/\Omega)\delta\dot{\underline{I}})$$

($K \simeq 0.94$ is the secular Love number, $k \simeq 0.28$ the volumic Love number, I_{33} is the polar moment of inertia, I_{11} the equatorial moment of inertia). And the equation of the pole motion (e.g. Lambeck, 1980) is :

$$i(\dot{\underline{m}}/\sigma_0) + m = \underline{\Psi}_D \qquad (7)$$

with $\sigma_0 = 2\pi/435$ rad/day ; $\underline{m} = m_1 + im_2$ represents the pole coordinates in radians, m_1 along the Greenwich meridian, m_2 along the 90° West meridian, counted from a conventional origin (CIO, close to the mean pole position - c.f. Figure 6).

Π_{2s}, as given by the inversion of the 1980 SV field, is of the order of 300 Pa (Π_{2c} happens to be smaller at this epoch). And a variation of this coefficient of the same order of magnitude is observed between 1970 and 1980. With $|\delta\underline{\Pi}_2| = \Delta\Pi_2 \simeq 300$ Pa, it comes, from (6) :

$$\Psi_D = 1.2\ 10^{-7} \text{ rad.}$$

which is to be compared to the value of $|\underline{m}|$ itself ($\simeq 7\ 10^{-7}$ rad.). It can then be concluded that the mechanism we propose gives an excitation function of an interesting order of magnitude and should be considered as possibly contributing to maintain the wobble.

But the efficiency of the mechanism depends on the time constants of the pressure (or velocity) field. Let us write explicitly the solution of (7) :

$$\underline{m}(t) = \exp(i\sigma_0 t)\{\underline{m}_0 - i\sigma_0 \int_0^t \underline{\Psi}(\tau)\exp(-i\sigma_0\tau)d\tau\} \quad (8)$$

($\underline{m}_0 = \underline{m}(0)$ being the initial ($t = 0$) position of the pole).

If the variation $\delta\underline{\Pi}_2$ is linear (from 1970 to 1980, with $|\Pi_2(1980) - \Pi_2(1970)| \simeq \Delta\Pi_2$, or from 1915 to 1970 ; see Figure 7), that is to say if $\delta\underline{I}$ is a ramp function, the mechanism we propose will not be very efficient to excite the wobble, as seen in a straightforward way from (8) (but it could contribute significantly to the drift of the mean pole).

Let us now assume that the total variation $\Delta\underline{\Pi}_2$ is reached through a succession of steps $\delta\underline{\Pi}_{2,n}H(t-t_n)$ at times $t_n > 0$ ($\sum \delta\underline{\Pi}_{2,n} = \Delta\underline{\Pi}_2$). Then (8) becomes :

$$\underline{m}(t) = \exp(i\sigma_0 t)\{\underline{m}_0 - \chi\sum_n \delta\underline{I}_n(1+(\sigma_0/\Omega)H(t-t_n))\exp(-i\sigma_0 t_n)\} + \chi\sum_n \delta\underline{I}_n H(t-t_n)$$

with :

$$\delta\underline{I}_n = 6\ 10^{25}\delta\underline{\Pi}_{2,n} \qquad \chi = K(K-k)^{-1}(I_{33}-I_{11})^{-1}$$

The effect of the pressure variations on the displacement of the mean pole is nearly unchanged (compared to the case of the ramp function), but the effect on the wobble magnitude depends essentially on the sequence of the t_n ($\Delta\Pi_2$ being given).

Is there any evidence of such steps $\delta\Pi_{2,n}H(t - t_n)$? Figure 7 represents the variation of the first time derivative $\delta X/\delta t$ of the North component X of the geomagnetic field in European observatories, after removing an estimate of the external (ionospheric and magnetospheric) field contribution (Gavoret et al., 1985). Some step-like variations do appear on the curve, superimposed on a clear linear decreasing trend till 1970 (on $-\delta X/\delta t$), a clear linear increasing trend after 1970. Some of these step variations are believed to be real and of internal origin. If true, taking into account the diffusion of the magnetic signal through the conducting mantle from the CMB, those variations must indeed be generated by quite rapid step-like variations in p (and u).

CONCLUSION

We have first given some evidence of a correlation between magnetic variations and variations of the Earth's rotation rate. This correlation certainly needs some further confirmation. A better determination of the SV field and a better correction of rotation data from external effects should render this confirmation possible in a near future. In particular, a clear test could be available in the end of the nineties if the 1978 magnetic event is comparable to the 1969 one : a decrease in the rotation rate should then occur if the correlation we propose is genuine. The mechanism permitting angular momentum transfert from the core to the mantle also needs further consideration.

We have then envisionned a possible contribution of the over-pressure associated with the varying fluid motion at the core-mantle boundary to the excitation of the Chandler wobble. The order of magnitude of the corresponding excitation function is such that this mechanism should be considered. But its efficiency depends on the time constant of the secular variation. And a detailed examination of the SV series in different observatories, in particular of the rapid events, is to be carried on, and the results confronted to the wobble data.

A permanent magnetic satellites program, which would provide an improved knowledge of the SV field, would greatly facilitate these studies.

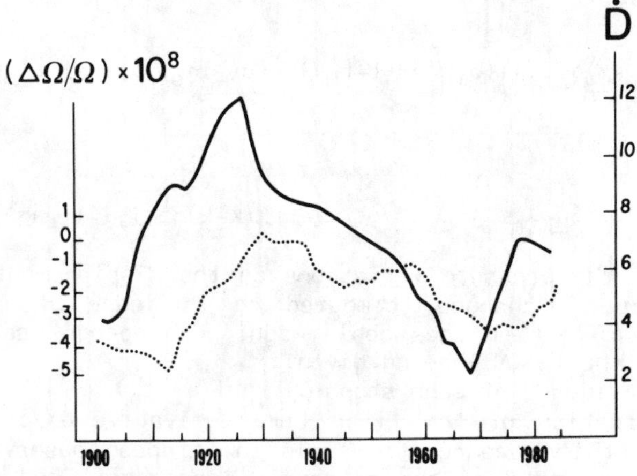

Fig. 1 The secular variation of declination, $\dot{D}$ (solid curve and right hand scale in arcmin. yr^{-1}), and the relative change in the Earth's rotation rate $\Delta\Omega/\Omega$ (dashed curve).

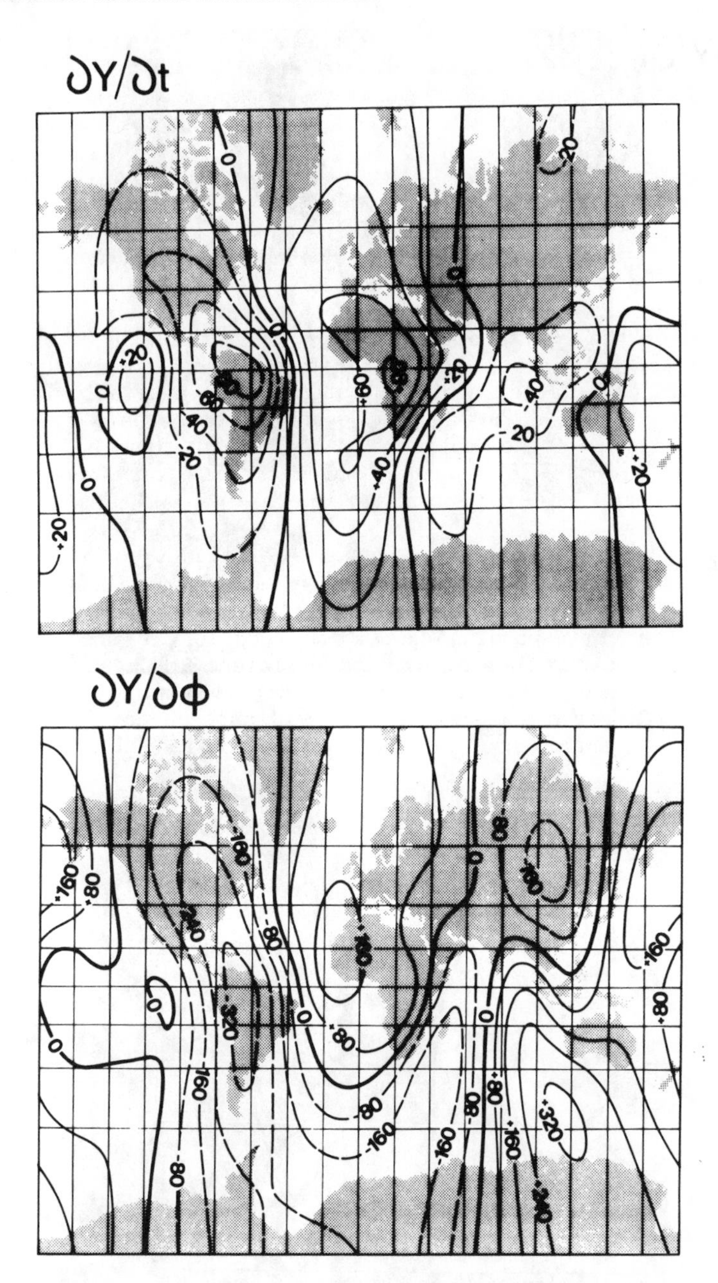

Fig. 2 a) the secular variation of the East component of the Earth magnetic field, $\delta Y/\delta t$ computed from IGRF80 model. Contour interval is 20 $nT.yr^{-1}$
b) azimutal change of East component, $\partial Y/\partial\phi$, computed from IGRF80 model. Contour interval is 80 $nT.rad^{-1}$.

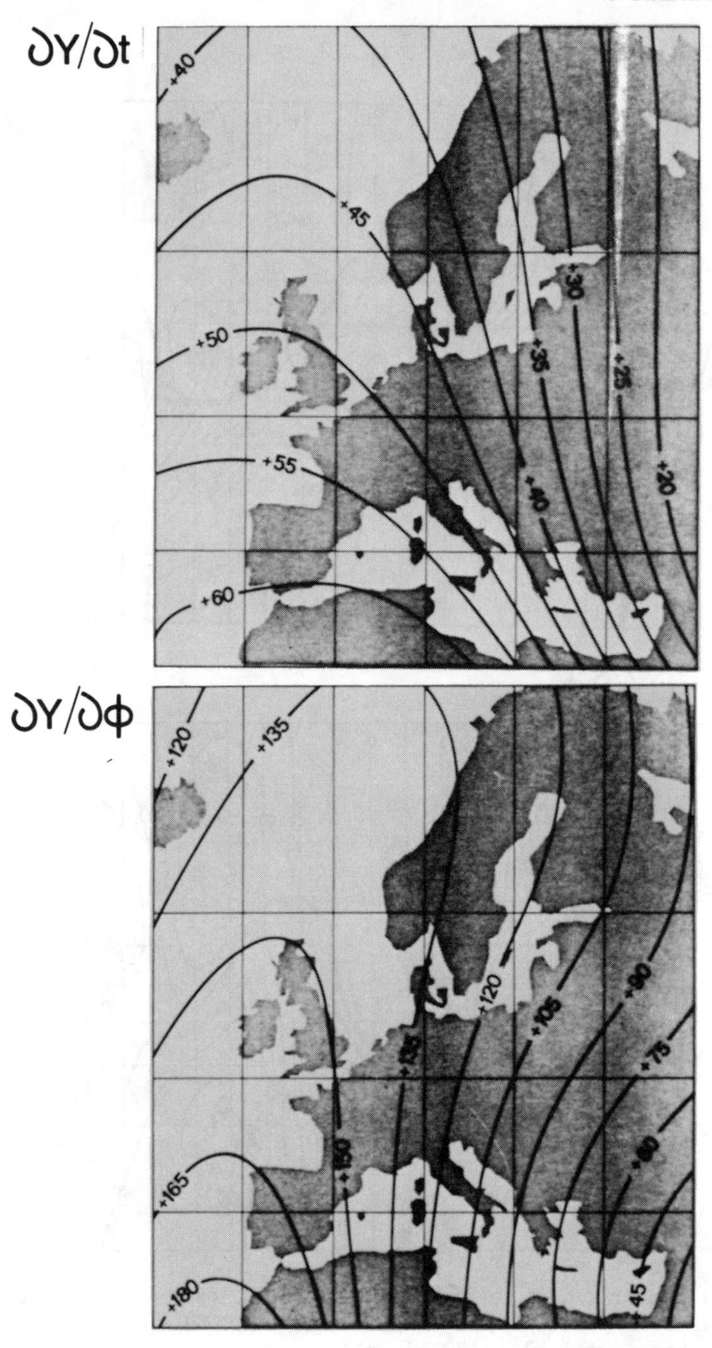

Fig. 3 a) The secular variation $\delta Y/\delta t$ in Europe. Units : $nT.yr^{-1}$.
b) Azimutal change $\partial Y/\partial\phi$ in Europe. Units : $nT.rad^{-1}$.

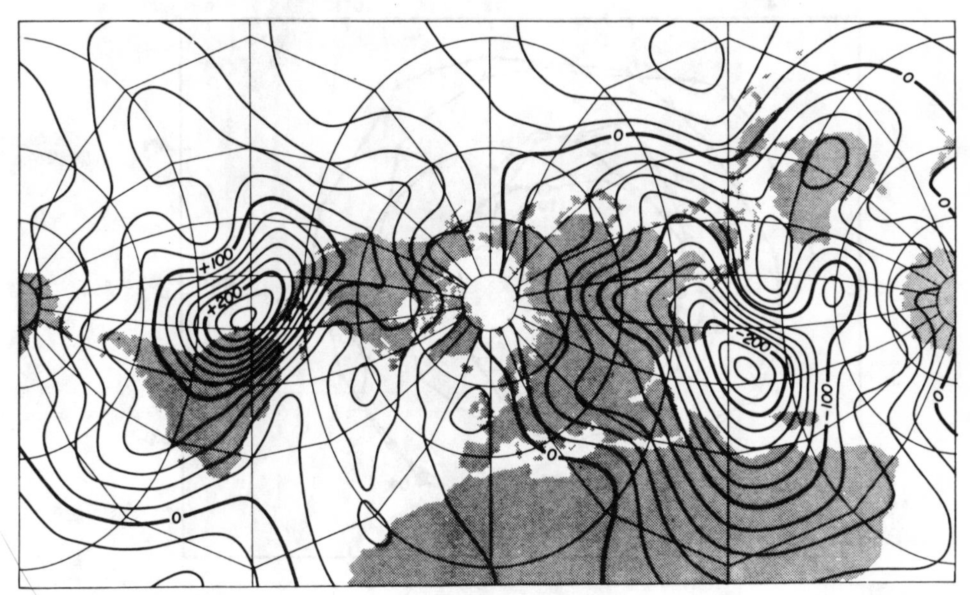

Fig. 4 Poloidal component $\underline{s}$ (= $\underline{\nabla}_S S$) of the geostrophic motion at the CMB in 1980. The contours are the poloidal scalar S contours. The poloidal flow is orthogonal to these contours. Contour interval : 2 10^{-4} rad^2 yr^{-1}.

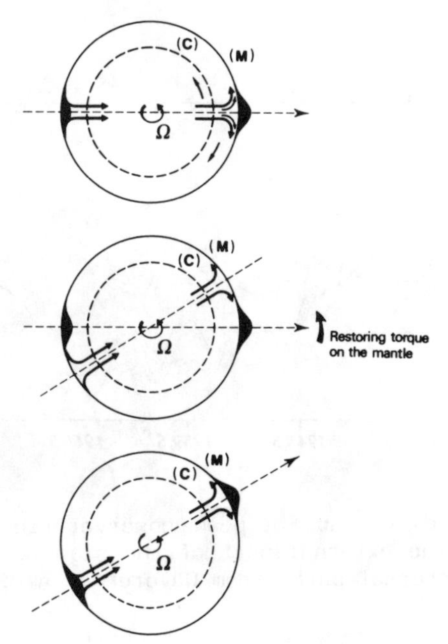

Fig. 5 Sketch of a possible coupling mechanism between core (C) and mantle (M).

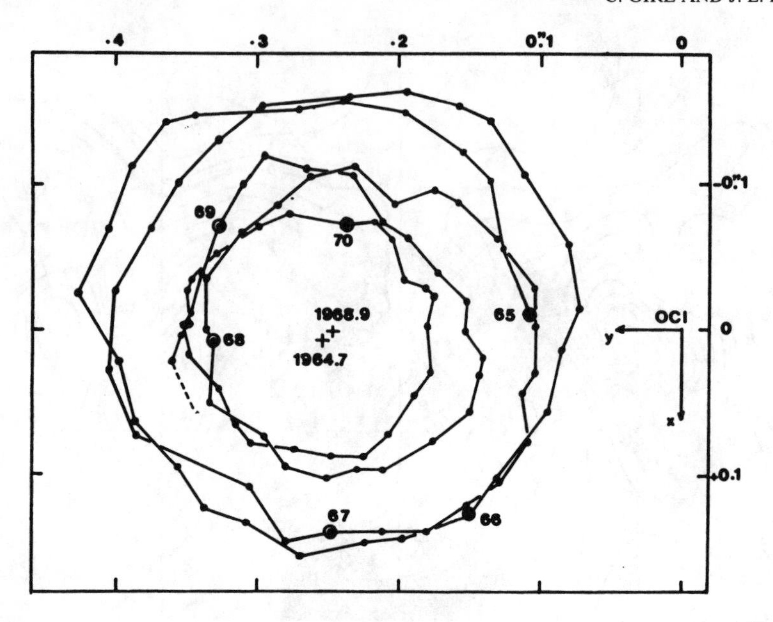

Fig. 6 Chandler wobble referred to the Conventional International Origin (CIO).

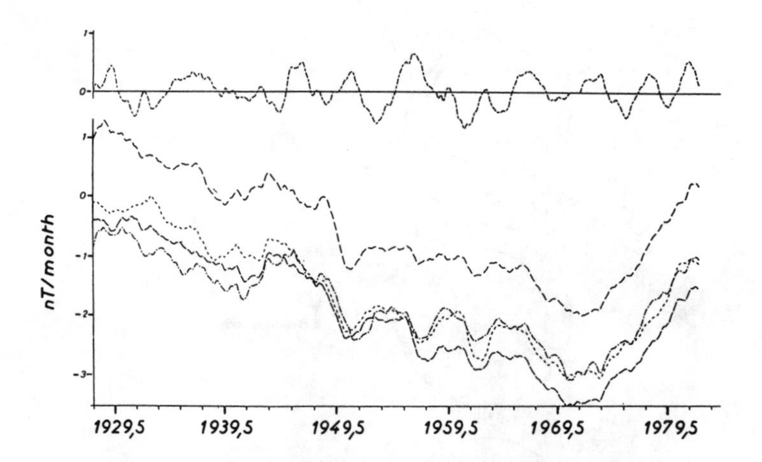

Fig. 7 $-\delta X/\delta t$ in different European observatories. The top curve represents the external part of the signal, the lower curves represent internal part (from Gavoret et al., 1985).

REFERENCES

Achache, J., Courtillot, V., Ducruix, J., Le Mouel, J.L., 1980 : 'The late 1960's secular variation impulse : Further constraint on deep mantle conductivity', Phys. Earth Planet. Inter., **23** ,72-75.

Allègre, C., Jaupart, C., Le Mouël, J.L., 1985 : 'Global geodynamics', Nature, (submitted).

Anderson, D., 1974 : 'Earthquakes and the station of the Earth'. Science, **182** , 49-50.

Backus, G., 1968 : 'Kinematics of geomagnetic secular variation in a perfectly conducting core.' Phil. Trans. Roy. Soc. Lond., A, **263** , 239-266.

Backus, G., 1983 : 'Application of mantle filter theory to the magnetic jerk of 1969.' Geophys. J.R. Astron. Soc., **74** , 713-746.

Backus, G. and Le Mouël, J.L., 1985 : 'The region on the core mantle boundary where a geostrophic velocity field can be determined from frozen-flux magnetic data', Nature, (submitted).

Cazenave, A., 1975 : ' Interactions entre les irrégularités de la vitesse de rotation de la Terre et les phénomènes météorologiques et climatiques.', Thesis, Univ. Toulouse, France, 258pp.

Cecchini, G., 1928 : 'Il problema della variazione della latudine'. Publicazioni del Reale observatorio Astronomico di Brero in Milano, **61** , 7.

Chao,b., 1985 :' On the excitation of the Earth's polar motion.' Geophys. Res. Lett., **12** ,526-529.

Eubanks, T.M., Steppe, J.A., Dickey, J.O. and Callabah, P.S., 1985 : 'A spectral analysis of the Earth's angular momentum budget'. J. Geophys. Res., **90** , 5385-5404.

Gavoret, J., Gibert, D., Menvielle, M., Le Mouël, J.L., 1985 : 'Long term variations of the internal and external components of the Earth's magnetic field', J. Geophys. Res. (submitted).

Gire, C., Le Mouël, J.L., Madden, T., 1984 : 'The recent westward drift rate of the geomagnetic field and the body drift of external layers of the core', Annales Geophysicae, **2** , 1, 37-46.

Haubrich, R., 1970 : 'An examination of the data relating pole motion to earthquakes', in Earthquake displacement fields and the rotation of the Earth, L. Manshina et al., Eds. Reidel, Dordrecht, 149-157.

Hide, R., 1984 : 'Rotation of the atmospheres of the Earth and Planets', Philos. Trans. R. Soc. London, A, **313** , 107-121.

Hide, R. and Horai, K, 1968 : 'On the topography of the core mantle interface', Phys. Earth Planet. Inter., **1** , 305-308.

Hide, R., Birch, N.T., Morrisson, L.V., Shea, D.J. and White A., 1980 : 'Atmospheric angular momentum fluctuations and changes in the length of the day', Nature, **286** , 114-117.

Hide, R., 1977 : 'Towards a theory of irregular variations in the length of day and core mantle coupling', Philos. Trans. R. Soc. London, A, **284** , 547-554.

Kanamori, H., 1977 : 'The energy release in great earthquakes', J. Geophys. Res., **82** , 2981-2987.

Jeffreys, H., 1940 : 'The variation of latitude', Monthly Notice of the Royal Astronomical Society, **100** , 139-155.

Lambeck, K., 1980 : 'The Earth's variable rotation : Geophysical cause and consequences', Cambridge University Press, 449 p.

Legros, H., Amavict, M., Hinderer, J., 1985 : 'Love numbers and planetary dynamics'. (in preparation)

Le Mouël, J.L., Madden, T., Ducruix, J. and Courtillot, V., 1981 : 'Decade fluctuations in geomagnetic westward drift and Earth rotation', Nature, **290** , 763-765.

Le Mouël, J.L., and Courtillot, V., 1981 : 'Core motions, electromagnetic core-mantle coupling and variations in the earth's rotation : new constraints from geomagnetic secular variation impulses', Phys. Earth Planet. Inter., **24** , 236-241.

Le Mouël, J.L., and Courtillot, V., 1982 : 'On the outer layers of the core and geomagnetic secular variation', J. Geophys. Res., **87** , 4103-4108.

Le Mouël, J.L., 1984 : 'Outer core geostrophic flow and secular variation of Earth's geomagnetic field', Nature, **311** , 734-735.

Le Mouël, J.L., Gire, C., Madden, T., 1985 : 'Motions at the core surface in the geostrophic approximation', Phys. Earth Planet. Inter.,

Le Mouël, J.L., Gire, C., Hinderer, J., 1985 : 'The flow of the core fluid at the core mantle boundary as a possible mechanism to excite the chandler's wobble (in french)', C.R. Acad.Sc. Paris, II **301** , 27-32.

Malin, S.R. and Hide, R., 1982 : 'Bumps on the core-mantle boundary : geomagnetic and gravitational evidence revisited', Philos. Trans. R. Soc. London, A, **306** , 281-289.

Morisson, L., 1979 : 'Redetermination of the decade fluctuations in the rotation of the Earth in the period 1861-1978', Geophys. J.R. Astron. Soc., 58 , 349-360.

Munk, W., and Hassan, S., 1961 : 'Atmospheric excitation of the Earth wobble', Geophys. Res. Astr. Soc., 4 , 339-358.

Nagata, T., 1965 : 'Main characteristics of recent geomagnetic secular variation', J. Geomag. Geoelec., **17** , 263-276.

O'Connel, R. and Dziewonski, A., 1976 : 'Excitation of the Chandler wobble by large earthquakes', Nature, **262** , 259-262.

Roberts, P.H. and Scott, S., 1965 : 'On analysis of the secular variation. A hydrodynamic constraint theory', J. Geomag. Geoelec., **17** , 137-151.

Rudnick, P., 1956 : 'The spectrum of the variation in latitude', Transaction of the American Geophysical Union, **37** , 137-142.

Runcorn, K., 1970 : 'A possible cause of the correlation between earthquakes and polar motion', Earthquake displacements fields and the rotation of the Earth, L. Manshina et al., Eds. Reidel, Dordrecht, 181-190.

Runcorn, K., 1982 : 'The role of the core in irregular fluctuations of the Earth's rotation and the excitation of the Chandler wobble', Phil. Trans. Soc. Lond., A**306** , 261-270.

Sasao, T., Okamoto, I. and Sakaï, S., 1977 : 'Dissipative core mantle coupling and nutational motion of the Earth', Publ. Astron. Soc. Jap., **20** , 83-105.

Sasao, T., Okubo, S. and Saïto, M., 1980 : 'A simple theory on the dynamical effects of a stratified fluid core upon nutational motion of the earth', Nutation and the Earth's rotation',I.A.V., Reidel Publishing cmpany, London.

Souriau, A. and Cazenave, A., 1985 : 'Reevaluation of the Chandler wobble seismic excitation from recent data', Earth Planet. Sc. Let., (submitted).

Stoyko, A. and Stoyko, N., 1966 : 'L'échelle de temps rotationnel et sa représentation', Bulletin de l'Académie royale de Belgique, **52** , 279-285.

Smylie, D. and Manshina, L., 1968 : 'Earthquakes and the observed motion of the rotation pole', J. Geophys. Res., **73** , 7661-7673.

Vestine, E., 1952 : 'On the variations of the geomagnetic field, fluid motions and the rate of the earth's rotation', Proc. Natn. Acad. Sci., USA, **38** , 1030-1038.

Vestine, E. and Kahle, A., 1968 : 'The westward drift and geomagnetic secular change', Geophys. J. R. Astr. Soc., **15** , 29-37.

Voorhies, C.V. and Benton, E.R., 1982 : 'Pole strength of the Earth from MAGSAT and magnetic determination of the core radius', Geophys. Res. Lett., 9 , 258-261.

Yukutake, T. and Tachinaka, H., 1969 : 'Separation of the Earth's magnetic field into the drifting and the standing parts', Bull. Earthq. Res. Inst., 47 , 65-97.

ELECTROMAGNETIC CORE-MANTLE COUPLING

M. Paulus and M. Stix
Kiepenheuer-Institut für Sonnenphysik
Schöneckstr. 6
D-7800 Freiburg
F.R.G.

ABSTRACT. Electromagnetic core-mantle coupling can account for the decade fluctuations of the Earth's rotation. The essential physical quantity is the electrical conductivity of the mantle, which must be large for a large couple - but not too large because the short period (several years) secular variation would then not be detectable at the surface of the Earth. Both requirements are best satisfied by a conductivity which steeply drops off outwards from the core-mantle boundary. Results based on such a conductivity profile are presented for the period 1903.5 to 1975.5, in two-year intervals, and compared with the decade fluctuations of the Earth' rotation.

1. INTRODUCTION. THE CONDUCTIVITY OF THE MANTLE

Electromagnetic core-mantle coupling was first considered by Bullard et al. (1950) in the context of the westward drift of the geomagnetic field. Adopting a conductivity, σ_m, of $10\,\Omega^{-1}\text{m}^{-1}$, these authors concluded that the observed fluctuations in the rate of rotation of the Earth could not be explained by electromagnetic coupling of the mantle to turbulent motions in the core "since the time constant of such processes will be at least as great as 300 years ...". This (mechanical) time scale is proportional to I_m/Γ, where Γ is the total torque and $I_m = 7.2\cdot10^{37}$ kg m² is the mantle's moment of intertia; it can be reduced only by increasing Γ to about 10^{18} Nm, or, equivalently, by a corresponding increase of the horizontal electromagnetic stress

$$F = \frac{B_r B_\varphi}{\mu} \approx \frac{B_r^2 \tau_m U}{\mu L_m} , \qquad (1)$$

where B_r and B_φ are the radial and azimuthal field components, μ is the permeability ($=4\pi\cdot10^{-7}$ Vs/Am), U is the velocity of the core relative to the mantle, and L_m is the thickness of the conducting layer at the base of the mantle. The electromagnetic time scale of the mantle, τ_m, is determined by

A. Cazenave (ed.), Earth Rotation: Solved and Unsolved Problems, 259–267.

$$\tau_m = \mu \sigma_m L_m^2 \quad ; \tag{2}$$

it should not exceed a few years because geomagnetic variations with time scales of years are still observed at the surface of the Earth.

In contrast to Bullard et al. (1950) Munk and Revelle (1952) concluded that core motions should in fact be responsible for the decade fluctuations in the length of the day (l.o.d.), because other sources, e.g. the oceans or the atmosphere, were found inadequate. With a value of σ_m ten times that of Bullard et al. they found electromagnetic coupling feasible, and McDonald (1957) indeed found this consistent with the observed secular variation. A detailed calculation of Rochester (1960), using $\sigma_m = 100\ \Omega^{-1}m^{-1}$, demonstrated that the couple could be further increased by a factor 1.6 through the effect of higher (than the dipole) field harmonics, and Roden (1963) showed that a high electrical conductivity is required only in the lowest layers of the mantle. 500 $\Omega^{-1}m^{-1}$ in the first 500 km above the core-mantle boundary, and 100 $\Omega^{-1}m^{-1}$ in the following 1500 km yielded nearly the same factor of 5 as if the conductivity in the whole layer of 2000 km thickness was increased by this factor. The relative contribution of higher field harmonics (a factor 2.1) could also be improved by such a two-layer model.

A conductivity profile of the form

$$\sigma_m(r) = \sigma_a (r/a)^{-\alpha} \tag{3}$$

is of particular convenience (a is the radius of the core). First, the induction equation in the mantle admits an analytical solution (e.g. Lahiri and Price, 1939) in the form of spherical surface harmonics multiplied by radial coefficients which are powers of r. Second, from (3) we obtain $L_m = a/\alpha$, and - for given τ_m - the second of Eqs. (1) immediately shows that $F \sim \alpha$, i.e. that the coupling is the stronger the more concentrated to the base of the mantle the conducting layer is. Braginsky and Fishman (1976) and Stix (1982) calculated the electromagnetic couple using a σ_m of form (3).

An apparently sudden change ("jerk") in the secular variation rate occured in 1969/1970 (Le Mouël et al., 1982; Malin et al, 1983), and was used in attempts to model the electrical conductivity of the Earth's mantle. Achache et al. (1980) found that a layer of 2000 km thickness could not have a (constant) conductivity of more than 150 $\Omega^{-1}m^{-1}$; otherwise the jerk could not have been seen with the suddenness which was in fact observed. On the other hand, Backus (1983) investigated the filter properties of an electrically conducting mantle separately for each spherical harmonic; he found in particular that $\sigma_a = 3000\ \Omega^{-1}m^{-1}$ and $\alpha = 16$ in Eq. (3) would still be consistent with the observed jerk of 1969/1970. The results which we report below are based on the same σ_a, and $\alpha = 30$, i.e. an even smaller average mantle conductivity.

The above-mentioned calculations of the electromagnetic couple all adopt a core velocity, U, of order, $3 \cdot 10^{-4}$ m/s, which is inferred by the westward drift of the observed geomagnetic field. In the subsequent section we shall derive such core velocities from the secular variation separately for each considered epoch; we shall see that their values strongly vary in time and so generate large fluctuations of the ensuing electromagnetic couple.

2. TIME-DEPENDENT ELECTROMAGNETIC CORE-MANTLE COUPLING

We now report some results obtained by Stix and Roberts (1984), who used geomagnetic field models, given in two-year intervals between 1902 and 1976 (D.R. Barraclough, 1982, private communication). The models were available in form of harmonic coefficients g_n^m and h_n^m, up to degree N=5 for 1902 to 1976, and N=6 for 1926 to 1976. The authors made the approximation of slow core motions, i.e. $U \cdot \tau_m / L_m \ll 1$ (Roberts, 1972; Benton and Whaler, 1983). In the 0-th order the mantle is then insulating, and the geomagnetic field $\mathbf{B}^{(0)}$ and its secular variation $\dot{\mathbf{B}}^{(0)}$ (obtained by differentiating the g_n^m and h_n^m) are readily extrapolated down to the core-mantle boundary. As a first-order perturbation, an electric current, with density $\mathbf{j}^{(1)}$, appears, and the associated first-order electromagnetic couple is

$$\mathbf{\Gamma}^{(1)} = \int \mathbf{r} \times (\mathbf{j}^{(1)} \times \mathbf{B}^{(0)})\, dV . \tag{4}$$

The field perturbation corresponding to $\mathbf{j}^{(1)}$ can be divided into its poloidal and toroidal parts, and we therefore also speak of the "poloidal" and "toroidal" first-order couple. The former contributes only about 10% to the total couple, and we refer to the original paper by Stix and Roberts for details. In the present brief presentation we shall only discuss the "toroidal" couple, which depends on mantle currents induced by core motions. We assume that these motions are axisymmetric and toroidal, i.e. a differential rotation of the form

$$u(\theta, t) = a \sin\theta \sum_{n=0}^{\infty} \zeta_n(t)\, P_n(\theta) . \tag{5}$$

The coefficients, ζ_n, are determined for each epoch separately through minimization of the expression

$$\int \left[\dot{\mathbf{B}}^{(0)} - \operatorname{curl}(\mathbf{u} \times \mathbf{B}^{(0)}) \right]_r^2 dS , \tag{6}$$

where the integral is taken over the core surface. Figures 1 and 2 show the results obtained in this way for the first two symmetric terms of (5), the mean westward drift, $-\zeta_0(t)$, and the symmetric torsional oscillation, $\zeta_2(t)P_2(\theta)$. For the model with truncation index N=6 these results are identical with those in Figures 2 and 4 of Stix and Roberts (1984), but for the N=5 model differences occur, in particular in the period 1926 to 1930. These are due to an incorrect treatment of the N=5 model in the earlier calculations: instead of using Barraclough's original N=5 model for the whole period, it had been replaced by the N=6 model, truncated at N=5, for all epochs since 1926. As spherical harmonics are orthogonal on the sphere, but not on a given set of observatory locations, differences must arise. We take this opportunity to present the corrected calculations here.

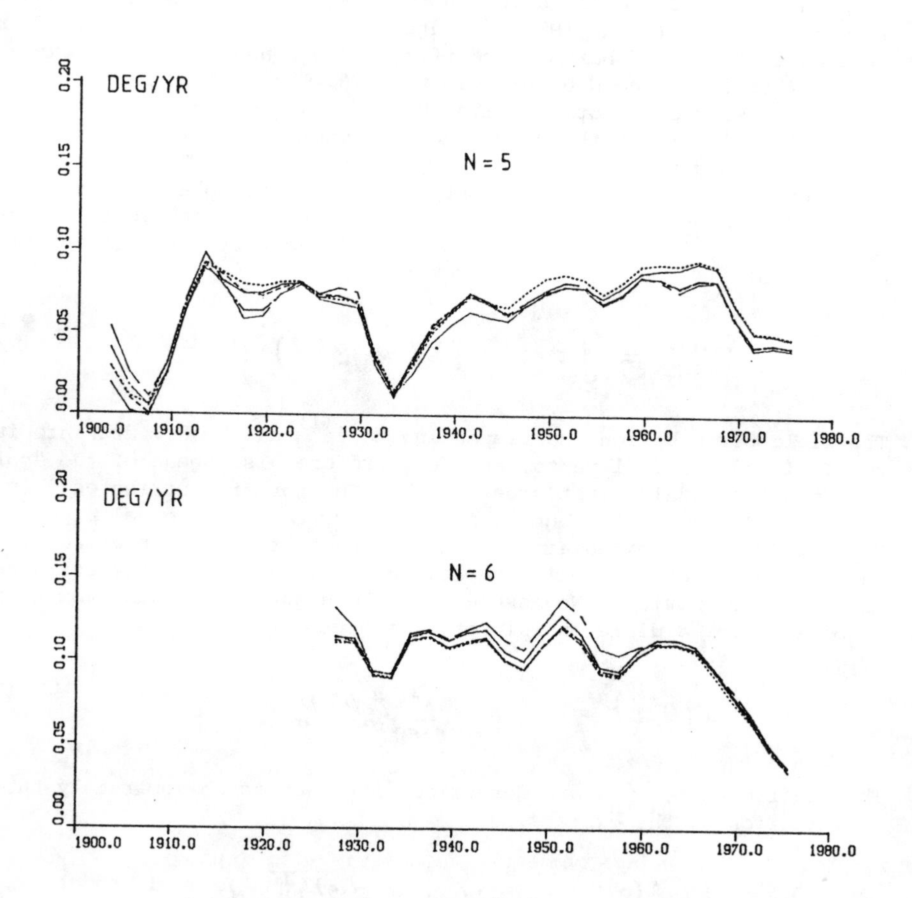

Figure 1. The westward drift, $-\zeta_o$, as a function of time, for geomagentic field models with harmonics up to degree N=5 and N=6. The various curves arise as 1 ... 5 motion harmonics are retained in the expansion (5).

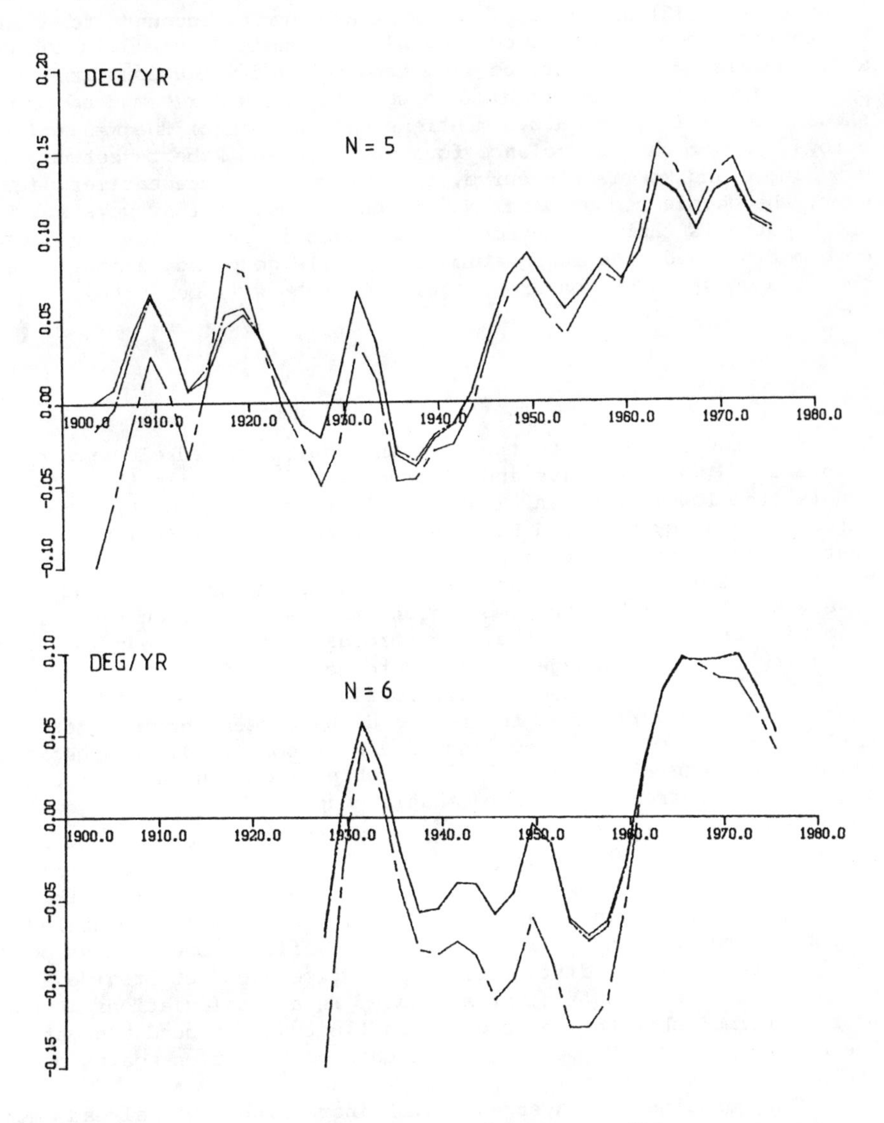

Figure 2. The second symmetric velocity coefficient, ζ_2, of the expansion (5), as a function of time, for geomagnetic field models with harmonics up to degree N=5 and N=6. The various curves arise as 3 ... 5 motion harmonics are retained in the expansion (5).

For details of the calculation of the toroidal couple we again refer to Stix and Roberts (1984), in particular their Eq. (57). For the N=5 (corrected) and N=6 models the result is shown in Figure 3. The similarity of these curves with Figure 1 immediately shows that the first term of (5) alone, i.e. the westward drift, accounts for almost the entire electromagnetic couple. Also shown in Figure 3 is the curve which Morrison (1979) derived from the observed variations in the length of the day. This is the required couple to which our computed couple should be similar. The above-mentioned correction of the N=5 model destroys some of the resemblance found by Stix and Roberts between the N=5 couple and Morrison's curve, in particular in the earlier part. However, the couple computed for N=6 is unchanged, so that we still find the hypothesis that the decade fluctuations in the l.o.d. are caused by core motions and electromagnetic core-mantle coupling, a most plausible one. The question of convergence with $N \rightarrow \infty$ will be further discussed in the last section.

3. DISCUSSION

We first draw attention to the fact that our computed electromagnetic couple is always negative and of order -10^{18} Nm, while the required couple (the lower curve in Figure 3) fluctuates around zero. The explanation offered by Stix and Roberts is based on Robert's (1972) idea that, in addition to the couple considered so far, i.e. the one induced by core motions, there is another couple, of opposite sign and comparatively slowly variable in time, which originates from currents leaking from the core into the mantle. The toroidal field gradient in the core necessary for such leakage was found to be - 0.5 G/km, - an indication for a "strong-field" dynamo in the core.

Long-term variations in the l.o.d. have been obtained for periods earlier than 1900 (e.g. Stephenson and Morrison, 1984), and geomagnetic field models are also available for such periods. In an attempt to calculate the electromagnetic core-mantle coupling before 1900 we have used 6 field models, with N=4, of Barraclough (1978) for epochs beween 1855 and 1902.5, and 8 models, also with N=4, of Thompson and Barraclough (1982), for epochs between 1600 and 1950. The result is that the torque required for the l.o.d. changes cannot in any detail be modelled. The reason is probably that, although the early field models might be reasonable, their simple differentiation in time does not provide secular variation models of sufficient accuracy. In all calculations it was however confirmed that the induced couple fluctuated around a negative value of order -10^{18} Nm, and that the main contribution again comes from the westward drift term in (5).

The question of convergence with increasing N was already mentioned. The concentration of the mantle current to a narrow layer above the core-mantle boundary (i.e. large α) makes the higher field harmonics more influential. In a model with N=12, for example, the harmonics with n=7, ... 12 still substantially contribute to the couple (Stix, 1982, Table 1). New methods to compute geomagnetic field models, i.e. the method of harmonic splines (Shure et al., 1982) and the stochastic inversion method (Gubbins, 1983) give harmonic coefficients which nicely converge as N is increased. Test calculations with such models, kindly provided by K.A. Whaler (1983, private communication) show that the

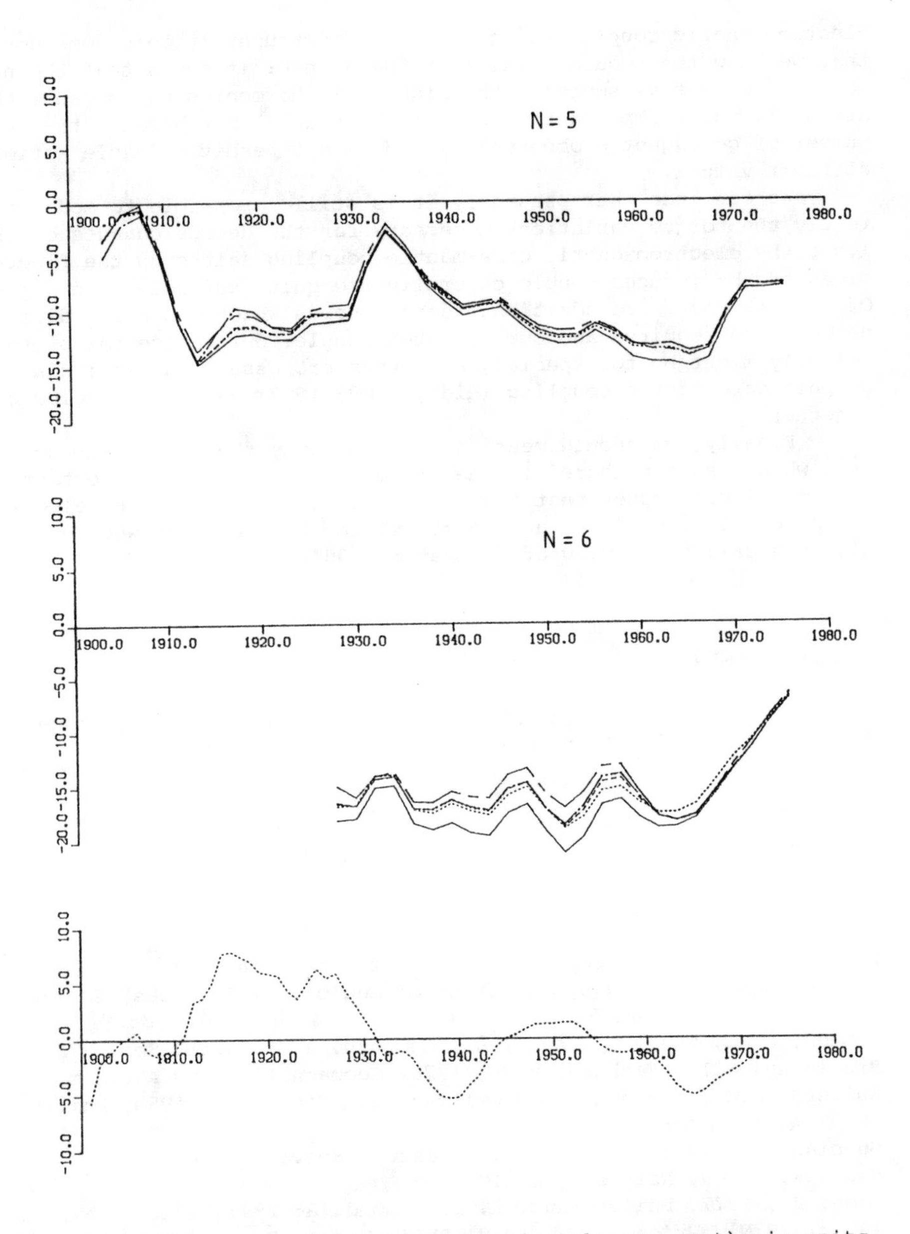

Figure 3. The total electromagnetic couple (axial component), in units of 10^{17} Nm, as a function of time, for geomagnetic field models with harmonics up to degree N=5 (top) and N=6 (middle). The various curves arise as 1...5 terms are retained in the expansion (5). The lower curve is the required torque, obtained by Morrison (1979) from observed fluctuations in the l.o.d.

electromagnetic couple also converges. Unfortunately this does not mean that we know the accurate value of the couple: it seems that the new models (correctly) suppress the high field harmonics not because they are small and unimportant (e.g. for the couple) but because the limited number of geomagnetic observatories does not permit reliable estimates at their values.

We conclude that presently it is still impossible to model perfectly the torque variations necessary for the decade changes of the l.o.d. by electromagnetic core-mantle coupling, although the fluctuations of the induced couple generally are quite sufficient in magnitude. Of course, there are additional deficiencies which can cause differences between the required and our computed couple: Our choice (5) of the core velocity might be too special, and other processes, in particular topographic core-mantle coupling (Hide, 1969; 1977) have been neglected altogether.

Finally, we should mention that not only the axial component of Γ, which was considered in the present work, might be important. Runcorn (1982) argues that the equatorial components of the electromagnetic couple should be capable of exiting the Chandler wobble - see also the critical review of Rochester (1984).

ACKNOWLEDGEMENT

We thank D.R. Barraclough and K.A. Whaler who made their geomagentic field models available to us.

REFERENCES

Achache, J., Courtillot, V., Ducruix, J., Le Mouël, J.-L., 1980, Phys. Earth Planet. Inter., 23: 72
Backus, G.E., 1983, Geophys. J. R. astr. Soc., 74: 713
Barraclough, D.R., 1978, Institute of Geological Sciences, Geomagnetic Bulletin 8, London
Benton, E.R., Whaler, K.A., 1983, Geophys. J. R. astr. Soc., 75: 77
Braginsky, S.I., Fishman, V.M., 1976, Geomagnetism and Aeronomy, 16: 443
Bullard, E.C., Freedman, C., Gellman, H., Nixon, J., 1950, Phil. Trans. R. Soc. Lond. A: 243: 67
Gubbins, D., 1983, Geophys. J. R. astr. Soc., 73: 641
Hide, R., 1969, Nature, 222: 1055
Hide, R., 1977, Phil. Trans. R. Soc. Lond. A: 284: 547
Lahiri, B.N., Price, A.T., 1939, Phil. Trans. R. Soc. Lond. A: 237: 509
Le Mouel, J.-L., Ducruix, J., Ha Duyen, C., 1982, Phys. Earth Planet. Inter., 28: 337
Malin, S.R.C., Hodder, B.M., Barraclough, D.R., 1983, Publ. Vol. Conmem. 75 aniversario Observatorio del Ebro, Roquetes, p. 239
Mc Donald, K.L., 1957, J. Geophys. Res., 62: 117
Morrison, L.V., 1979, Geophys. J.R. astr. Soc., 58: 349

Munk, W., Revelle, R., 1952, Mon. Not. R. Astron. Soc., Geophys. Suppl., 6: 331
Roberts, P.H., 1972, J. Geomagn. Geoelectr., 24: 231
Rochester, M.G., 1960, Phil. Trans. R. Soc. Lond. A: 252: 531
Rochester, M.G., 1984, Phil. Trans. R. Soc. Lond. A: 313: 95
Roden, R.B., 1963, Geophys. J. R. astr. Soc., 7: 361
Runcorn, S.K., 1982, Phil. Trans. R. Soc. Lond. A: 306: 261
Shure, L., Parker, R.L., Backus, G.E., 1982, Phys. Earth Planet. Int. 28: 215
Stephenson, F.R., Morrison, L.V., 1980, Phil. Trans. R. Soc. Lond. A: 313: 47
Stix, M., 1982, Geophys. Astrophys. Fluid Dynamics, 21, 303
Stix, M., Roberts, P.H., 1984, Phys. Earth Planet. Int. 36: 49
Thompson, R., Barraclough, D.R., 1982, J. Geomag. Geoelectr., 34: 245

[illegible], R. [illegible] 1972. [illegible]
[illegible]
Roberts, [illegible]
Lorenz [illegible]
Keating [illegible]
[illegible] 1962 [illegible]
[illegible] 198[illegible]
Spence [illegible]
[illegible]
Stephenson, F.R., [illegible] 1980 [illegible]
[illegible]
[illegible], M., 1984, [illegible]
[illegible], Roberts, P.H., 1984 [illegible]
Thom[illegible] [illegible]

GLOBAL EARTH MODELS AND EARTH ROTATION

V. DEHANT
Institut d'Astronomie et de Géophysique G. Lemaître
Université Catholique de Louvain
2, Chemin du Cyclotron
B1348 Louvain-La-Neuve
Belgium

ABSTRACT. In this study we consider the existing Earth models and supply them with an inelastic mantle. Results showing the dispersion effects expressed in terms of frequency dependant Lamé parameters are presented. Using Zschau's inelastic parameters the gravimetric factor increases by about .3 percent.

1. INTRODUCTION

In order to study the theoretical response of the Earth to an external potential, a model of the Earth is adopted. A short review of the existing models will be presented and Wahr's model will be more detailed. The equations governing the displacement field are based on the Earth physical properties. In the first part of this work, the Earth is considered as spherical and non rotating, with an elastic inner core, a fluid outer core and an elastic mantle. In the second part, the Earth is rotating and slightly elliptical. In order to extend the last existing model (Wahr's model), the inelasticity in the mantle will be considered (paragraph 3). In the last part, the effects of dispersion on the gravimetric factor will be presented.

In order to understand the philosophy of this work, we only have to remind the obvious difference between the observed response of the Earth to the luni-solar attraction and the theoretical one. This difference is prominent either in the nutation results or in the gravimetric factor, δ, which is the proportionality coefficient between the external potential and the acceleration of the instrument which mesures it. Dickey (1983) and Gwinn et al. (1984) pointed out possible discrepancies between the observed nutations and the nutation series adopted by the I.A.U. in 1981.

Melchior et al. (1983) pointed out a discrepancy of about 1 percent between the observed gravimetric factor and the computed one. This difference can be related to an experimental and/or a theoretical effect. Recent results (Edge, 1985, Van Ruymbeke, 1985) could explain about .7% difference due to a calibration effect. The remaining gap of .3 to .5% should be filled by considering a physical phenomenon which

A. Cazenave (ed.), Earth Rotation: Solved and Unsolved Problems, 269–275.

has not yet been taken into account and the inelasticity of the mantle is considered as a possible explanation.

The inelasticity of the Earth is accounted for by complex Lame parameters depending on the frequency as it will be explained later. Those parameters describe the structure of the Earth and are necessary for the integration of the equations of motion through the Earth. In a first step and to check the sensibility on the gravimetric factor and the 18.6 years nutation, arbitrary values of the Lame parameters have been selected. The values adopted in the PREM (Dziewonski and Anderson, 1981) and in the 1066A model (Gilbert and Dziewonski, 1974) have been arbitrarily decreased down to 81%. The corresponding changes of G, the part of δ which does not depend on latitude (Wahr, 1979), are presented in Table I for tide O_1. Those values even overreach the real observation. Table II presents, in percent, the variations of the nutations with respect to the PREM model. The very important sensitivity of to the variations of the Lame parameters shows of the importance of the mantle inelasticity.

TABLE I

percent of the PREM values	constant part of the gravimetric factor
81%	1.168
85%	1.164
90%	1.159
93%	1.156
95%	1.154
97%	1.153
99%	1.151
100%	1.150

TABLE II

percent of the PREM values	variation of the nutations with respect to the Prem nutation
85%	.05%
90%	.03%
95%	.02%
97%	.01%
100%	.0%

2. SPHERICAL NON-ROTATING EARTH MODEL

The response of a spherical non-rotating Earth, is deduced from a system of partial differential equations composed of the equations of motion, the Poisson's equation and the stress-strain relationship. In order to integrate it, the equations are converted in the frequency domain and the components of scalars, vectors and tensors are expanded using the Generalized Spherical Harmonic functions (GSH). Finally, ordinary first order differential equations are deduced and solved in the Earth interior using the boundary conditions and a seismical Earth model (i.e. values for the parameters such as density, shear modulus, bulk modulus at each point of the Earth interior).

3. ELLIPTICAL ROTATING EARTH MODEL

The authors studying the response of a slightly elliptical rotating Earth are splitted in two groups according the way they proceed.

The first group (including Sasao, Okubo and Saito, Molodensky, Jeffreys, Vicente, Capitaine, Hinderer...) use a reference frame tied to the Earth and rotating at a speed equal to the real Earth rotation. Ellipticity is considered as a perturbation which implies an additional torque in the right member of the angular momentum equation. The nutations and the variations of the length of day are computed using the angular momentum conservation equation.

The second group (Smith and Wahr) considers a reference frame in uniform rotation at a speed equal to the mean Earth rotation. In this case, the nutations and the variations of the length of day appear in the displacement field. They are deduced using the equations of motion. The ellipticity is considered directly in the equations : the surfaces of equal shear modulus, equal density, equal bulk modulus and equal initial potential are considered as ellipsoids. The Lamé parameters, λ and μ, the density, ρ, and the initial potential, W_{in} are not only functions of radius (r) as in the spherical case, but also functions of the colatitude (θ); they have the general form :

$$\rho(r) = \rho(r,\theta) = \rho_0(r) + \frac{2}{3}\,\varepsilon(r)\,r\,\frac{d}{dr}\rho_0(r)\,D^2_{00}(\theta)$$

$$\lambda(r) = \lambda(r,\theta) = \lambda_0(r) + \frac{2}{3}\,\varepsilon(r)\,r\,\frac{d}{dr}\lambda_0(r)\,D^2_{00}(\theta)$$

$$\mu(r) = \mu(r,\theta) = \mu_0(r) + \frac{2}{3}\,\varepsilon(r)\,r\,\frac{d}{dr}\mu_0(r)\,D^2_{00}(\theta)$$

$$W_{in}(r) = W_{in}(r,\theta) = W_{in0}(r) + \frac{2}{3}\,\varepsilon(r)\,r\,\frac{d}{dr}W_{in0}(r)\,D^2_{00}(\theta)$$

where $\varepsilon(r)$ is the ellipticity of the ellipsoid corresponding to an equivolume sphere of radius r and $D^2_{00}(\theta)$ is a GSH function which is, in this case, the Legendre polynomial of order 2.

As in the spherical case, the partial differential equations are transformed in a system of ordinary first order differential equations which then contains products of GSH functions. This leads to a coupling between spheroidal displacements of order ℓ, toroidal displacements of order $\ell+1$, $\ell-1$, $\ell+3$, $\ell-3$,... and spheroidal displacements of order $\ell+2$, $\ell-2$, $\ell+4$, $\ell-4$,... The ordinary first order differential equations provide an infinite number of systems (one for each value of ℓ as in the spherical non rotating case) with an infinite number of variables and equations (Smith, 1974). To solve such a system, truncation is requested and at the first order of the ellipticity, the integration is done in two steps (Wahr, 1979). In the first one, the solution of a supertruncated system allows to find spheroidal displacement of order ℓ and toroidal displacement of order $\ell+1$ and $\ell-1$. These displacements are then used to solve the second step and to find out the corresponding spheroidal displacement of order $\ell+2$ and $\ell-2$. The summation of all these displacements gives the spheroidal response of the Earth to the external potential of order ℓ(Wahr, 1981a and 1981b).

4. INELASTICAL EARTH MANTLE

In this part, the inelasticity of the mantle is introduced in the equations describing the displacement field of an uniformly rotating elliptical Earth.

The inelasticity can be expressed by the linear relationship between the strain and its derivatives and the stress and its derivatives. The correspondence principle of Biot allows to transform the time domain to the frequency domain in which a simple relationship between the Fourier transform of the strain and the Fourier transform of the stress is obtained with a complex proportionality coefficient. At this stage, the stress-strain relationship has exactly the form of an elastic Hooke law with complex Lamé parameters. With this principle we can account for the mantle inelasticity in the frequency domain equations. In the first step of the integration, the ten complex equations are splitted into twenty equations with twenty variables.

To integrate inside the Earth, the new inputs to the ordinary differential equations are values of the complex Lamé parameters as functions of radius. The different inelasticity models provide different profiles for the shear modulus inside the Earth. Either a description of stress or of strain is possible, using a distribution of stress relaxation time or of strain retardation time. In the computations, the real and the imaginary part of the shear modulus are taken as functions of frequency. The distributions used here are, for the strain retardation time, the constant distribution of Liu et al. (1976) and the power α distribution of Anderson and Minster (1979). We use also the truncated Gaussian stress relaxation distribution (Zschau's model) which supposes diffusion controlled processes (Zschau, personal communication, 1985). The first two distributions were constructed only in accordance with the observed quality factors of seismic and free oscillations data inside the considered frequency band while Zschau's distribution is based on a larger set of data and frequencies.

5. RESULTS

One aspect produced by inelasticity is dispersion which is accounted for by the frequency dependence of the Lamé parameters. Table III shows the gravimetric factor values for the raw data of PREM (computed at 1 second) for Liu et al.'s distribution (PREM freq), Anderson and Minster's distribution and Zschau's distribution. An increase of the gravimetric factor for O_1 of .2% appears in Liu et al.'s inelastic model, of .5% in Anderson and Minster's model and of .3% in Zschau's model. We conclude then that the dispersion in the mantle can increase the gravimetric factor by an amount corresponding to the expected value.

TABLE III

tides	Constant part of the gravimetric factor for			
	PREM	PREM freq (Liu et al.)	Anderson and Minster	Zschau
O_1	1.150	1.152	1.155	1.153
M_1	1.150	1.152	1.155	1.152
Q_1	1.150	1.152	1.155	1.153
π_1	1.149	1.149	1.152	1.149
P_1	1.145	1.147	1.150	1.148
S_1	1.142	1.144	1.147	1.144
K_1	1.130	1.133	1.136	1.133
ψ_1	1.233	1.233	1.233	1.232
ϕ_1	1.165	1.167	1.169	1.167
J_1	1.152	1.155	1.158	1.155
OO_1	1.152	1.154	1.157	1.154
M_2	1.158	1.160	1.162	1.160
M_f	1.152	1.155	1.160	1.157
M_3	1.605	1.606	1.609	1.607

ACKNOWLEDGEMENTS

Prof. Zschau is greatfully acknowledged for the advices in numerous useful discussions. Dr. Wahr is also acknowledged for providing us the

programs which allowed this work. The author wants to thank them both also for their hospitality.
We are thankful to Prof. Melchior for his helpful comments and encouragement.
This work was financially supported by a grant from the Fonds National de la Recherche Scientifique of Belgium. The computations were done at the Royal Observatory of Belgium.

REFERENCES

Anderson D.L. & Minster B.J., 1979, 'The frequency dependence of Q in the Earth and implications for mantle rheology and Chandler Wobble.', Geophys. J. Roy. astr. Soc. , 58, pp 431-440.

Dickey J.O., Williams J.G., Newhall XX & Yoder C.F., 1983, 'Geophysical applications of Lunar Laser Ranging.', JPL Geodesy and Geophysics publication , n°104.

Dziewonski A.M. & Anderson D.L., 1981, 'Preliminary Reference Earth Model.', Phys. Earth Planet. Int. , 25, pp 297-356.

Edge B.J., 1985, 'The calibration procedure of Lacoste-Romberg tidal gravimeters: Initial results from Bidston, U.K., Bad Homberg, F.R.G. and Brussels, Belgium.', communication at 58th Journees Luxembourgeoises de Geodynamique, Working Group of the Europeen Concil.

Gilbert F. & Dziewonski A.M., 1975, 'An application of normal mode theory to the retrieval of structural parameters and source mechanism for seismic spectra.', Phil. Trans. Roy. Soc. A , 278, pp 187-269.

Gwinn C.R., Herring T.A. & Shapiro L.I., 1984, 'Geodesy by Radio Interferometry: corrections to the IAU 1980 nutation series.', E.O.S. , 65, n°45.

Liu H.P., Anderson D.A. & Kanamori H., 1976, 'Velocity dispersion due to anelasticity; implications for seismology and mantle composition.', Geophys. J. Roy. astr. Soc. , 47, pp 41-58.

Melchior P. & De Becker M., 1983, 'A discussion of the world-wide measurements of tidal gravity with respect to oceanic interactions, lithosphere heterogeneities, Earth's flattening and inertial forces.', Phys. Earth Planet. Int. , 31, pp 27-53.

Smith M.L., 1974, 'The scalar equations of infinitesimal elastic-gravitational motion for a rotating, slightly elliptical Earth.', Geophys. J. R. astr. Soc. , 37, pp 491-526.

Van Ruymbeke M., 1985, 'Calibration of Lacoste-Romberg gravimeters by the inertial force resulting from a vertical periodic movement.', communication at 59th Journees Luxembourgeoises de Geodynamique, Working Group of Europeen Concil.

Wahr J.M., 1979, 'The tidal motions of a rotating, elliptical, elastical and oceanless Earth.', Ph. D. thesis , University of Colorado.

Wahr J.M., 1981a, 'A normal mode expansion for the forced response of a rotating Earth.', Geophys. J. Roy. astr. Soc. , 64, pp 651-675.

Wahr J.M., 1981b, 'Body tides on an elliptical, rotating, elastical and oceanless Earth.', Geophys. J. Roy. astr. Soc. , 64, pp 677-703.

RESONANCE EFFECTS OF THE EARTH'S FLUID CORE

J. Hinderer
Institut de Physique du Globe (UA 323 du CNRS)
5, rue René Descartes
67084 Strasbourg Cedex
France

ABSTRACT. The elastic deformation of a self-gravitating Earth's model with a liquid core submitted to volume potentials and pressure fields at the core-mantle interface is expressed by a Love number formalism. The resonance processes and the consequences on the deformation (and the resulting gravity change) are investigated. The classical nearly diurnal 'core resonance' is precised with the help of the Euler equations for conservation of angular momentum. The resonance effects on the tidally forced rotational motion and elastic deformation are shown. Other kinds of resonances, mainly due to possible core internal oscillations, are considered. A simple model expressing the gravity changes due to these modes is proposed. The experimental results concerning these resonances are finally discussed.

1. INTRODUCTION

The presence of a liquid core within an elastic mantle is known to affect the geophysical and astronomical behaviour of the Earth: appearance of a new eigenmode (usually called 'nearly diurnal free wobble' or 'free core nutation') in the rotational spectrum and modification of the Euler period relative to a rigid model, perturbations in the forced luni-solar nutations and tidal deformation; we don't consider here the influence of the fluid core in the changes of the length-of-day.

The first study on the dynamical effects of a fluid-filled elliptic cavity within a rigid rotating mantle is attributed to Hopkins (1839). This subject was then developed by other theoreticians (Hough 1895; Sludskii 1896; Poincaré 1910), who showed that a second eigenmode (of nearly diurnal period) appears in addition to the Euler wobble first observed by Chandler in 1891.

The introduction of elastic models with a more realistic distribution of the physical parameters within the Earth was first achieved by Jeffreys & Vicente (1957a, b) and Molodensky (1961).

These theoretical investigations were then extended by many authors (e.g. Shen & Mansinha 1976 ; Smith 1974, 1977;

A. Cazenave (ed.), Earth Rotation: Solved and Unsolved Problems, 277–296.

Sasao et al. 1977, 1980; Wahr 1981a,b; Lanzano 1984). As a consequence, the tidal elastic deformation was found to be resonant at diurnal frequencies. Moreover, other free modes, which are of completely different origin and dependent on the nature of the core stability, can theoretically appear and lead to resonant effects for other specific tidal waves (e.g. Pekeris & Accad 1972; Mansinha & Shen 1976).

Section 2 is devoted to the equations governing the deformation of a self-gravitating elastic body. The solutions (displacement, potential of mass redistribution) are expressed with a Love number formalism in order to take into account, in particular, pressure effects acting at the core-mantle boundary.

The resonance mechanisms due to excitation sources (volume potential, core pressure) of the free modes relative to our model and their consequences, especially on the gravity, are investigated.

The usually called 'fluid core resonance' caused by the proximity of some luni-solar diurnal waves with the nearly diurnal free wobble is precised in section 3 with the help of the Euler equations describing the rotational response of an elliptical Earth's model with liquid core and elastic mantle, especially when a special attention is paid to the core dynamics.

On one side, these equations give the classical expressions of the eigenperiods relative to the Chandler and nearly diurnal free rotational modes in the lack of tidal excitation.

On the other side, they allow to obtain the tidally forced rotational motions and elastic deformation. The resulting gravity change at the Earth's surface is then expressed in terms of a generalized (resonant) gravimetric factor.

Other kinds of resonances due to possible gravity-inertia internal oscillations in the fluid core are investigated in section 4.

A model expressing the gravity changes at the Earth's surface involving the gravito-elastic deformation of the mantle caused by the pressure field associated with these oscillatory modes is proposed.

The experimental results concerning the geophysical observations (e.g. gravity, tilt, strain), mainly relative to the nearly diurnal resonance, are considered in section 5. The astronomical observations (e.g. ellipticity of the luni-solar nutations) are also briefly considered.

2. GRAVITO-ELASTIC DEFORMATION

2.1. Basic formulation

The elastic deformation of a gravitating Earth is usually considered in the static and linearized case (to the first-order with respect to the displacement $\vec{u}$ and potential of mass redistribution V') assuming spherical symmetry and hydrostatic prestress.

The resulting perturbations due to the ellipticity and rotation (e.g. Coriolis force) are found to be small of the order of 1 % (Wahr 1981a).

Navier's elasto-static equations are (Alterman et al. 1959):

$$\vec{\nabla}.T + (\rho_o \vec{u}.\vec{\nabla}V_o) + \rho'\vec{\nabla}V_o + \rho_o\vec{\nabla}V' + \rho_o\vec{\nabla}W = 0 \qquad (1)$$

where T is the stress tensor related to the displacement $\vec{u}$ by Hooke's law in the perfect linear elastic case.

The gravific potential V_o before deformation is related to the density ρ_o before deformation by Poisson's law, and similarly for the perturbations after deformation V' and ρ'; besides, ρ' must satisfy the continuity equation.

When the effects of rotation and ellipticity are neglected, there is no coupling between the toroidal and spheroidal modes of deformation, and, in the last case, we must solve a linear differential system of six first-order equations within the mantle.

The boundary conditions at the core-mantle limit r = b are the usual ones, modified, in our case, by a hydrodynamical pressure that we note P_n (coefficient of the spherical harmonic development in $(r/b)^n$, where n is the order);we don't consider tangential tractions of viscomagnetic origin for instance.

The solutions of the gravito-elastic problem with this type of appropriate boundary conditions are given by generalized Love numbers (Legros & Amalvict 1986); in our specific case:

$$u_r(a) = h_n \frac{V_n}{g} + \bar{h}_n \frac{P_n}{\rho g}$$

$$u_t(a) = l_n \frac{V_n}{g} + \bar{l}_n \frac{P_n}{\rho g} \qquad (2)$$

$$V'(a) = k_n V_n + \bar{k}_n \frac{P_n}{\rho}$$

where V_n is the coefficient of the spherical harmonic development into $(r/a)^n$ of the volume potential (e.g. tidal, centrifugal), which is acting on the whole Earth, and expressed at r = a; $u_r(a)$, $u_t(a)$, the radial and tangential components of the displacement at the outer surface; V'(a), the potential of mass redistribution at the same limit; h_n, l_n, k_n, classical volume Love numbers and $\bar{h}_n$, $\bar{l}_n$, $\bar{k}_n$, internal pressure Love numbers of order n (independent on the degree m for a given order n for symmetry reasons); with the definition of the Love numbers that we will use hereafter, ρ is the mean density of the mantle and g the mean outer surface gravity.

Similarly, the radial displacement of the core limit $u_r(b)$ becomes:

$$u_r(b) = h_n(b) \frac{V_n}{g} + \bar{h}_n(b) \frac{P_n}{\rho g} \qquad (3)$$

with the help of the quantities (akin to Love numbers) $h_n(b)$ and $\bar{h}_n(b)$.

A more complete expression taking into account the existence of a superficial fluid layer (oceanic or atmospheric) and additional acting phenomena can be found in Legros et al. (1985 a).

2.2. Resonances

For a fluid layer, a whole class of eigenmodes can, in principle, exist and be excited by various kinds of processes. It is conceivable that oscillatory modes due to magnetic, inertial or buoyancy restoring forces may exist within the fluid core (e.g. Smylie & Rochester 1981; Crossley 1984; Friedlander 1985) in addition to eigenmodes of rotational origin relative to an Earth's model composed by a fluid core and solid shell.

If we note Ψ_n and Π_n the inertial potentials and σ_i, σ_j the respective eigenfrequencies, we can write, to the main order:

$$\Psi_n = a_{in} \frac{V_n}{(\sigma - \sigma_i)} + b_{in} \frac{P_n}{\rho(\sigma - \sigma_i)}$$

$$\Pi_n = c_{jn} \frac{V_n}{(\sigma - \sigma_j)} + d_{jn} \frac{P_n}{\rho(\sigma - \sigma_j)} \qquad (4)$$

when supposing that the only acting sources are here a volume potential V_n and a core pressure P_n (other sources can, of course, exist, but are neglected); the coefficients a_{in}, b_{in}, c_{jn}, d_{jn} are 'resolving' parameters which are relative to the specific excitation equation.

If the sources of excitation are of frequency close to one of these eigenfrequencies, there is a possibility of resonant deformation of the Earth.

When substituting (4) into (2), then we get for the radial displacement $u_r(a)$, for instance:

$$u_r(a) = \left[h_n\left(1 + \frac{a_{in}}{(\sigma - \sigma_i)}\right) + \bar{h}_n \frac{c_{jn}}{(\sigma - \sigma_j)}\right] \frac{V_n}{g}$$

$$+ \left[h_n \frac{b_{in}}{(\sigma - \sigma_i)} + \bar{h}_n\left(1 + \frac{d_{jn}}{(\sigma - \sigma_j)}\right)\right] \frac{P_n}{\rho g} \qquad (5)$$

and similar expressions for $u_t(a)$ and $V'(a)$ by changing h_n, $\bar{h}_n$ into l_n, $\bar{l}_n$ and k_n, $\bar{k}_n$, respectively; each coefficient (e.g. a_{in}) is relative to a given harmonic term of order n and degree m, and eigenfrequency (e.g. σ_i).

The knowledge of $u_r(a)$, $u_t(a)$ and $V'(a)$ enables to express many geophysical phenomena: the relative change in gravity, the variation of latitude, the deflection of vertical, the Earth's shape, the displacement of equipotential surface, etc...

Moreover, as the changes in the Earth's and core inertia tensors of tidal or rotational origin are also dependent on the gravito-elastic deformation, some peculiar combinations of Love numbers will play an important role in the Euler equations (cf. section 3.1) and, consequently, for the rotational dynamics.

We don't intend to study all these combinations, but rather consider now a special case, the relative variation of gravity.

If we are interested in the resonance phenomena occuring in the gravity changes at the outer surface, we obtain a formula similar to (5), when changing $gu_r(a)$ into $- a\Delta g/n$ and h_n, $\bar{h}_n$ into coefficients δ_n, $\bar{\delta}_n$:

$$\Delta g(a) = \left[\delta_n\left(1 + \frac{a_{in}}{(\sigma - \sigma_i)}\right) + \bar{\delta}_n \frac{c_{jn}}{(\sigma - \sigma_j)}\right] \frac{(-nV_n)}{a} + \left[\delta_n \frac{b_{in}}{(\sigma - \sigma_i)} + \bar{\delta}_n\left(1 + \frac{d_{jn}}{(\sigma - \sigma_j)}\right)\right] \frac{(-nP_n)}{\rho a} \tag{6}$$

with:

$$\delta_n = 1 + 2h_n/n - (n+1)k_n/n$$

$$\bar{\delta}_n = 2\bar{h}_n/n - (n+1)\bar{k}_n/n$$

3. NEARLY DIURNAL RESONANCE

We further suppose now that the sole periodical forcing mechanism is the diurnal part of the luni-solar tidal potential W , which has a tesseral distribution of order n = 2 and degree m = 1; we neglect for the moment the excitation in pressure ($P_n = 0$).

The rotational motion of an Earth's model composed by a liquid core and elastic mantle is described by the Euler equations for conservation of angular momentum. These equations will allow us to get the expressions of the rotational eigenperiods and the 'resolving' coefficient (like c_{jn}) relative to the forced rotational motion of tidal origin.

3.1. Euler equations

Correct to the first-order, the equatorial components of these equations become in the Tisserand frame of the mantle mean axes (e.g. Hinderer et al., 1982):

$$\begin{aligned} A\dot{\omega} - iA\alpha\Omega\omega + \Omega\dot{c} + i\Omega^2 c + \dot{l}^c + i\Omega l^c &= G \\ A^c\dot{\omega} - iA^c\alpha^c\Omega\omega + \Omega\dot{c}^c + i\Omega^2 c^c + \dot{l}^c + i\Omega l^c &= G^c + N + \Gamma \end{aligned} \tag{7}$$

using a complex notation for the Earth's wobble components $\omega = \omega_1 + i\omega_2$, Earth's and core tensors of inertia $c = c_{13} + ic_{23}$, $c^c = c^c_{13} + ic^c_{23}$, core relative angular momentum $l^c = l^c_1 + il^c_2$, passive core-mantle interaction torques due to the fluid pressure $N = N_1 + iN_2$ or of viscomagnetic origin $\Gamma = \Gamma_1 + i\Gamma_2$, and gravitational torques of tidal origin $G = G_1 + iG_2$, $G^c = G^c_1 + iG^c_2$ applied to the whole Earth and core respectively.

A, α are the mean Earth's equatorial moment of inertia and dynamical ellipticity; A^c, α^c, the corresponding quantities for the core; Ω is the uniform axial rotation rate.

The angular momentum l^c of the core depends on the fluid velocity field $\vec{v}$ with respect to the mantle which necessarely must satisfy the Navier-Stokes hydrodynamical equations.

These equations are often considered in a restricted linear and steady-state form without buoyancy forces, assuming that the viscous and magnetic terms are confined within small boundary layers near the core-mantle interface and thus don't play any role within the deeper core:

$$\dot{\vec{\omega}} \times \vec{r} + 2\,\vec{\omega} \times \vec{v} = -\frac{\vec{\nabla}p}{\rho^c} + \vec{\nabla}\Psi + \vec{\nabla}V \qquad (8)$$

where $2\vec{\omega} \times \vec{v}$ is the classical Coriolis term, $\dot{\vec{\omega}} \times \vec{r}$ the more unfamiliar Poincaré term, p, ρ^c the fluid pressure (non-hydrostatic part) and density , $\Psi = (1/2)(\vec{\omega} \times \vec{r})^2 - (\Omega^2 r^2)/3$ the centrifugal potential and V any other potential (e.g. tidal).

When the core is homogeneous, as the right hand side of (8) is the gradient of a scalar potential, the left hand side must be too a gradient and its curl is identically zero; such a conclusion is also valid for a stratified core in a barotropic approximation.

To the dominant order, $\dot{\vec{\omega}} \times \vec{r} \simeq (\vec{\omega} \times \vec{\Omega}) \times \vec{r}$ (e.g. Malkus 1968), and the curl of (8) implies a relation between the mantle rotation and the velocity of the fluid particles.

When the fluid motion is of uniform vorticity (Poincaré 1910), satisfying $\vec{\nabla}.\vec{v} = 0$ and $\vec{v}.\vec{n} = 0$ at the elliptical core-mantle limit of unit outward normal $\vec{n}$, we obtain then a compatibility condition between ω and the core rotation ω^c (Legros & Amalvict 1986), which means that the core has an inertial response to the rotational motion of the container and, consequently, that one more degree of freedom in wobble appears in addition to the classical wobble and nutation relative to a wholly rigid Earth's model (e.g. Smith 1977; Chao 1983; Dickman 1983).

The knowledge of the velocity field leads to the expression of the angular momentum by integration over the core volume.

The core-mantle interaction torques N and Γ are also dependent on the relative core motion and on some other parameters like the geometry of the core limit, core viscosity, lower mantle and core electrical conductivities, amplitude of the Earth's magnetic field. The forcing torques of tidal origin G, G^c are functions of the dynamical ellipticities of the Earth ($\alpha = (C-A)/A$) and core ($\alpha^c = (C^c-A^c)/A^c$) respectively, and of the luni-solar gravitational potential W of order

$n = 2$.

The gravito-elastic response under the influence of the various potentials (e.g. tidal, centrifugal, inertial) acts to change the inertia tensors c^c, c which can be easily expressed with the help of a 'Love number' formalism previously described.

The development of all the terms involved in (7) leads to the following set of equations (for a detailed derivation, see e.g. Legros et al. 1985b):

$$\dot{\omega}(1 + \alpha k/k_s) - i\Omega\omega\alpha(1 - k/k_s) + (\dot{\omega}^c + i\Omega\omega^c)(A^c/A - \alpha k_1/k_s) = (3\alpha k)(k_s a^2)^{-1}(\dot{W}/\Omega + iW) - 3i\alpha W/a^2$$

$$\dot{\omega}(1 + q_o h^c/2) + \dot{\omega}^c(1 - q_o h_1^c/2) + i\Omega\omega^c(1 + \alpha^c + K' - iK) = (3q_o h^c)(2a^2)^{-1}(\dot{W}/\Omega) \tag{9}$$

with k, k_s, k_1, h^c, h_1^c, Love numbers (or combinations) of various kind: k, k_s, volume and secular Love numbers ; $k_1 = \bar{k}(\rho^c/\rho)(b/a)^2$, where $\bar{k}$ is defined in (2); $h^c = h(b)(a/b)$ and $h_1^c = \bar{h}(b)(b/a)(\rho^c/\rho)$ where $h(b)$, $\bar{h}(b)$ are defined in (3).

The parameter $q_o = \Omega^2 a/g$ is the geodynamical constant expressing the ratio of the centrifugal force to the gravity at the outer surface; K, K' are viscomagnetic coupling constants (here in a dimensionless notation) involving, among others, the outer core Eckman number and the lower mantle magnetic Eckman number (e.g. Loper 1975; Rochester 1976).

3.2. Free wobbles

The eigenfrequencies of this system in the lack of tidal excitation ($W = 0$) are:

$$\tilde{\sigma}_{nd} = \sigma_{nd} + i\eta_{nd} = \sigma_{nd}(1 - i(2Q_{nd})^{-1}) = -\Omega[1 + (A/A^m)(\alpha^c + q_o h_1^c/2 + K' - iK)] \tag{10}$$

$$\sigma_{cw} = (A/A^m)\alpha\Omega(1 - k/k_s)$$

where η_{nd}, Q_{nd} are the damping coefficient and 'quality factor' of the nearly diurnal free wobble.

Because of the small values of the coupling constants K and K', probably of order 10^{-7} with the usual estimates of the physical parameters near the core-mantle boundary, there is no perturbation of the Chandler frequency σ_{cw}, to the chosen order of approximation, due to the viscomagnetic torque and, thus, no damping of this mode.

It is clear that the consideration of anelastic properties within the mantle (e.g. Zschau 1978; Anderson & Minster 1979; Smith & Dahlen 1981; Okubo 1982) would transform the Love numbers k, h_1^c into complex terms and induce a damping mechanism for both eigenmodes.

3.3. Forced rotational motion

With the help of the Euler equations (9), it is easy to obtain the amplitude of the Earth's and core rotations ω, ω^c as a function of the tidal potential W. Setting $\omega = \omega_o \exp(i\sigma t)$, $\omega^c = \omega_o^c \exp(i\sigma t)$ and $W = W_o \exp(i\sigma t)$, the rotational responses ω_o and ω_o^c (expressed here in the rotating Tisserand frame) become, for $(\sigma + \Omega)/\Omega \ll 1$, to the main order (see Hinderer et al. 1985b):

$$\omega_o = \frac{A\left[(\sigma + \Omega)\left(1 - \frac{A^c q_o h^c}{2\alpha A}\right) + \Omega\left(\alpha^c + \frac{q_o h_1^c}{2}\right)\right]}{A^m\,(\sigma - \tilde{\sigma}_{nd})} \; \frac{3\alpha W_o}{\Omega a^2}$$

$$\omega_o^c = \frac{A\left(1 - \frac{q_o h^c}{2\alpha}\right)}{A^m\,(\sigma - \tilde{\sigma}_{nd})} \; \frac{3\alpha W_o}{a^2} \tag{11}$$

We immediately see the resonance effect in $(\sigma - \tilde{\sigma}_{nd})^{-1}$ appearing in the rotational motions for a given potential of amplitude W_o and frequency σ.

The slight damping introduced by the imaginary part of the eigenfrequency is able to cause a phase lead for some luni-solar nutations as pointed out by Toomre (1966).

Notice that, because of the assumption $(\sigma + \Omega)/\Omega \ll 1$, the core response $\omega_o^c \gg \omega_o$ by one order of magnitude.

Similar formulae expressing the tidally forced rotations in a quasi-inertial frame linked to the vernal equinox can be found in Amalvict et al. (1985a, b).

3.4. Tidal deformation

The predominant core rotation will lead to a fluid pressure $\rho^c b^2 \omega_o^c \Omega$ at the core-mantle limit r = b and, substituting (11) into (5) when $P_n = a_{in} = 0$, the resulting elasto-gravitational deformation is then specified by the following formulae:

$$u_r(a) = \left(h + h_1 \frac{c_{nd}}{(\sigma - \tilde{\sigma}_{nd})}\right)(W_2/g) = h^*(W_2/g)$$

$$u_t(a) = \left(l + l_1 \frac{c_{nd}}{(\sigma - \tilde{\sigma}_{nd})}\right)(W_2/g) = l^*(W_2/g) \tag{12}$$

$$V'(a) = \left(k + k_1 \frac{c_{nd}}{(\sigma - \tilde{\sigma}_{nd})}\right) W_2 = k^*\, W_2$$

where $c_{nd} = \Omega(A/A^m)(\alpha - q_o h^c/2)$ is the resolving parameter relative to the present problem (proportional to the quantity noted c_{in} in equation (5); h_1, l_1, k_1 are $\bar{h}$, $\bar{l}$, $\bar{k}$ respectively multiplied by the constant quantity $(\rho^c/\rho)(b/a)^2$;; we have lighten the notation by omitting in (12) the order n = 2 of the Love numbers.

We don't intend to consider now all the combinations of these Love numbers expressing any observable parameter at the surface of the Earth, but it is clear that the nearly diurnal resonance commonly called 'fluid core resonance' will be present in every combination.

For instance, the gravity change at the outer surface becomes:

$$\Delta g(a) = - \delta^* (2W_2/a) \tag{13}$$

with:

$$\delta^* = (1 + h - 3k/2) + \frac{A\,(h_1 - 3k_1/2)(\alpha - q_o h^c/2)\Omega}{A^m\,(\sigma - \tilde{\sigma}_{nd})}$$

It should be noted that the theoretical gravity variation involves a gravimetric factor δ^* which consists of two terms: the usual (static) gravimetric factor $\delta = 1 + h - 3k/2$ and a 'dynamical' contribution showing the core resonance for some tidal waves in the diurnal band of frequency σ close to the nearly diurnal wobble eigenfrequency σ_{nd}.

The frequency dependence of the Love numbers shown above expresses the resonant behaviour of the deformation and is, of course, completely independent on the presence of inertial (e.g. Coriolis term due to the rotation) or anelastic terms (e.g. visco-elastic rheology) in equation (1) that we supposed in the elasto-static approximation.

Another important combination of Love number is the tilt diminishing factor $\gamma^* = 1 + k^* - h^*$ (deflection of the vertical relatively to the deformed surface), which also occurs in the expression of the displacement of the equipotential surface with respect to the deformed one (e.g. amplitude of oceanic tides).

The deflection of the vertical with respect to fixed stars as observed by astronomical methods involves a different coefficient $\lambda^* = 1 + k^* - l^*$.

In addition to these phenomena, it is possible to observe the strain at the Earth's surface which again can be written with the help of various combinations of the Love numbers h^* and l^*).

4. OTHER KINDS OF RESONANCE

4.1. Core oscillations

The detection of core oscillations (especially the identification of the eigenperiods) with surface data is of fundamental interest for the knowledge of the stability of the density stratification in the fluid core and its thermal state.

The most adequate possibility of detection of these core modes is probably offered by the sensitive and stable superconducting gravimeters which are now available.

The equations of motion relative to the dynamics of these core waves become (e.g. Friedlander 1985):

$$\partial^2\vec{u}/\partial t^2 + 2\,\vec{\Omega} \times \partial\vec{u}/\partial t = -\vec{\nabla}\chi - c^2N^2(\vec{\nabla}.\vec{u})(\vec{g}/g) \qquad (14)$$

neglecting the advective, magnetic and viscous terms from a scaling analysis (Crossley & Rochester 1980; Smylie & Rochester 1981; Crossley 1984). $\vec{u}$ is the displacement vector, $\vec{g}$ and g the unit vector and amplitude of the equilibrium gravity respectively; $\chi = p'/\rho - V'$, where p' and V' are the perturbed Eulerian pressure and gravitational potential, and $\rho(r)$ the local equilibrium density of the core; c, the local P-wave velocity and N, the local buoyancy frequency related to the density stratification in the core:

$$\begin{aligned} N^2(r) &= -\,g(r)(d\rho(r)/dr)\rho^{-1}(r) - g^2(r)c^{-2}(r) \\ &= \alpha_o(r)g(r)(dT/dr - dT_a/dr) \qquad (15) \\ &= -\,g^2(r)c^{-2}(r)\beta(r) \end{aligned}$$

where α_o is the thermal expansion coefficient, dT/dr and dT_a/dr, the actual and adiabatic temperature gradients respectively; β is the stability parameter introduced by Pekeris & Accad (1972).

To form a closed system, the equations of motion (14) must be implemented by the law of mass conservation, Poisson's equation for the perturbed density and gravific potential and an equation of state.

The interpretation of a given oscillatory regime observed at the surface as a function of the Brunt-Vaisala buoyancy frequency N(r) is a difficult task as pointed out by many authors (e.g. Crossley 1984; Friedlander 1985).

Even with the (most unlikely) assumption that N is constant throughout the core volume, the problem is complex, mainly because of the influence of the Earth's rotation (Coriolis force), but also because of the Earth's elasticity and self-gravitation (e.g. Shen 1983).

The damping of the core oscillations due to ohmic losses and viscosity, although supposed very small (Crossley & Smylie 1975), introduces additional difficulties for the observational detection (Crossley 1984).

We modelize now the eventual gravity effect due to a core oscillation assuming that it results from the elastostatic deformation induced by a pressure field acting at the core-mantle limit; we don't consider the direct gravitational effect (existing even for a rigid mantle) due to the mass redistribution within the liquid core as supposed elsewhere (Gubbins 1975; Crossley 1984).

From equation (6), it yields:

$$\Delta g(a) = -\left(\delta_n + \bar{\delta}_n \frac{c_{jn}}{(\sigma - \sigma_j)}\right)\frac{nW_n}{a} - \bar{\delta}_n \frac{nP_n}{\rho a} \qquad (16)$$

If the 'ultra-slow' modes are supposed to be confined within the core (e.g. Pekeris & Accad 1972), the vibrating radial displacement tends to vanish at the core-mantle boundary and it appears reasonable to consider that this limit is not a nodal surface for the normal stress that will act to deform the mantle in the elasto-static approximation justified for phenomena of period larger than the one of the classical free oscillations (say a few hours).

The term in $(\sigma - \sigma_j)^{-1}$ shows the resonant behaviour of the tidal deformation due to the tidal potential W_n of frequency σ in the proximity of a core mode of frequency σ_i and is, actually, very similar to the diurnal resonance studied before; the only differences are in the values of the eigenfrequency and resolving parameter c_{jn}.

It means that the study of the tidal deformation can (in principle) be useful for the discussion of the state of stability of the core (in case of resonance).

Consider now the last term at the right hand side of (16) which is relative to a core mode oscillating at its own eigenfrequency, because of some excitation process that we don't precise (e.g. due to an earthquake), and giving rise to a pressure field P of degree n; notice that the core oscillations are rather expressed with the help of Hough functions (Olson 1977) as a sum of various associated Legendre polynomials than only with a single harmonic component. For n = 2 and no tidal potential, equation (16) becomes in the non-resonant case:

$$\Delta g(a) = -(\bar{h} - 3\bar{k}/2)(2P_2/\rho a) \qquad (17)$$

Setting $\rho = 5.10^3\,\mathrm{kg.m^{-3}}$, $a = 6.4\ 10^6$ m, $\bar{h} = -0.23$ and $\bar{k} = -0.13$ (Legros et al. 1986) leads to:

$$\Delta g\ (\text{nanogals}) \simeq 20\ P\ (\text{mbars}) \qquad (18)$$

A gravity change at the surface of about 10 nanogals could be produced by a pressure at the core surface of order 0.5 mbars; such a small value is physically compatible with the order of magnitude of the velocity fields involved in the (theoretical) core oscillations (e.g. Crossley & Rochester 1980; Crossley 1984).

4.2. Hydrospheric oscillations

It is conceivable to think of other kinds of resonance mechanisms which are, on the contrary to the last ones, located in the atmosphere or in the oceans.

The world ocean oscillations are mainly divided in vorticity and gravity modes (Platzman et al. 1981). If the possible resonance of the vorticity modes with tidal waves needs further investigation, it is quite certain that, since the classical range of periods (8-30 hours) relative to the gravity modes contains the semi-diurnal and diurnal tidal bands (Sundermann1982; Wahr 1984), some near-resonant conditions should exist.

As the atmosphere itself can exhibit free vibrations either of gravity-inertial (e.g. Chapman & Lindzen 1970) or rotational origin (Legros & Amalvict 1985), additional possibilities of resonance can be taken into account; a methodology similar to the previous one for the resonances originating within the core can be proposed if one uses the appropriate combinations of Love numbers (external pressure, load) involved in each specific case.

5. Observational results

5.1. Nearly diurnal resonance

The first experimental results for the tilt factor γ^* of the three principal diurnal waves O_1, P_1 and K_1 were obtained with quartz horizontal pendulums (Melchior 1966) and showed a good agreement with one of the first theoretical study on the dynamical effects of the fluid core within an elastic mantle (Molodensky 1961) .

These results were latter confirmed by a very large number of pendulums observations; besides, similar resonant effects were seen in the gravimetric factor δ^* (of O_1 , P_1 , K_1) with the help of gravimeters measuring the vertical tidal component.

This first kind of observations could essentially provide the relative amplitude ratio of the previous principal waves, but could not allow a precise determination of the asymtote (corresponding to the location of the NDFW frequency) of the hyperbolic resonance curves given by equation (13).

A major contribution to obtain more information on the fine structure of the core resonance appeared with the possibility of determining the small tidal waves ψ_1 and ϕ_1; unfortunately, the critical wave ψ_1, which is the closest to the eigenfrequency, is of very small amplitude (less than one hundredth of K_1 at latitude 50°).

Moreover, the separation of closer tidal frequencies needs of course longer durations of observations (at least one year in order to separate K_1 and ψ_1).

Such a determination could be first achieved principally with gravity data (e.g. Lecolazet & Steinmetz 1974; Abours & Lecolazet 1979; Lecolazet & Melchior 1977; Warburton & Goodkind 1978), but also with clinometric (e.g. Blum et al. 1973; Lecolazet & Melchior 1977) and strain recordings (Melchior 1979; Levine 1978).

An illustration of the observed resonance effect on the tilt γ^* and gravimetric δ^* factors is given by figure 1; the amplitudes of these factors for the tidal waves P_1 , K_1, ψ_1,ϕ_1 are divided by $\gamma(O_1)$ and $\delta(O_1)$, respectively, in order to get 'normalized ratios', which are independent on the calibration of the instruments; the continuous curves are relative to theoretical values computed by Wahr (1981a) for the neutrally stratified variant of 1066A Earth's model (Gilbert & Dziewonski 1975) .

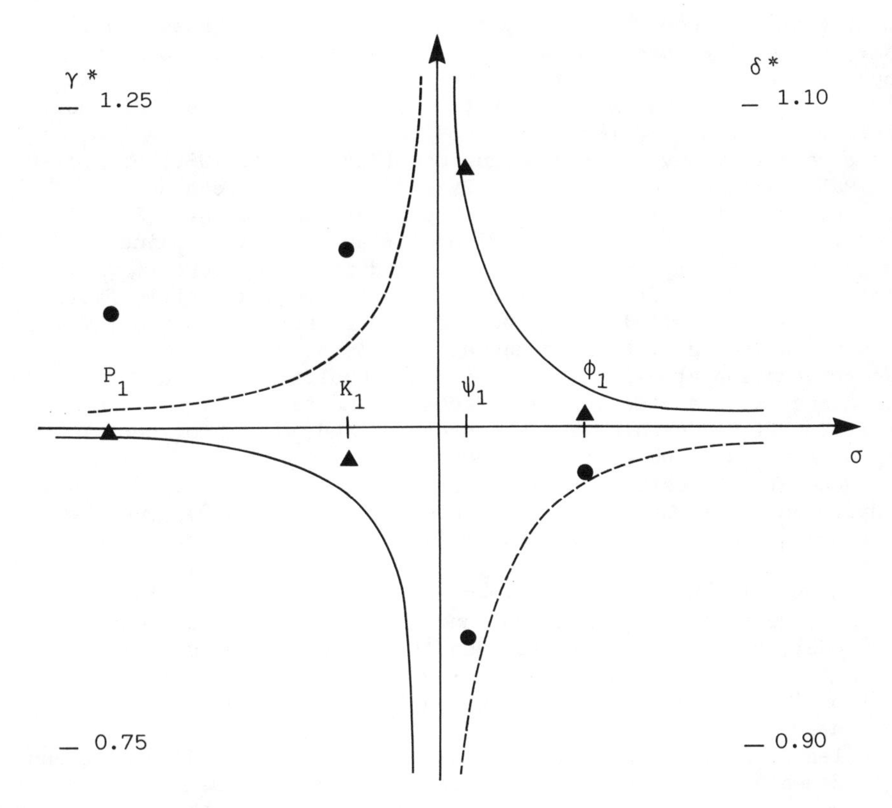

Figure 1. Illustration of the nearly diurnal resonance effect on the tilt and gravimetric factors. The closed circles, ●, are observed mean values obtained from the analysis of long tiltmeter series (about 250 000 readings) according to Melchior (1980). The closed triangles, ▲, are measured gravimetric factors after correction for atmospheric effects using 10 years (1973-1983) of gravity registration from Strasbourg. The dashed and solid lines are theoretical clinometric and gravimetric (normalized) factors respectively according to Wahr's (1981 a)computations.

The improved precision of the measurements (especially in gravity) could then allow to determine not only the amplitudes of the tidal waves but also their phase lags (or leads) and therefore provide new constraints for the damping of the nearly diurnal free wobble (see equations 10 and 13), which is only dependent on the viscomagnetic friction torque between the fluid core and the mantle in the frame of the model discussed in section 3.

The determination of the 'quality factor' Q_{nd} was initiated by Warburton & Goodkind (1978) and precised by Goodkind (1983), who first took into account the diurnal resonance in the oceanic tide itself and its loading effect upon the gravity predicted by Wahr & Sasao (1981).

Notice that the core resonance effect (with a damping coefficient of about 1/year) was indeed observed in the oceanic tide around Japan from the analysis of the amplitude ratio and phase difference of the principal waves K_1 and P_1 (Ooe & Tamura 1985).

More recently, several investigations (Zurn et al. 1985; Hinderer et al. 1985a) allowed to simultaneously find the eigenfrequency, damping and strength (given by the term proportional to $(h_1 - 3k_1/2)$ in equation 13) from the study of the 'loads' (Warburton & Goodkind 1978) of the waves P_1, K_1, ψ_1, ϕ_1(after oceanic and pressure corrections); the 'load' for a given frequency corresponds to the gravimetric factor of this wave given by (13) minus the δ factor relative to a tidal wave which is not affected by the resonance (in general O_1).

An interesting attempt to determine the quality factor and the eigenfrequency with a stacking method using simultaneous long records of gravity, tilt and strain was also proposed and could provide meaningful results (Neuberg & Zurn 1985).

In general, the estimated eigenfrequencies ($\simeq - \Omega(1 + 1/430)$) are quite different from the theoretical ones ($\simeq - \Omega(1 + 1/460)$) and the quality factors are of order 3000-4000 (with relatively large uncertainties).

The spread in the strength values seems to be more important; notice that, because the geometrical and inertia properties of the Earth's model are known from independent techniques, the estimate of the strength of the resonance leads to an observed value of the combination $(h_1 - 3k_1/2)$ involving internal pressure Love numbers defined in (2).

It clearly appears that the accuracy of the ocean load corrections is of fundamental importance, even in Central Europe where the above corrections are small (Zurn et al. 1985), especially for the determination of the quality factor Q_{nd}.

Notice, at last, that, before interpreting the core resonance parameters in terms of core-mantle coupling properties or liquid core dynamics, one must take care of the fact that other physical mechanisms are probably operating like mantle anelasticity or unmodelled oceanic effects (ocean bottom friction, dynamical processes, etc...).

The direct astronomical observation of the nearly diurnal free wobble is up to now controversial and the recent studies suggest a very small upper bound for its amplitude (e.g. Sasao & Wahr 1981; Capitaine 1982; Wahr & Larden 1983).

Therefore, once again, it is the indirect effect (on the forced luni-solar nutations) that is investigated by the astronomers in order to infer some constraints on the resonance process.

As there is a strong relation between the amplitudes of some diurnal tidal waves and the amplitudes of the associated luni-solar nutations (Melchior & Georis 1968), there is of course an expected resonance effect for some of these nutations (cf. equation 11); from a theoretical point of view, the corrections due to the existence of a fluid core within an elastic mantle can reach a value as high as 0.02 arcseconds for the principal (18.6 years) and semi-annual nutations (Jeffreys & Vicente 1957 a, b; Molodensky 1961; Wahr 1981b;

Amalvict et al. 1985b) and are consistent with astronomical observations (e.g. Sasao et al. 1977; Kinoshita et al. 1979; Melchior 1980; Wahr 1981b).

5.2. Detection of core modes

As we have seen it previously, the gravito-elastic deformation of the Earth can be dependent on the existence of core internal oscillations and can show a resonant behaviour or not according to the excitation process.

On one hand, the possibility of resonant deformation was proposed by Mansinha & Shen (1976), who found that, for a stable outer core with a stability parameter $\beta(r) = -0.2$ (see equation 15), a theoretical free oscillation occurs near a period of 12 hours between the L_2 and λ_2 sectorial semi-diurnal waves (see their figure 3).

This resonance, although there is some indication in tidal gravity records (see e.g. Warburton & Goodkind 1978; Richter 1985) suggesting an abnormal amplification of these waves (may be of oceanic origin), needs to be experimentally confirmed (or infirmed).

On the other hand, a spectral peak of 10 to 15 nanogals around a period of 13.9 hours was detected very recently with a superconducting gravimeter at Brussels (Melchior & Ducarme 1985) and attributed to an inertial gravity oscillation in the Earth's liquid core excited by a deep earthquake.

This second kind of non-resonant gravity change (at least without any tidal contribution), if caused by the mechanism mentioned in section 4, would require pressure fields at the core-mantle interface of reasonable amplitude in case of low degree ($n = 2$) distribution (cf. equation 18).

In order to check the reality of such periodical phenomena (of resonant nature or not) and to extract more information (e.g. the spatial characteristics of the core modes), it would be useful to achieve similar treatments on combined high quality data relative to the same time periods of different stations.

In addition to the removal of local effects, it should, in principle, provide a higher ratio of signal with respect to stochastic noise and allow a better detection of these modes of very small amplitude.

6. Conclusion

The gravito-elastic deformation of the Earth is dependent on the acting phenomena (e.g. volume potential) and the suitable boundary conditions (e.g. overpressure).

The existence of oscillatory flows within the fluid core that complete the classical spectrum of the eigenmodes relative to a solid model leads to complementary conditions at the core -mantle interface. These core oscillations are either of rotational origin (e.g. core wobble) or caused by gravity-inertial restoring forces (core modes). When the frequency of a given excitation source is close to one of these eigenfrequencies, a resonant deformation appears and implies that the generalized Love numbers (h*, l*, k*) introduced to express the gravito-elastic solutions (radial and tangential displacement, potential of mass redistribution) exhibit a frequency-dependent 'dynamical' part in addition to the usual (non-resonant) part.

As a consequence, all the observable phenomena (e.g. gravity, tilt, strain) that involve various combinations of Love numbers are resonant at the same frequencies.

On one hand, the existence of a core flow of uniform vorticity (Poincaré flow) well represented by a quasi-rigid rotation of the core with respect to the mantle adds to the Euler wobble a new rotational mode of nearly diurnal period called nearly diurnal free wobble (NDFW). Therefore, when the frequency of a luni-solar tidal wave is in the proximity of this eigenfrequency, the rotational responses of the Earth and, especially, of the core become resonant.

The enhanced core rotation leads to an overpressure condition at the core-mantle limit and, consequently, the tidal deformation of the Earth shows a resonant behaviour in the vicinity of the nearly diurnal eigenfrequency: this is the classical 'fluid core resonance' effect.

On the other hand, when buoyancy effects related to the stability of the density distribution in the core are taken into account, a completely different kind of free oscillations is theoretically possible.

These 'ultra-slow' modes have eigenperiods larger than the classical (elastic) ones (bounded by $\simeq$ 1 hour) and some tidal waves (e.g. semi-diurnal) could find resonance and, once again, lead to a resonant deformation of the Earth.

When assuming that the main effects caused by these modes are due to pressure actions at the core-mantle boundary, the resulting deformation (in the elasto-static approximation probably justified for periods of several hours) due to a tidal periodical action (near-resonant case) or the excited core eigenmode itself is obtained.

The experimental results of geophysical relevance (gravity, tilt, strain) concerning the 'free core resonance' are numerous and clearly confirm the predicted effect. Even some evidence for the same effect in the ocean tide has recently appeared. Moreover, the observational results relative to the amplitude of some luni-solar nutations bring further evidence.

The problem of fundamental interest concerning the experimental detection and identification (e.g. by surface gravimeters) of core internal oscillations of resonant nature or not is a difficult one that is still open and needs further work.

Acknowledgments

The author wishes to thank H. Legros for helpful discussions and constructive suggestions that improved this manuscript.

References

Abours, S. & Lecolazet, R., 1979. 'New results about the dynamical effects of the liquid outer core as observed at Strasbourg', Proc. 8th Int. Symp. Earth Tides (eds M. Bonatz & P. Melchior), 689-697.
Alterman, Z., Jarosch, H. & Pekeris, C.L., 1959. 'Oscillations of the Earth', Proc. R. Soc. A, **252**, 80-95.
Amalvict, M., Hinderer, J. & Legros, H., 1985a. 'Influence of elasticity on the precessions of the earth and its fluid core', Proc. 10th Int. Symp. Earth Tides, in press.
Amalvict, M., Legros, H. & Hinderer, J., 1985b. 'Core dynamics and the Earth's rotation. Part III: Precession and nutations', Geophys. J. R. astr. Soc., submitted.
Anderson, D.L. & Minster, J.B., 1979. 'The frequency dependence of Q in the Earth and implications for mantle rheology and Chandler wobble', Geophys. J. R. astr. Soc., **58**, 431-440.
Blum, P.A., Hatzfeld, D. & Wittlinger, G., 1973. 'Résultats expérimentaux sur la fréquence de résonance due à l'effet dynamique du noyau liquide', C.R. Acad. Sc. Paris, **277**, B, 241-244.
Capitaine, N., 1982. 'Effets de la non-rigidité de la Terre sur son mouvement de rotation: Etude théorique et utilisation d'observations', Thesis, Université Pierre et Marie Curie, Paris, 205 pp.
Chao, B.F., 1983. 'Normal mode study of the Earth's rigid body motions', J. geophys. Res., **88**, 9437-9442.
Chapman, S. & Lindzen, R.S., 1970. Atmospheric tides, D. Reidel, Dordrecht, 200pp.
Crossley, D.J. & Smylie, D.E., 1975. 'Electromagnetic and viscous damping of core oscillations', Geophys. J. R. astr. Soc., **42**, 1011-1033.
Crossley, D.J. & Rochester, M.G., 1980. 'Simple core undertones', Geophys. J. R. astr. Soc., **60**, 129-161.
Crossley, D.J., 1984. 'Oscillatory flow in the liquid core', Phys. Earth Planet. Int., **36**, 1-16.
Dickman, S.R., 1983. 'The rotation of the ocean-solid Earth system', J. geophys. Res., **88**, 6373-6394.
Friedlander, S., 1985. 'Internal oscillations in the Earth's fluid core', Geophys. J. R. astr. Soc., **80**, 345-361.
Gilbert, F. & Dziewonski, A.M., 1975. 'An application of normal mode theory to the retrieval of structural parameters and source mechanisms from seismic spectra', Phil. Trans. R. Soc. A, **278**, 187-269.
Goodkind, J.M., 1983. 'Q of the nearly diurnal free wobble', Proc. 9th Int. Symp. Earth Tides (ed. J. Kuo), Schweizerbart, Stuttgart, 569-575.
Gubbins, D., 1975. 'Can the Earth's magnetic field be sustained by core oscillations?', Geophys. Res. Lett., **2**, 409-412.
Hinderer, J., Legros, H. & Amalvict, M., 1982. 'A search for Chandler and nearly diurnal free wobbles using Liouville equations', Geophys. J. R. astr. Soc., **71**, 303-332.

Hinderer, J., Legros, H., Amalvict, M. & Lecolazet, R., 1985a. 'Theoretical and observational search for the gravimetric factor: diurnal resonance and long-period perturbations', Proc. 10th Int. Symp. Earth tides, in press.

Hinderer, J., Amalvict, M. & Legros, H., 1985b. 'Core dynamics and the Earth's rotation. Part II: wobble and axial rotation', Geophys. J. R. astr. Soc., submitted.

Hopkins, W., 1839. 'Researches in physical geology', Phil. Trans. R. Soc. A, **129**, 381-423.

Hough, S.S., 1895. 'The oscillations of a rotating ellipsoidal shell containing fluid', Phil. Trans. R. Soc. A, **186**, 469-506.

Jeffreys, H. & Vicente, R.O., 1957a. 'The theory of nutation and variation of latitude', Mon. Not. R. astr. Soc., **117**, 142-162.

Jeffreys, H. & Vicente, R.O., 1957b. 'The theory of nutation and variation of latitude: the Roche core model', Mon. Not. R. astr. Soc., **117**, 162-173.

Kinoshita, H., Nakajima, K., Kubo, Y., Nakagawa, I., Sasao, T. & Yokoyama, K., 1979. 'Note on nutation in ephemerides', Publ. int. Latit. Obs. Mizusawa, **12**, 71-108.

Lanzano, P., 1984. 'Earth's tides and polar motion', J. Geodyn., **1**, 121-142.

Lecolazet, R. & Steinmetz, L., 1974. 'Sur les ondes diurnes de la marée gravimétrique observée à Strasbourg', C.R. Acad. Sci. Paris, **277**, B, 295-297.

Lecolazet, R. & Melchior, M., 1977. 'Experimental determination of the dynamical effects of the liquid core', Ann. Geophys., **33**, 11 -22.

Legros, H. & Amalvict, M., 1985. 'Rotation of a deformable earth with with dynamical superficial fluid layer and liquid core- Part I: Fundamental equations', Ann. Geophysicae, **3**, 5, 655-670.

Legros, H. & Amalvict, M., 1986. 'The Rotation', in Physics and Evolution of the Earth's interior, vol. 4, Low frequency geodynamics, ed. R. Teisseyre, Elsevier Publ. Co. Polish Sci. Publ.

Legros, H., Amalvict, M. & Hinderer, J., 1985a. 'Love numbers and planetary dynamics', Proc. 10th Int. Symp. Earth Tides, in press.

Legros, H., Hinderer, J. & Amalvict, M., 1985b. 'Core dynamics and the Earth's rotation, Part I: Fundamental equations', Geophys. J. R. astr. Soc., submitted.

Legros, H., Amalvict, M. & Hinderer, J., 1986. 'Love numbers and planetary dynamics', in preparation.

Levine, J., 1978. 'Strain-tide spectroscopy', Geophys. J. R. astr. Soc., **54**, 27-41.

Loper, D.E., 1975. 'Torque balance and energy budget for the precessional driven dynamo', Phys. Earth planet. Int., **11**, 43-60.

Malkus, W.V.R., 1968. 'Precession of the Earth as the cause of geomagnetism', Science, **160**, 259-264.

Mansinha, L. & Shen, P.Y., 1976. 'Core stability and Earth tides', Tectonophysics, **35**, 285-293.

Melchior, P., 1966. 'Diurnal Earth tides and the Earth's liquid core', Geophys. J. R. astr. Soc., **12**, 15-21.

Melchior, P., 1979. 'Report on the activities of the International Centre for Earth Tides (ICET)', Proc. 8th Int. Symp. Earth Tides (eds

M. Bonatz & P. Melchior), 30-43.
Melchior, P., 1980. 'Luni-solar nutation tables and the liquid core of the Earth', Astr. Astrophys., **1**, 1-4.
Melchior, P. & Georis, B., 1968. 'Earth tides, precession-nutation and the secular retardation of the Earth's rotation', Phys. Earth planet. Int., **1**, 267-287.
Melchior, P. & Ducarme, B., 1985. 'Detection of inertial gravity oscillations in the Earth's core with a superconducting gravimeter at Brussels', Phys. Earth planet. Int., in press.
Molodensky, M.S., 1961. 'The theory of nutation and diurnal Earth tides', Comm. Obs. R. Belg., **288**, 25-56.
Neuberg, J. & Zurn, W., 1985. 'Investigation of the nearly diurnal resonance using gravity, tilt and strain data simultaneously', Proc. 10th Int. Symp. Earth Tides, in press.
Okubo, S., 1982. 'Theoretical and observed Q of the Chandler wobble-Love number approach', Geophys. J. R. astr. Soc., **71**, 647-657.
Olson, P., 1977. 'Internal waves in the Earth's core', Geophys. J. R. astr. Soc., **51**, 183-215.
Ooe, M. & Tamura, Y., 1985. 'Fine structures of tidal admittance and the fluid core resonance effect in the ocean tide around Japan', Manuscripta Geodaetica, **10**, 1, 37-49.
Pekeris, C.L. & Accad, Y., 1972. 'Dynamics of the liquid core of the Earth', Phil. Trans. R. Soc. A, **273**, 237-260.
Platzman, G.W., Curtis, G.A., Hansen, K.S. & Slater, R.D., 1981. 'Normal modes of the world ocean, II, Description of modes in the period range 8 to 80 hours', J. Phys. Oceanogr., **11**, 579-603.
Poincaré, H., 1910. 'Sur la précession des corps déformables', Bull. astr., **27**, 321-356.
Richter, B., 1985. 'The spectrum of a registration with a superconducting gravimeter', Proc. 10th Int. Symp. Earth Tides, in press.
Rochester, M.G., 1976. 'The secular decrease of obliquity due to dissipative core-mantle coupling', Geophys. J. R. astr. Soc., **46**, 109-126.
Sasao, T., Okamoto, J. & Sakai, S., 1977. 'Dissipative core-mantle coupling and nutational motion of the Earth', Publ. astr. Soc. Japan, **29**, 83-105.
Sasao, T., Okubo, S. & Saito, M., 1980. 'A simple theory on dynamical effects of a stratified fluid core upon nutational motion of the Earth', in Proc. IAU Symp. No 78, Nutation and the Earth's rotation, Kiev, May 1977, 165-183, eds. E.P. Fedorov, M.L. Smith & P.L. Bender, D. Reidel, Dordrecht.
Sasao, T. & Wahr, J.M., 1981. 'An excitation mechanism for the free 'core nutation'', Geophys. J. R. astr. Soc., **64**, 729-746.
Shen, P.Y., 1983. 'On oscillations of the Earth's fluid core', Geophys. J. R. astr. Soc., **75**, 737-757.
Shen, P.Y. & Mansinha, L., 1976. 'Oscillation, nutation and wobble of an elliptical rotating Earth with liquid outer core', Geophys. J. R. astr. Soc., **46**, 467-496.
Sludskii, F., 1896. 'De la rotation de la Terre supposée fluide en son intérieur', Bull. Soc. Nat. Moscou, **9**, 285-318.

Smith, M.L., 1974. 'The scalar equations of infinitesimal elastic gravitational motion for a rotating, slightly elliptical Earth', Geophys. J. R. astr. Soc., **37**, 491-526.

Smith, M.L., 1977. 'Wobble and nutation of the Earth', Geophys. J. R. astr. Soc., **50**, 103-140.

Smith, M.L. & Dahlen, F.A., 1981. 'The period and Q of the Chandler wobble', Geophys. J. R. astr. Soc., **64**, 223-281.

Smylie, D.E. & Rochester, M.G., 1981. 'Compressibility, core dynamics and the subseismic wave equation', Phys. Earth planet. Int., **24**, 308-319.

Sundermann, J., 1982. 'The resonance behaviour of the world ocean', in Tidal friction and the Earth's rotation, II, 165-174, eds. P. Brosche & J. Sundermann, Springer-Verlag, Berlin.

Toomre, A., 1966. 'On the coupling of the core and mantle during the 26000 yr precession', in The Earth-Moon system, 33-45, eds. B.G. Mardsen & A.G.W. Cameron, Plenum Press, New-York.

Wahr, J.M., 1981a. 'Body tides of an elliptical, rotating, elastic and oceanless earth', Geophys. J. R. astr. Soc., **64**, 677-703.

Wahr, J.M., 1981b. 'The forced nutations of an elliptical, rotating, elastic and oceanless earth', Geophys. J. R. astr. Soc., **64**, 705 -727.

Wahr, J.M. & Sasao, T., 1981. 'A diurnal resonance in the ocean tide and in the Earth's load response due to the resonant free 'core nutation'', Geophys. J. R. astr. Soc., **64**, 747-765.

Wahr, J.M. & Larden, D.R., 1983. 'An analysis of lunar laser ranging data for the Earth's free core nutation', Proc. 9th Int. Symp. Earth Tides (ed. J. Kuo), Schweizerbart, Stuttgart, 547-553.

Wahr, J.M., 1984. 'Normal modes of the coupled Earth and ocean system', J. geophys. Res., **89**, 7621-7630.

Warburton, R.J. & Goodkind, J.M., 1978. 'Detailed gravity-tide spectrum between one and four cycles per day', Geophys. J. R. astr. Soc., **52**, 117-136.

Zschau, J., 1978. 'Tidal friction in the solid Earth: loading tides versus body tides', in Tidal friction and the Earth's rotation, I, 62-93, eds. P. Brosche & J. Sundermann, Springer-Verlag, Berlin.

Zurn, W., Rydelek, P.A. & Richter, B., 1985. 'The core-resonance effect in the record from the superconducting gravimeter at Bad Homburg', Proc. 10th Int. Symp. Earth Tides, in press.

LONG PERIOD CORE DYNAMICS

D. E. Smylie
Department of Earth and Atmospheric Science
York University
Downsview, Ontario, M3J 1P3
Canada

and

M. G. Rochester
Department of Earth Sciences
Memorial University of Newfoundland
St. John's, Newfoundland
Canada

ABSTRACT. Fundamental to the determination of the dynamical behaviour of the whole Earth is the computation of the dynamical response of the fluid core. It is a rotating, stratified, self-gravitating, contained, compressible fluid exhibiting a rich array of possible behaviours. At periods beyond a substantial fraction of a day, Coriolis coupling between terms in traditional spherical harmonic representations of the motions is very strong and the series are poorly convergent. To overcome this problem a single scalar second order governing equation has been developed, called the subseismic equation, which is valid at frequencies below the elasto-gravitational mode spectrum. This equation has been separated and new eigenfunctions which are uncoupled in the body of the fluid core are being investigated as representations of the motions. In general, the boundary conditions are not separable and superpositions of the eigenfunctions of the subseismic equation are required for their satisfaction. Successful computation of the normal modes of the fluid core will allow the complete determination of its contribution to the body tide response, free and forced nutations, changes in the length of day and polar motion as well as its long period free mode spectrum.

1. INTRODUCTION

The problem of calculating the effects of the fluid core on Earth's dynamics has attracted the attention of scientists at least since the study of Hopkins (1839) who sought its effect on precession. Interest in the stability of such rotating self-gravitating fluids arising from planetary theories, and the stimulation of the discovery of the Chandler wobble with a period lengthened by 40% compared to the expected Euler

A. Cazenave (ed.), Earth Rotation: Solved and Unsolved Problems, 297–324.

period for a rigid Earth, led to a flurry of activity in the late nineteenth century with major contributions by Poincaré (1885), Bryan (1889), Hough (1895) and Sludskii (1896). These early studies were concerned with the dynamics of a rotating, uniform, incompressible fluid, either contained in a rigid envelope or assumed to be self-gravitating with a free bounding surface. The governing equation has come to be known as the Poincaré equation and remarkably, although a variety of analytical solutions for it have been obtained for a single bounding surface, beginning with that of Poincaré, difficulties have been encountered in constructing solutions when, as in the Earth's liquid core, there is an inner boundary (Stewartson, 1978; Stewartson and Rickard, 1969).

Modern theories of Earth's dynamics require a description of the dynamical response of the liquid core in a wide variety of problems, including calculation of the body ties, the free and forced nutations, the response to surface loading by the oceans and atmosphere and changes in the length of day and polar motion (Wahr et al. 1981). The solutions of each of these problems in turn can be represented as a superposition of normal modes, a result which might have been anticipated on general physical grounds for such a linearized system, but which was not formally demonstrated until very recently (Wahr, 1981).

The normal modes of the liquid core themselves are of intrinsic interest because of the information on core structure and the magnetic field which they might provide if they were detected (see Smylie, Szeto and Rochester, 1984 for a recent review). Owing to the strong coupling between terms in a conventional spherical harmonic representation provided at long periods by the Coriolis acceleration, ellipticity and centrifugal acceleration (Smylie, 1974; Smith, 1974; Crossley, 1975), these modes are extremely difficult to compute using such a representation, and in all studies to date where they have been used in Earth dynamics calculations, solutions truncated to two or three terms have been employed (Shen and Mansinha, 1976; Smith, 1977; Wahr, et al., 1981) Concerns about convergence of these series led Johnson and Smylie (1977) to develop a fast computational method based on the implementation of a variational principle via piecewise cubic splines. While they were able to double the number of terms included, they found the truncation error to still be very serious. A similar conclusion was drawn by Crossley and Rochester (1980), who made a number of simplifications to the physical problem which allowed them to carry the numerics even further.

Poor convergence of solutions in spherical harmonic series prompted Smylie and Rochester (1981) to investigate the basic governing equations in more detail with a view to obtaining an appropriate description valid at long periods. They rejected the solenoidal flow approximation commonly made in core dynamo theory on the grounds that while the square of the flow speed is much less than the square of the speed of sound and the square of the frequencies of possible oscillatory motions is much less than the square of acoustic frequencies, the scale-height of the density stratification is not significantly larger than the vertical scale of core motions (see for example, Batchelor, 1967, pp 167-177). Instead, they made what they called the 'subseismic approximation' which neglects the effect of flow pressure fluctuations on the density but retains in full the effects of radial non-homogeneity, compressibility, self-gravitation, hydrostatic prestress and rotation. A second order governing partial differential equation in a single scalar dependent variable was obtained called the 'subseismic equation' in recognition of the fact that it is based on the form the system of equations governing the seismically-excited, elasto-gravitational oscillations take at frequencies below seismic frequencies (< 300 μHz).

The character of the subseismic equation in the space variables (elliptic, parabolic, hyperbolic) can be related directly to four possible solution régimes depending on the density structure of the outer core; (i) the strongly-stable case where the period of oscillation of a fluid element about its equilibrium position (the Brunt-Väisälä period) is less than half a sidereal day, (ii) the weakly-stable case where the Brunt-Väisälä period is longer than half a sidereal day, (iii) the weakly-unstable case where the square of the Brunt-Väisälä period is negative but its magnitude is greater than half a sidereal day and (iv) the strongly-unstable case where the square of the Brunt-Väisälä period is negative and its magnitude is less than half a sidereal day.

We find here a separation of variables solution to the subseismic equation in spherical geometry in terms of new eigenfunctions of colatitude which are governed by a second order, ordinary differential equation with six regular singular points in the finite plane and one at infinity. The locations of the singular points are related to the density structure of the outer core and to the character of the subseismic equation. Numerical techniques are devised for the fast computation of eigenfunctions representing solutions finite over the whole outer core (the corresponding eigenvalues are the allowed values of the separation constant for finite solutions). The resulting eigenfunctions are then superimposed to satisfy the boundary conditions giving a secular determinant for the calculation of eigenperiods. We show results for a strongly stably-stratified core.

2. THE SUBSEISMIC EQUATION AND PROPERTIES OF ITS SEPARATED SOLUTIONS

In its neglect of the effect of flow pressure on the density at periods beyond the acoustic range or at flow speeds below the speed of sound, the subseismic approximation is seen to be the planetary analogue of the low-speed flow assumption of incompressibility in ordinary laboratory fluid dynamics. However, in planetary scale bodies the density variation arising from the depth-dependent pressures induced by self-gravitation make the assumption of incompressibility untenable except for small scale phenomena.

After Fourier transformation of the time dependence, the complete governing system of subseismic equations is:

$$- \omega^2 \vec{u} + 2i\,\omega\,\vec{\Omega} \times \vec{u} = - \nabla \chi - \vec{g}_o \psi, \tag{1}$$

$$\psi = \beta\, \nabla \cdot \vec{u} = - \frac{\beta}{\alpha^2} \vec{u} \cdot \vec{g}_o \,, \tag{2}$$

and

$$\nabla^2 V_1 = 4\pi\, G\, \rho_o\, \psi. \tag{3}$$

Here ω is the angular frequency, $\vec{u}$ is the vector displacement field, $\vec{\Omega}$ is the constant spin vector of the Earth frame, $\vec{g}_o$ is the equilibrium gravity vector and α is the P wave velocity. The variable χ is defined as

$$\chi = \frac{1}{\rho_o} p_1 - V_1 \tag{4}$$

with ρ_o the equilibrium mass density, p_1 the flow pressure perturbation and V_1 the decrease in gravitational potential associated with the motion. The dimensionless stability factor β is defined by

$$\beta = 1 - \frac{\alpha^2}{\rho_o g_o^2} \vec{g}_o \cdot \nabla \rho_o \;. \tag{5}$$

It is related to the square of the Brunt-Väisälä angular frequency $4N^2\Omega^2$ through

$$4N^2 \Omega^2 = - \beta (g_o/\alpha)^2 \;. \tag{6}$$

Once χ and ψ are determined from equations (1) and (2), (3) and (4) can be used to find V_1 and p_1. Alternatively, following procedures outlined in Smylie and Rochester (1981) and in Smylie, Szeto and Rochester (1984), a single scalar equation in χ can be found, called the subseismic equation, which can be written

$$\{\sigma^2 \nabla^2 - (\hat{k} \cdot \nabla)^2\} \chi = \frac{A}{B} \vec{C} \cdot \nabla \chi + \vec{C}^* \cdot \nabla \left(\frac{\vec{C} \cdot \nabla \chi}{B}\right) \tag{7}$$

with

$$A = \frac{\omega^2}{\beta^2} (\sigma^2 - 1) + \sigma^2 (4\pi\, G\, \rho_o - 2\Omega^2) + (\hat{k} \cdot \nabla) \hat{k} \cdot \vec{g}_o, \tag{8}$$

$$B = \frac{\alpha^2 \omega^2}{\beta} (\sigma^2 - 1) + \sigma^2\, g_o^2 - (\hat{k} \cdot \vec{g}_o)^2 \,, \tag{9}$$

and

$$\vec{C} = (\hat{k} \cdot \vec{g}_o) \hat{k} + i\sigma\, \hat{k} \times \vec{g}_o - \sigma^2\, \vec{g}_o \;. \tag{10}$$

$\hat{k}$ is the unit vector in the direction of the rotation, $\vec{C}^*$ is the complex conjugate of the vector $\vec{C}$ and

$$\sigma = \omega/2\Omega. \tag{11}$$

Because of its great generality in describing fully the effects of rotation, stratification, non-homogeneity, compressibility and gravity, the construction of solutions to the subseismic equation (7) is a formidable task and it is useful to first consider how it degenerates into some of the simpler descriptions of rotating fluids obtained previously.

The operator on the LHS of the subseismic equation (7) is the same as that arising in the Poincaré description so that the terms on the RHS describe physical effects beyond pure inertial waves in an incompressible, uniform fluid. The second of the terms on the RHS of (7) is easily shown to vanish in the case of a neutrally-stratified fluid ($\beta \to 0$) and since it is the only additional term involving second derivatives, it is seen that the spatial character of solutions in the case of a neutrally-stratified fluid will be the same as that of solutions to the Poincaré equation. Even in the neutrally-stratified case, though, the governing equation is distinct from the Poincaré equation since the first term on the RHS of (7) then takes the limiting form $\vec{C}\cdot\nabla\chi/\alpha^2$. The RHS of (7) only vanishes completely when the further restriction of absolute incompressibility is applied ($\alpha^2 \to \infty$). There is, of course, a physical contradiction in considering fluids which are both perfectly incompressible and stratified, but a number of investigators have nonetheless examined the behaviour of such implausible fluids. For example, Needler and LeBlond (1973) obtain a governing equation for the pressure perturbation which is equivalent to equation (7) with $A \to 0$.

To examine the relationship of the mathematical structure of solutions to the subseismic equation to physical conditions in the outer core, we first adopt a spherically-stratified Earth model and study the form the equation takes in spherical polar coordinates. The density and gravity are then functions of radius alone and

$$\nabla\cdot\vec{g}_o = -\frac{dg_o}{dr} - \frac{2g_o}{r} = -4\pi G + 2\Omega^2 . \tag{12}$$

Replacing the cosine of the colatitude by the variable $z = \cos\theta$ we find

$$A\frac{r}{g_o} = 3z^2 - 1 + M^2(\zeta_1^2-\sigma^2) + \frac{r}{g_o}(4\pi G\rho_o - 2\Omega^2)(\sigma^2 - z^2) \tag{13}$$

$$B = g_o^2(\zeta_1^2 - z^2), \tag{14}$$

$$\vec{C}/g_o = \hat{r}(\sigma^2 - z^2) + \hat{\theta}z(1-z^2)^{\frac{1}{2}} - \hat{\phi}\, i\sigma(1-z^2)^{\frac{1}{2}}, \tag{15}$$

where $\hat{r}$, $\hat{\theta}$, $\hat{\phi}$ are the unit vectors along the coordinate directions. We have defined the positive definite quantity

$$M^2 = \frac{g_o r}{\alpha^2} \tag{16}$$

as the local compressibility number and we use the shorthand

$$\zeta_1^2 = 1 + (\sigma^2 - 1)(1 - \frac{\sigma^2}{N^2}) = \frac{\sigma^2}{N^2}\zeta_2^2 \tag{17}$$

with

$$\zeta_2^2 = 1 + (N^2 - \sigma^2) = \frac{N^2}{\sigma^2}\zeta_1^2 , \tag{18}$$

for parameters which we shall see shortly provide important characteristics of the solutions. Only solutions, single-valued in the longitude ϕ are of interest, hence, without loss of generality, we may assume that χ depends on longitude as $e^{im\phi}$, with m an integer or zero. The subseismic equation in spherical polar coordinates then becomes

$$(\zeta_1^2-\sigma^2)\{(\sigma^2-z^2)r^2 \frac{\partial^2\chi}{\partial r^2} - 2(1-z^2)zr \frac{\partial^2\chi}{\partial r\partial z}$$

$$-(\zeta_2^2-z^2)(1-z^2) \frac{\partial^2\chi}{\partial z^2}\}$$

$$+ \{(\zeta_1^2-z^2)(2\sigma^2-(1-z^2))-2z^2(1-z^2)$$

$$- (\sigma^2-z^2)(2\sigma^2-(1-z^2) + M^2(\zeta_1^2-\sigma^2))$$

$$+ 2\frac{(\sigma^2-z^2)}{(\zeta_1^2-z^2)} (z^2(1-z^2)-(\zeta_1^2-\sigma^2)(\sigma^2-z^2) \frac{r}{N}\frac{dN}{dr})\} r \frac{\partial\chi}{\partial r}$$

$$+ \{(\zeta_1^2-z^2)(3(1-z^2)-2\sigma^2) + (1-z^2)(2(3z^2-1)$$

$$+ \sigma^2 - z^2 + M^2(\zeta_1^2-\sigma^2)) - 2\frac{(1-z^2)}{(\zeta_1^2-z^2)} (z^2(1-z^2)$$

$$- (\zeta_1^2-\sigma^2)(\sigma^2-z^2) \frac{r}{N}\frac{dN}{dr})\} z \frac{\partial\chi}{\partial z}$$

$$- m\sigma\{m\sigma \frac{(\zeta_1^2-z^2)}{1-z^2} + 3z^2-1 + \sigma^2-z^2-m\sigma + M^2(\zeta_1^2-\sigma^2)$$

$$- \frac{2}{\zeta_1^2-z^2} (z^2(1-z^2)-(\zeta_1^2-\sigma^2)(\sigma^2-z^2) \frac{r}{N}\frac{dN}{dr})\} \chi = 0 . \qquad (19)$$

At the boundaries, the radial displacements are to be continuous. These are given by (2) in terms of ψ which is found to be related to χ in the variables elimination leading to the subseismic equation (7) (Smylie, Szeto and Rochester, 1984) by

$$\psi = \frac{\vec{C}\cdot\nabla\chi}{B} . \qquad (20)$$

In the spherically-stratified Earth model adopted here we have

$$-4N^2\Omega^2(\zeta_1^2-z^2)\vec{u}\cdot\hat{r} = \hat{r}(\sigma^2-z^2) \frac{\partial\chi}{\partial r} -\hat{\theta} \frac{z(1-z^2)}{r} \frac{\partial\chi}{\partial z} +\hat{\phi} \frac{m\sigma}{r} \chi. \quad (21)$$

The problem is thus seen to depend on the azimuthal number m, the dimensionless angular frequency σ and the physical properties of the core (specified by the compressibility number M^2 and the square of the dimensionless Brunt-Väisälä angular frequency N^2 and its radial derivative). In particular, m enters the problem only as a product with σ, σ otherwise appearing as a square. Standing wave solutions are therefore possible, in general, only in the axisymmetric case. In the non-axisymmetric case, we can restrict m to the positive integers without loss of generality. Then solutions to the problem for positive values of σ move eastward at the angular rate σ/m, while solutions for negative values of σ move westward at the angular rate $-\sigma/m$. Such travelling wave solutions are common to a wide variety of problems in the dynamics of laboratory rotating fluids (Greenspan, 1969). A further property of the solutions to (19) and (21), which follows directly from the fact that z appears only as a square in the coefficients of its even order derivative terms (with respect to z) and as an odd power in the coefficients of its odd derivative terms, is that they separate into two groups, one of which is even in the equatorial plane, the other odd. Again, this is a common property of solutions of rotating fluids problems and even extends to certain mean-field dynamos (Roberts and Stix, 1972).

The spatial character of the solutions to the subseismic equation (19) is determined by the sign of its discriminant. The discriminant is

$$\frac{1}{r^2}\frac{\sigma^4}{N^2}(\sigma^2-1)^2(\zeta_1^2-z^2), \tag{22}$$

with sign turning on that of $N^2(\zeta_1^2-z^2)$. If the sign is positive, the solutions are spatially hyperbolic, if it is negative they are elliptic, and on the bounding curves ($\zeta_1^2 = z^2$) between the two, they are parabolic. There are four distinct régimes of core structure which determine the the nature of the bounding curves; (i) the strongly-stable case in which $N^2 > 1$, with bounding curves given by

$$2\sigma^2 = N^2 + 1 \pm \sqrt{(N^2+1)^2 - 4N^2\cos^2\theta} \quad , \tag{23}$$

(ii) the weakly-stable case in which $0 < N^2 < 1$, with bounding curves given by

$$2\sigma^2 = 1 + N^2 \pm \sqrt{(1+N^2)^2 - 4N^2\cos^2\theta} \quad , \tag{24}$$

(iii) the weakly-unstable case in which $0 < \mathcal{N}^2 = -N^2 < 1$, with bounding curve given by

$$2\sigma^2 = 1 - \mathcal{N}^2 + \sqrt{(1-\mathcal{N}^2)^2 + 4\mathcal{N}^2\cos^2\theta} \quad , \tag{25}$$

(iv) the strongly - unstable case in which $\mathcal{N}^2 > 1$, with bounding curve given by

$$2\sigma^2 = -\mathcal{N}^2 + 1 + \sqrt{(\mathcal{N}^2-1)^2 + 4\mathcal{N}^2\cos^2\theta} \quad . \tag{26}$$

The bounding curves as functions of latitude for these four régimes are illustrated in Figure 1.

When the properties of the liquid outer core as given by N^2 and M^2 are supposed independent of radius, equation (19) is specified by m, σ and the two constants (with radius), ζ_1^2 and ζ_2^2.

It is then homogeneous in the radius and therefore has separated solutions expressible as

$$\chi = r^{\nu} Z(z) e^{im\phi} , \tag{27}$$

where ν is a separation constant which can be complex valued. Writing

$$Z(z) = (1-z^2)^{m/2}(\zeta_2^2-z^2)^{\lambda/2} t(z), \tag{28}$$

the differential equation satisfied by t(z) is found to be

$$\frac{d^2t}{dz^2} - 2\left\{\frac{m+1}{1-z^2} - \frac{1}{\zeta_1^2-z^2} + \frac{1}{\zeta_2^2-z^2}\right\} z \frac{dt}{dz}$$

$$+\left\{\frac{m^2 + 2(m+1) - m\sigma\left(\frac{N^2}{\sigma^2} - M^2\right) - (1+M^2) - 2\frac{m\sigma}{\sigma-1}\frac{N^2}{\sigma^2} \text{ ---}}{(N^2-\sigma^2)\ (1-z^2)}\right.$$

$$\text{--- } \frac{-m(m+1)(N^2-\sigma^2) + \frac{1}{4}(1-M^2)^2(\sigma^2-1)}{}$$

$$- \frac{2m + (1-M^2) - 2\frac{m\sigma}{\sigma-1}\frac{N^2}{\sigma^2}}{\left(\frac{N^2}{\sigma^2} - 1\right)\ (\zeta_1^2 - z^2)}$$

$$- \frac{m^2 + 2(m+1) + m\sigma\left(\frac{N^2}{\sigma^2} + M^2\right) + (\sigma^2-1)(1+M^2) \text{ ---}}{(N^2-\sigma^2)(\zeta_2^2-z^2)}$$

$$\text{--- } \frac{+\frac{1}{4}(N^2-\sigma^2)(8m + 1 - M^4) - 2\sigma^2 + \frac{1}{4}(1-M^2)^2(\sigma^2-1)}{}$$

$$\left. - \frac{\lambda^2 N^2(\zeta_1^2-z^2)}{(1-z^2)(\zeta_2^2-z^2)^2}\right\} t = 0\ , \tag{29}$$

$$\text{with} \quad \lambda^2 = (\nu-\eta)^2 = -\mu^2\ , \quad \eta = -\frac{1}{2}(1-M^2)\ . \tag{30}$$

This equation has six regular singular points in the finite z-plane at ± 1, $\pm\zeta_1$, $\pm\zeta_2$ and one at the point at infinity. In an obvious extension of the Riemann P-notation (Ince, 1926, p 389), we can write

$$t(z) = P\left\{\begin{matrix} 1 & -1 & \zeta_1 & -\zeta_1 & \zeta_2 & -\zeta_2 & \infty & \\ 0 & 0 & 0 & 0 & \lambda/2 & \lambda/2 & m+1+\eta & z \\ -m & -m & 2 & 2 & -\lambda/2 & -\lambda/2 & m-\eta & \end{matrix}\right\}. \tag{31}$$

Here the exponents of the solutions at the regular singular points are written in two rows below the row giving the point locations. Even when both exponents at a regular singular point are positive integers or zero, as is the case for the points $\pm\ \zeta_1$, the possibility exists that one of the two independent solutions may have a logarithmic branch point there (in fact, as is shown in Rochester and Smylie, 1985, this is not the case at $\pm\ \zeta_1$ for our solutions and these points are only apparent singularities of the equation).

On the physical portion (-1,1) of the real axis of the z-plane, all quantities in equation (29) are real. The parameter λ^2 has been introduced in the variables separation with allowed values which we shall see later are determined by the solutions of algebraic eigenvalue problems.

For each of the four distinct régimes of core structure shown in Figure 1, the points $\pm\ \zeta_1$, $\pm\ \zeta_2$ in the complex z-plane have distinct locations depending on whether one is studying the spatially hyperbolic case or elliptic case of the equation. These locations are illustrated in Figures 2 and 3.

3. COMPUTATION OF EIGENFUNCTIONS AND EIGENVALUES

When the physical portion (-1,1) of the real axis in the complex z-plane is free of singularities (except possibly at the end points), a more useful dependent variable for computational purposes than t(z) is found to be

$$s(z) = (\zeta_1^2-z^2)^{-3/2}\ (\zeta_2^2-z^2)^{-1/2}\ t(z). \tag{32}$$

The differential equation (29) governing t(z) is easily found to transform to

$$\frac{d}{dz}\left\{(1-z^2)^{m+1}(\zeta_1^2-z^2)^2(\zeta_2^2-z^2)^2\,\frac{ds}{dz}\right\}$$

$$+\left\{\frac{m^2 + 4(m+1) - m\sigma\ (\frac{N^2}{\sigma^2} - M^2)-(1+M^2)\ -2\ \frac{m\sigma-(2m+3)\sigma^2}{\sigma^2-1}\ \frac{N^2}{\sigma^2}}{(N^2-\sigma^2)(1-z^2)}\right. ---$$

$$--- \frac{-m(m+1)(N^2-\sigma^2) + \frac{1}{4}(1-M^2)^2(\sigma^2-1)}{}$$

$$+\ \frac{10 + 4(m+1) + (1+M^2) + 2\ \frac{m\sigma-(2m+3)\sigma^2}{\sigma^2-1}\ \frac{N^2}{\sigma^2}}{(\frac{N^2}{\sigma^2} - 1)\ \ (\zeta_1^2 - z^2)}$$

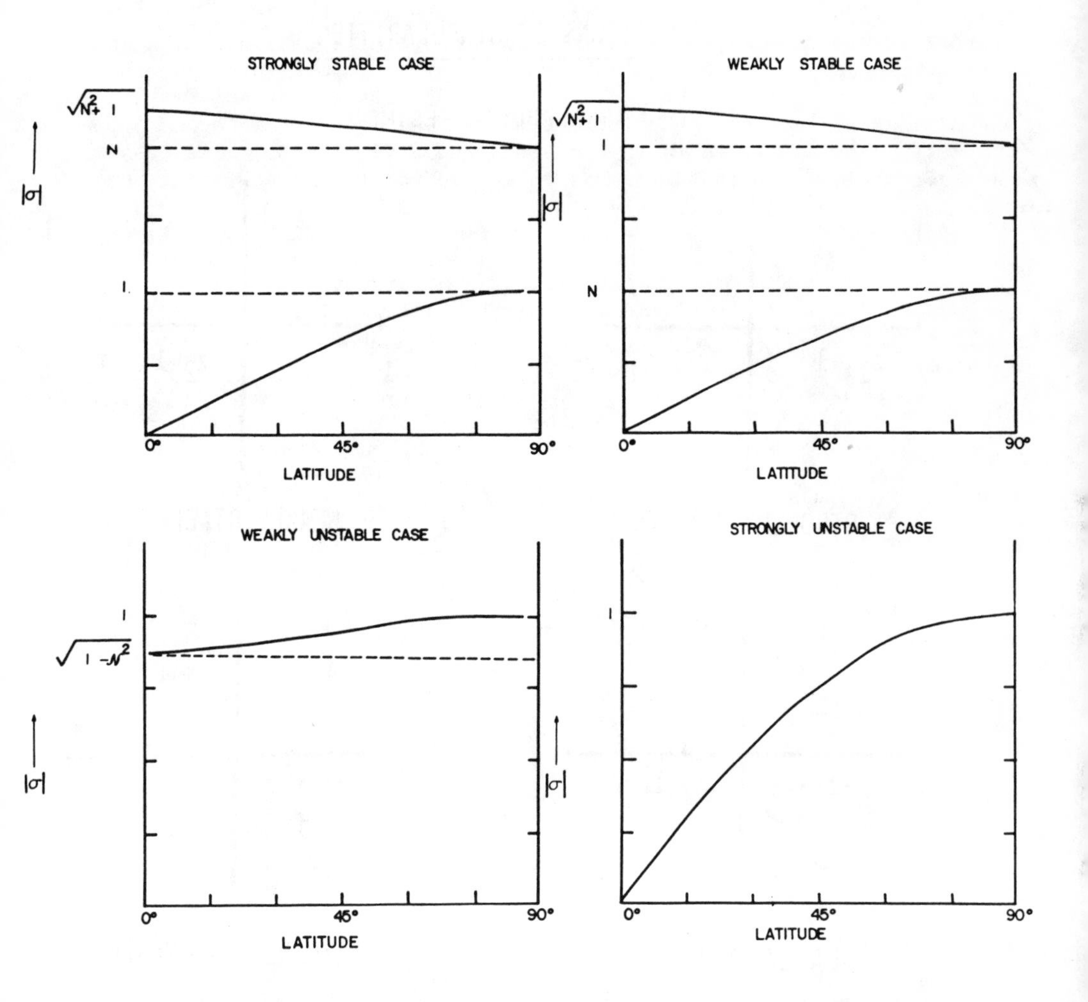

Figure 1. The four régimes of core structure which determine the nature of the bounding curves (on which the subseismic equation is parabolic) separating the elliptic and hyperbolic domains; (i) the strongly stable case in which $N^2 > 1$ and the equation is globally hyperbolic in the band of angular frequencies $1<|\sigma|<N$, (ii) the weakly stable case in which $0<N^2<1$ and the equation is globally hyperbolic in the band of angular frequencies $N<|\sigma|<1$, (iii) the weakly unstable case in which $0<N^2<1$ and the equation is globally hyperbolic in the band of angular frequencies $0<|\sigma|<(1-N^2)^{\frac{1}{2}}$, (iv) the strongly unstable case in which $0<1<N^2$ and there is no global hyperbolic domain.

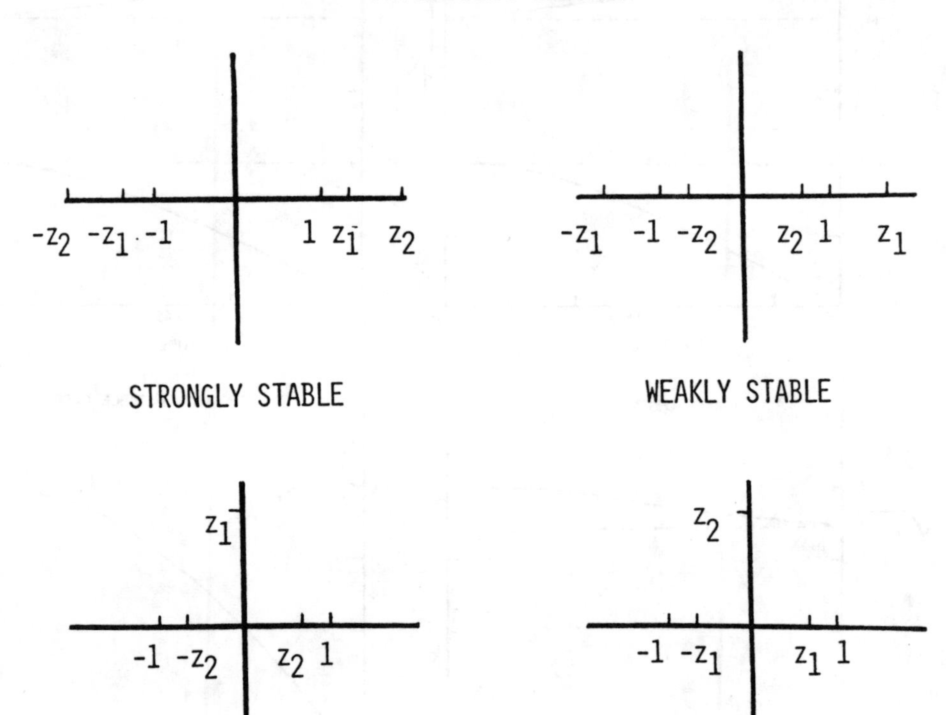

Figure 2. Singularity locations in the hyperbolic domain of the subseismic equation.

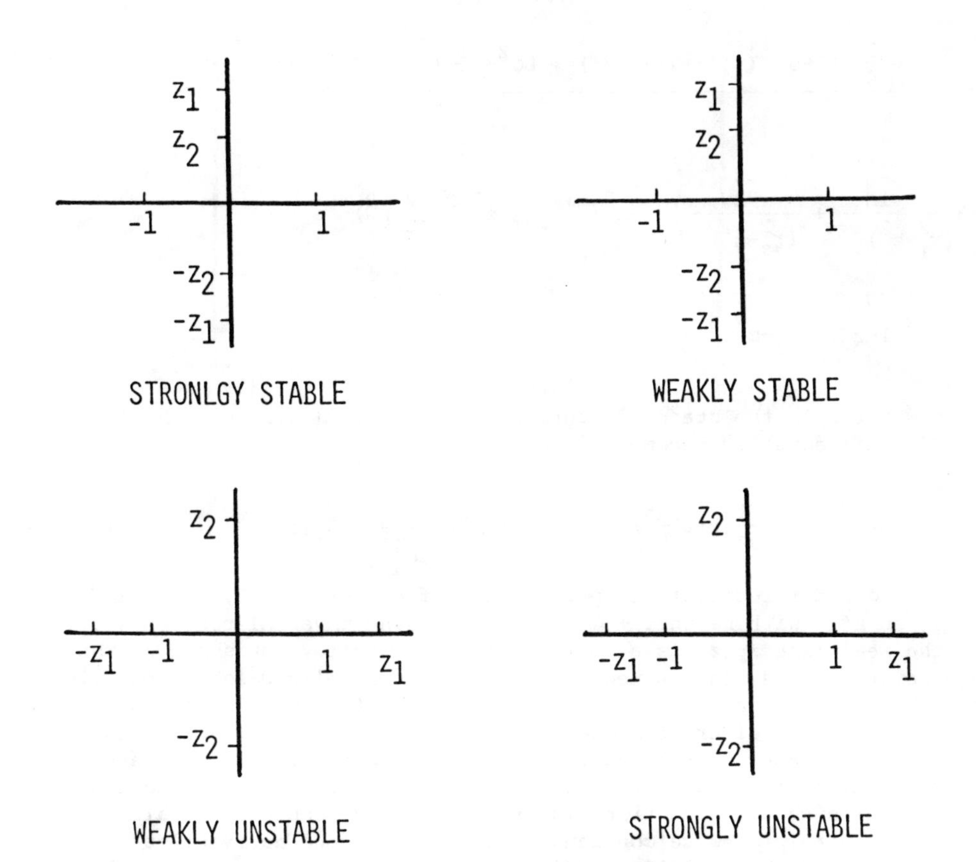

Figure 3. Singularity locations in the elliptic domain of the subseismic equation.

$$- \frac{m^2 + 4(m+1) + m\sigma(\frac{N^2}{\sigma^2} + M^2) + (\sigma^2-1)(1+M^2)}{(N^2-\sigma^2)(\zeta_2^2-z^2)} - - -$$

$$- - - \frac{+ \frac{1}{4}(N^2-\sigma^2)(8(2m+7)+1-M^4) + 8\sigma^2 + \frac{1}{4}(1-M^2)^2(\sigma^2-1)}{}$$

$$-3 \left\{ \frac{\zeta_1^2}{(\zeta_1^2-z^2)^2} + \frac{\zeta_2^2}{(\zeta_2^2-z^2)^2} \right\} (1-z^2)^{m+1}(\zeta_1^2-z^2)^2(\zeta_2^2-z^2)^2 \; s$$

$$- \lambda^2 N^2 (1-z^2)^m (\zeta_1^2-z^2)^3 \; s = 0 \; . \tag{33}$$

When written in the standard Sturm-Liouville form (Moiseiwitsch, 1966, p.147) this equation becomes,

$$\frac{d}{dz}(p(z)\frac{ds}{dz}) + (q(z) + \lambda^2 r(z) \; s = 0 \; , \tag{34}$$

with p, q and r respectively polynomials of degrees m + 5, m + 4 and m + 3 in z^2. With neither $\pm \zeta_1$ nor $\pm \zeta_2$ on the physical portion (-1,1) of the real line, p(z) is a positive function there, vanishing at the end points, while r(z) is negative for the hyperbolic case and positive for the elliptic case. It is well-known that this problem has solutions only for a real, discrete sequence of eigenvalues of λ^2 (increasing in the elliptic case, decreasing in the hyperbolic case), that to each allowed value of λ^2 there corresponds an eigenfunction s(z), and that the set of such eigenfunctions is orthogonal and complete for all functions with piecewise continuous first and second derivatives (Courant and Hilbert, 1953, p.293).

In extended Riemann P-notation we then have

$$s(z) = P \left\{ \begin{matrix} 1 & -1 & \zeta_1 & -\zeta_1 & \zeta_2 & -\zeta_2 & \infty & \\ 0 & 0 & -3/2 & -3/2 & -1/2+\lambda/2 & -1/2+\lambda/2 & m+5+\eta & z \\ -m & -m & 1/2 & 1/2 & -1/2-\lambda/2 & -1/2-\lambda/2 & m+4-\eta & \end{matrix} \right\} . \tag{35}$$

The change from dependent variable t(z) to s(z) has left the exponents at z = $\pm$ 1 unchanged but has had the desirable effect of producing polynomial coefficients p(z), q(z) and r(z). We shall see that coefficients of this form possess some advantages in implementing numerical procedures.

The exponents at $z = \pm 1$ differ by an integer and only one regular solution associated with the largest exponent (zero) exists there. The regular solution is distinguished by a particular value of the ratio of the derivative $\frac{ds}{dz}(z)$ to the function $s(z)$ at these points. This ratio is easily computed from the series solution itself as

$$\pm \frac{m^2+4(m+1)-m\sigma(\frac{N^2}{\sigma^2} - M^2)-(1+M^2) -2\frac{m\sigma-(2m+3)\sigma^2}{\sigma^2-1}\frac{N^2}{\sigma^2} - - - -m(m+1)(N^2-\sigma^2)+(\eta^2-\lambda^2)(\sigma^2-1)}{2(m+1)(N^2-\sigma^2)}, \qquad (36)$$

the positive sign applying to the ratio at $z = 1$ and the negative sign to the ratio at $z = -1$. Expression (36) is used to constrain numerical solutions to physically acceptable regular behaviour.

One of the most computationally efficient methods of solving for the eigenfunctions and eigenvalues of the system (34) is by a variational method implemented with piecewise cubic Hermite splines (Shultz, 1972; Johnson and Smylie, 1977). The appropriate functional is

$$L(s) = \int_0^1 \{p(z)(\frac{ds}{dz})^2 - (q(z) + \lambda^2 r(z))s^2\}\, dz. \qquad (37)$$

As already observed in Section 2, the solutions separate into those even about the equatorial plane ($z = 0$) and those odd about the equatorial plane. Since p, q and r are polynomials in z^2, the integral in the functional expression (37) need only be taken over the positive half of the interval (-1,1). Further, since the entire integrand is piecewise polynomial it can be integrated exactly by a numerical procedure of sufficient order. Solutions even in the equatorial plane have vanishing derivative at $z = 0$, while for those odd in the equatorial plane, the function itself must vanish there.

With the interval (0,1) divided into L-1 segments there are L segment end points or nodes and $s(z)$ is written

$$s(z) = \sum_{i=1}^{L} \{ \alpha_i \eta_i(z) + \beta_i \psi_i(z)\} \qquad (38)$$

where the α_i, β_i are expansion coefficients and $\eta_i(z)$, $\psi_i(z)$ are local cubics defined by

$$
\eta_i(z) = \begin{cases} \eta_o\left(\dfrac{z-z_i}{z_{i+1}-z_i}\right), & z_i \leqslant z \leqslant z_{i+1} \\[2ex] \eta_o\left(\dfrac{z-z_i}{z_i-z_{i-1}}\right), & z_{i-1} \leqslant z \leqslant z_i \\[2ex] 0, & \text{otherwise} \end{cases} \tag{39}
$$

and

$$
\psi_i(z) = \begin{cases} (z_{i+1}-z_i)\psi_o\left(\dfrac{z-z_i}{z_{i+1}-z_i}\right), & z_i \leqslant z \leqslant z_{i+1} \\[2ex] (z_i-z_{i-1})\psi_o\left(\dfrac{z-z_i}{z_i-z_{i-1}}\right), & z_{i-1} \leqslant z < z_i \\[2ex] 0, & \text{otherwise}. \end{cases} \tag{40}
$$

Of course η_I, and ψ_I, vanish for $z < 0$ and η_L and ψ_L vanish for $z > 1$. $\eta_o(x)$ and $\psi_o(x)$ are the canonical cubics

$$
\eta_o(x) = \begin{cases} 2x^3-3x^2+1, & 0 \leqslant x \leqslant 1 \\ -2x^3-3x^2+1, & -1 \leqslant x \leqslant 0 \\ 0, & \text{otherwise}, \end{cases} \tag{41}
$$

and

$$
\psi_o(x) = \begin{cases} x^3-2x^2+x, & 0 \leqslant x \leqslant 1 \\ x^3+2x^2+x, & -1 \leqslant x \leqslant 0 \\ 0, & \text{otherwise}. \end{cases} \tag{42}
$$

$\eta_o(x)$ is the cubic on $(-1,1)$ which has unit value and zero derivative at $x = 0$ and which has vanishing derivative and function values at the end points of the interval. $\psi_o(x)$ also has vanishing derivative and function values at $x = \pm 1$, but has unit derivative and zero function value at $x = 0$. Thus α_i represents the value of $s(z)$ at z_i and β_i is its derivative there. It is convenient to introduce matrix notation and write

$$\begin{pmatrix} s(z) \\ \frac{ds}{dz}(z) \end{pmatrix} \sum_{i=1}^{L} \begin{pmatrix} \eta_i(z) & \psi_i(z) \\ \frac{d\eta_i}{dz}(z) & \frac{d\psi_i}{dz}(z) \end{pmatrix} \begin{pmatrix} \alpha_i \\ \beta_i \end{pmatrix}$$

$$= \sum_{i=1}^{L} D_i(z)\underline{s}_i = \mathcal{D}(z)\underline{s} \tag{43}$$

where we have introduced the 2 x 2L partitioned matrix

$$\mathcal{D} = (D_1, D_2, \ldots D_L) \tag{44}$$

and the partitioned vector $\underline{s}$ with tranpose

$$\underline{s}^T = (\underline{s}_1, \underline{s}_2, \ldots, \underline{s}_L)^T . \tag{45}$$

The functional (37) is then easily shown to be expressible as the quadratic form

$$L(s) = \underline{s}^T P\underline{s} - \underline{s}^T Q\underline{s} - \lambda^2 \underline{s}^T R\underline{s} \tag{46}$$

with
$$P = \int_0^1 \mathcal{D}^T(z) \begin{pmatrix} 0 & 0 \\ 0 & p(z) \end{pmatrix} \mathcal{D}(z)dz , \tag{47}$$

$$Q = \int_0^1 \mathcal{D}^T(z) \begin{pmatrix} q(z) & 0 \\ 0 & 0 \end{pmatrix} \mathcal{D}(z)dz , \tag{48}$$

and
$$R = \int_0^1 \mathcal{D}^T(z) \begin{pmatrix} r(z) & 0 \\ 0 & 0 \end{pmatrix} \mathcal{D}(z)dz . \tag{49}$$

P, Q and R are 2L x 2L symmetric matrices which, because of the local nature of the Hermite splines, are block tri-diagonal with 2 x 2 blocks. Their ith diagonal blocks are respectively

$$\int_0^1 p(z) \begin{pmatrix} \eta_i'^2 & \eta_i' \psi_i' \\ \eta_i' \psi_i' & \psi_i'^2 \end{pmatrix} dz , \tag{50}$$

$$\int_0^1 q(z) \begin{pmatrix} \eta_i^2 & \eta_i \psi_i \\ \eta_i \psi_i & \psi_i^2 \end{pmatrix} dz , \tag{51}$$

and
$$\int_0^1 r(z)\begin{pmatrix} \eta_i^{\prime 2} & \eta_i\psi_i \\ \eta_i\psi_i & \psi_i^2 \end{pmatrix} dz \ , \tag{52}$$

while their sub-diagonal blocks on the ith row of blocks are

$$\int_0^1 p(z)\begin{pmatrix} \eta_i' \, \eta_{i-1}' & \eta_i' \, \psi_{i-1}' \\ \eta_{i-1} \, \psi_i' & \psi_i' \, \psi_{i-1}' \end{pmatrix} dz \ , \tag{53}$$

$$\int_0^1 q(z)\begin{pmatrix} \eta_i \, \eta_{i-1} & \eta_i \, \psi_{i-1} \\ \eta_{i-1} \, \psi_i & \psi_i \, \psi_{i-1} \end{pmatrix} dz \ , \tag{54}$$

and
$$\int_0^1 r(z)\begin{pmatrix} \eta_i \, \eta_{i-1} & \eta_i \, \psi_{i-1} \\ \eta_{i-1} \, \psi_i & \psi_i \, \psi_{i-1} \end{pmatrix} dz \ , \tag{55}$$

respectively. Primes are used to indicate derivatives.

The condition for stationarity of the quadratic form (46) is

$$P\underline{s} - Q\underline{s} - \lambda^2 R\underline{s} = 0 \ . \tag{56}$$

We note that

$$\underline{s}^T R\underline{s} = \int_0^1 r(z) \ s^2(z) \ dz \tag{57}$$

and hence that the matrix R is positive definite in the elliptic case and negative definite in the hyperbolic case. With A = P-Q, the eigenvalue-eigenvector problem becomes

$$A \ \underline{s} = \lambda^2 \ B \ \underline{s} \tag{58}$$

in the elliptic case with B = R positive definite, while in the hyperbolic case we take B = -R, again a positive definite matrix, and write the eigenvalue-eigenvector problem as

$$A\,\underline{s} = \mu^2\, B\,\underline{s}\ . \qquad (59)$$

Problems of the forms (58), (59) where A and B are band matrices have been considered by Peters and Wilkinson (1969) who describe an algorithm for their solution. Alternatively, an algorithm is outlined by Rochester and Smylie (1985) which takes specific advantage of the block tri-diagonal nature of A and B and which appears to be very efficient and stable in practice. As shown there, the condition (36) for a solution regular at $z = 1$ can be incorporated directly by replacing the usual inverse iteration scheme for the eigenvectors of (58) or (59) by

$$\{A_1 + \lambda_i^2\, A_2 - \lambda_i^2(B_1 + \lambda_i^2\, B_2)\}\ \underline{s}_{r+1}$$
$$= \{B_1 + 2\,\lambda_i^2\, B_2 - A_2\}\,\underline{s}_r\ , \qquad (60)$$

where we write $A = A_1 + \lambda_i^2\, A_2$, $B = B_1 + \lambda_i^2\, B_2$ in explicit recognition of the fact that condition (36) requires terms in A and B proportional to the particular eigenvalue λ_i^2 being considered. Finally, we observe that for solutions even in the equatorial plane β_1 vanishes, while for those odd in the equatorial plane α_1 vanishes. The systems (58), (59) are therefore of order $2L - 1$.

A further numerical advantage can be gained from the polynomial nature of $p(z)$, $q(z)$ and $r(z)$ in the integrands of the matrices of integrals P, Q and R given by (47), (48) and (49). These integrations can be carried out and stored with powers of z^2 replacing $p(z)$, $q(z)$ and $r(z)$ and then when σ is changed no further integrations are required since P, Q, and R can be formed as linear combinations of existing stored matrices.

In the hyperbolic case, for a strongly stably stratified core, solution of the eigenvalue-eigenvector problem (59) yields eigenfunctions (27) of the forms

$$\chi = r^{\underline{\eta+i\mu}}\left(1-z^2\right)^{\frac{m}{2}}\left(\zeta_1^2-z^2\right)^{3/2}\left(\zeta_2^2-z^2\right)^{\frac{1+i\mu}{2}} s(z)\ e^{im\phi}\ . \qquad (61)$$

Thus, the nth eigenfunction can be written as the linear combination

$$\chi_n = r^{-\eta}\left\{A_n \cos\left(\frac{\mu n}{2}\log\frac{r^2}{r_o^2}\left(\zeta_2^2-z^2\right)\right) + B_n \sin\left(\frac{\mu n}{2}\log\frac{r^2}{r_o^2}\left(\zeta_2^2-z^2\right)\right)\right\}\cdot \qquad (62)$$
$$\left(1-z^2\right)^{\frac{m}{2}}\left(\zeta_2^2-z^2\right)^{3/2}\left(\zeta_2^2-z^2\right)^{1/2} s(z)\ e^{im\phi},$$

where A_n, B_n are arbitrary constants and r_o is some fixed reference radius. Unfortunately, these individual eigenfunctions cannot be made to satisfy boundary conditions of the form (21) and they must be superimposed to meet boundary constraints.

Our technique is to form the residual function,

$$R(r,z) = \sum_{n=1}^{N} \left(s_n, \frac{ds_n}{dz}\right) \begin{pmatrix} \alpha & \gamma \\ \beta & 0 \end{pmatrix} \cdot$$

$$\cdot \begin{pmatrix} \cos\left(\frac{\mu_n}{2} \log \frac{r^2}{r_o^2}(\zeta_2^2-z^2)\right) & \sin\left(\frac{\mu_n}{2} \log \frac{r^2}{r_o^2}(\zeta_2^2-z^2)\right) \\ -\mu_n \sin\left(\frac{\mu_n}{2} \log \frac{r^2}{r_o^2}(\zeta_2^2-z^2)\right) & \mu_n \cos\left(\frac{\mu_n}{2} \log \frac{r^2}{r_o^2}(\zeta_2^2-z^2)\right) \end{pmatrix} \begin{pmatrix} A_n \\ B_n \end{pmatrix}, \tag{63}$$

where

$$\alpha(z) = m\sigma(\zeta_1^2-z^2)(\zeta_2^2-z^2) + \eta(\sigma^2-z^2)(\zeta_1^2-z^2)(\zeta_2^2-z^2)$$

$$+mz^2(\zeta_1^2-z^2)(\zeta_2^2-z^2) + 3\, z^2(1-z^2)(\zeta_2^2-z^2) + z^2(1-z^2)(\zeta_1^2-z^2) \tag{64}$$

$$\beta(z) = -z\,(1-z^2)(\zeta_1^2-z^2)(\zeta_2^2-z^2)\,, \tag{65}$$

and

$$\gamma(z) = (\sigma^2-z^2)(\zeta_1^2-z^2)(\zeta_2^2-z^2) + z^2(1-z^2)(\zeta_1^2-z^2)\,, \tag{66}$$

as a measure of how closely the condition (21) on the radial displacement is met. For the rigid boundaries assumed here our object is to make $R(r,z)$ as close to zero as possible on both boundaries of the fluid core. Substitution of expression (43) in (63) gives directly the values of R on both boundaries in terms of the nodal values of the eigenfunctions. Differentiation of (63) and the use of the differential equation (34) to replace $\frac{d^2 s_n}{dz^2}$ with

$$\frac{1}{p(z)}\left[-\frac{dp(z)}{dz}\frac{ds_n}{dz} - (q(z) - \mu_n^2\, r(z))s_n(z)\right] \tag{67}$$

yields a similar expression for $\frac{\partial R}{\partial z}$ at the nodes. Hence, in analogy to (43) we have the expression

$$\begin{pmatrix} R(r,z) \\ \frac{\partial R}{\partial z} \end{pmatrix} = \mathcal{D}(z)\ \underline{c}\ (r,\sigma)\,. \tag{68}$$

The eigenfunctions $s_n(z)$ are then themselves used to expand $R(r,z)$ on both boundaries as

$$R = \sum_{n=1}^{N} C_n(r)\, s_n(z) \tag{69}$$

and

$$\frac{\partial R}{\partial z} = \sum_{n=1}^{N} C_n(r)\, \frac{ds_n(z)}{dz} , \tag{70}$$

with the coefficients of the expansion for appropriately normalized eigenfunctions being given by

$$C_n = - \int_0^1 r(z)\, s_n(z)\, R(z)\, dz, \; n = 1, 2, \ldots, N . \tag{71}$$

For rigid boundaries, we set these coefficients to zero at both boundaries to ensure that both the residual function (69) and its derivative (70) vanish.

Expressions (43), (49) and (68) show that the conditional equations take the form

$$\begin{aligned} \underline{s}_n^T R\, \underline{c}\, (a,\sigma) &= 0 , \\ & \qquad n = 1, 2, \ldots, N \\ \underline{s}_n^T R\, \underline{c}\, (b,\sigma) &= 0 , \end{aligned} \tag{72}$$

where a is the radius of the inner boundary and b is that of the outer boundary. Note that in equations (71), (72) $r(z)$ is the coefficient introduced in the Sturm-Liouiville equation (34) and not the radius while R is the matrix (49) and not the residual function. The conditions (72) constitute a system of homogenous equations of order 2N and the determinant of the coefficient matrix becomes the secular determinant for the calculation of eignefrequencies.

Figures 4, 5, 6 and 7 show the eigenfunctions $s_1(z)$, $s_2(z)$, $s_5(z)$ and $s_{10}(z)$ for a core with $N^2 = 6$ and $M^2 = 6/35$. These values correspond to a Väisälä period of 4.9 hr and a compressibility number of 0.17. Eigenperiods computed using L = 50, N = 10 are shown in Figure 8. In all cases results are for solutions even in the equatorial plane with azimuthal number m = 1.

4. DISCUSSION

A separation of variables solution to the subseismic equation for constant core properties has been outlined. Numerical techniques have been developed and implemented for a stably-stratified core. While the

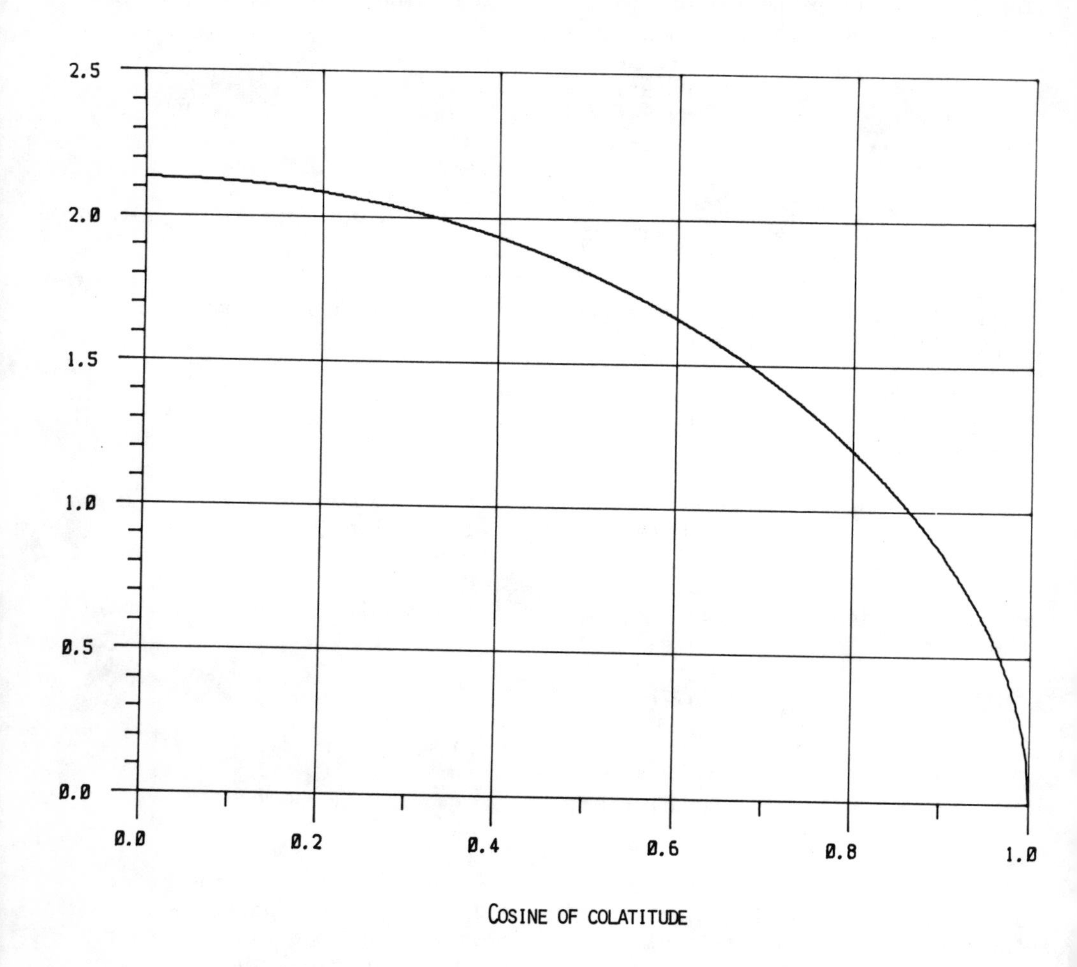

Figure 4. Eigenfunction $s_1(z)$, where z is the cosine of the colatitude, for a core with $N^2=6$, $M=6/35$.

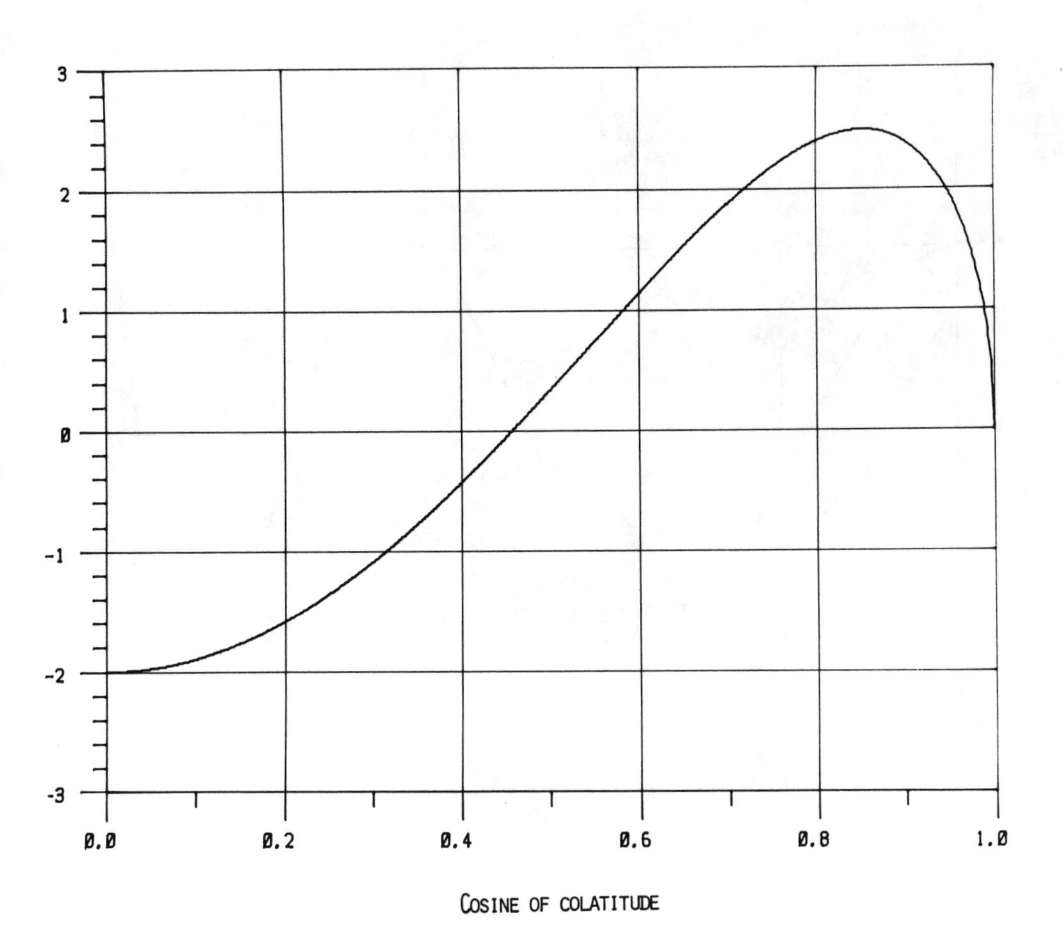

Figure 5. Eigenfunction $s_2(z)$, where z is the cosine of the colatitude, for a core with $N^2=6$, $M^2=6/35$.

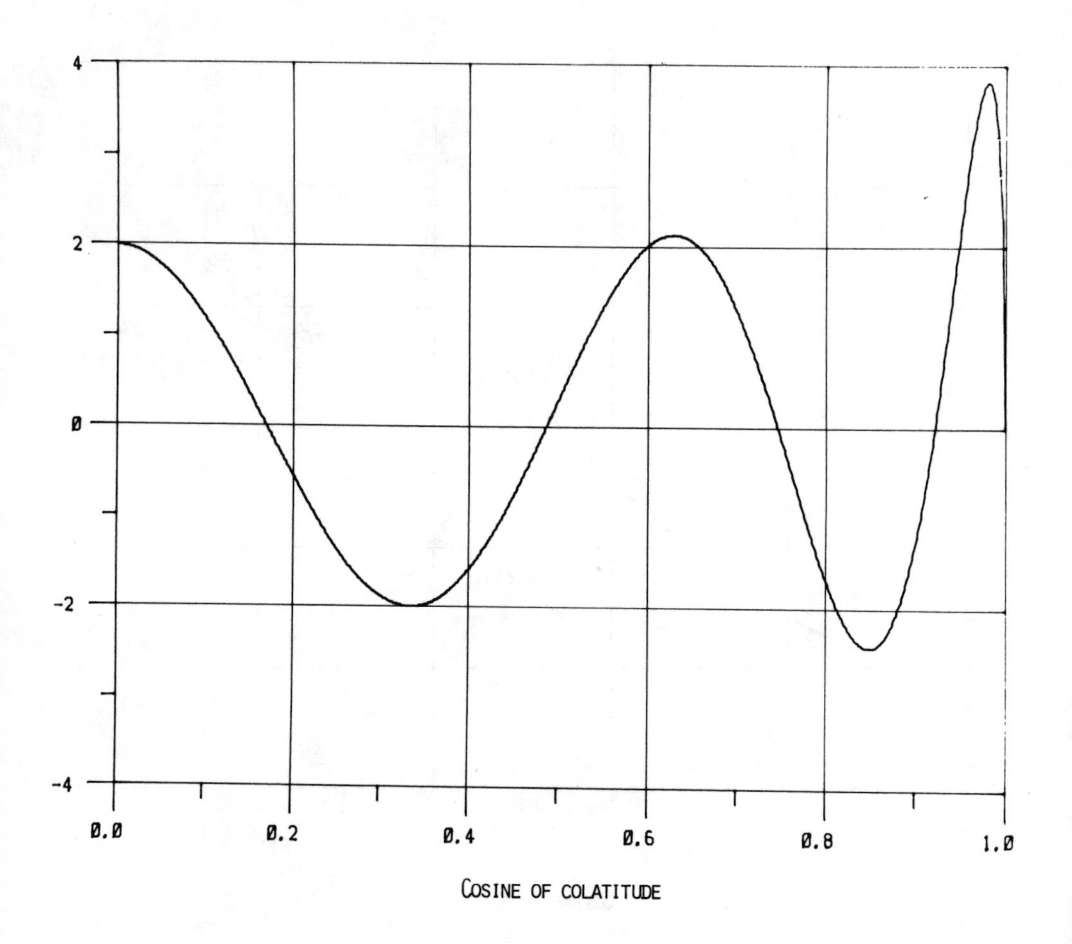

Figure 6. Eigenfunction $s_5(z)$, where z is the cosine of the colatitude, for a core with $N^2=6$, $M^2=6/35$.

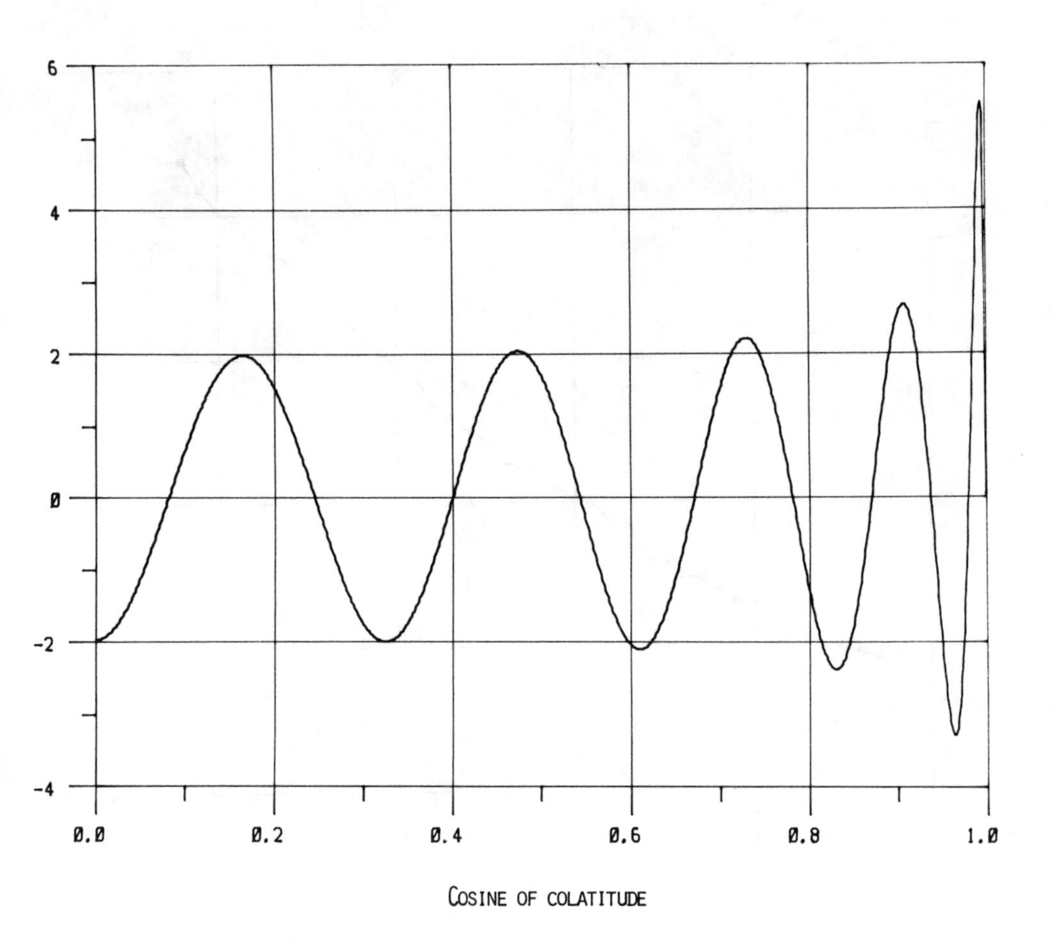

Figure 7. Eigenfunction $s_{10}(z)$, where z is the cosine of the colatitude, for a core with $N^2=6$, $M^2=6/35$.

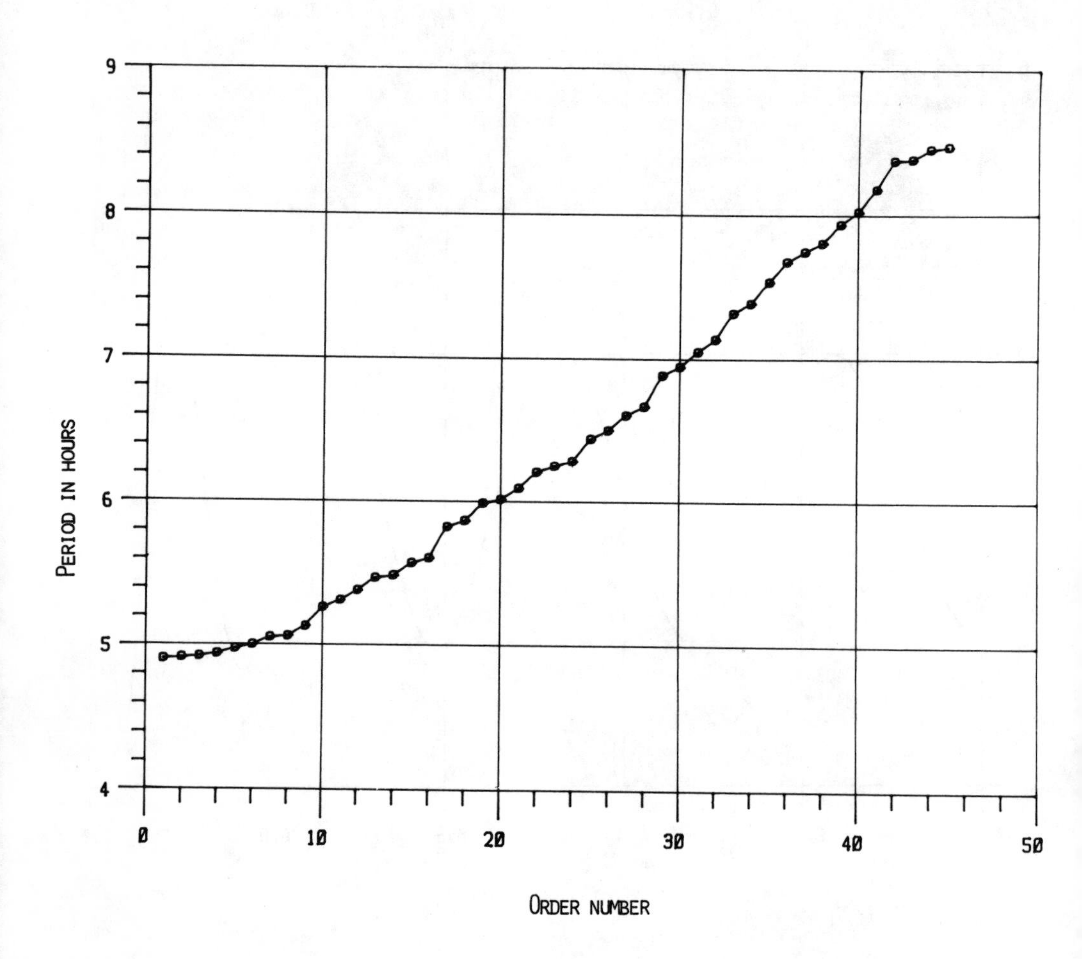

Figure 8. Eigenperiods computed using the first ten eigenfunctions of the subseismic equation for a core with $N^2=6$, $M^2=6/35$.

model is restricted to constant core properties, it serves the very useful purpose of displaying explicitly the relation between the analytical properties of solutions to the subseismic equation and the physical conditions prevailing in the liquid core. We expect this to be extremely useful in guiding solutions to the problem for variable core properties via a recently developed variational principle (Smylie and Rochester 1985).

While some further development remains to be completed, particularly in the connection between solutions of the subseismic equation and the classical Poincaré equation, a firm basis has now been established for a complete dynamical description of the liquid core in a wide variety of Earth dynamics calculations.

ACKNOWLEDGEMENTS

One of us (D.E.S.) thanks the National Research Council of Canada for the award of a six month visit to the Groupe de Recherches de Geodesie Spatiale in Toulouse under the France-Canada Exchange of Scientists and Engineers Programme where much of this work was carried out. He especially thanks Bernard Lago and Anny Cazenave for the many arrangements required to make this visit pleasant and productive. D.E.S. is also grateful to the Natural Sciences and Engineering Research Council (NSERC) for the award of time on the Cray vector processor which allowed definitive numerical results to be obtained. Both of us thank NSERC for continued generous support through operating grants.

REFERENCES

Batchelor, G. K., 1967. An Introduction to Fluid Dynamics, Cambridge University Press, Cambridge, 615 pp.

Bryan, G. H., 1889. The waves on a rotating liquid spheroid of finite ellipticity, Phil. Trans. Roy. Soc. London, A, 180, 187-219.

Courant, R. and Hilbert, D., 1953. Methods of Mathematical Physics I, Interscience, New York, 561 pp.

Crossley, D. J., 1975. Core Undertones with Rotation, Geophys. J.R. astr. Soc., 42, 477-488.

Crossley, D. J. and Rochester, M. G., 1980. Simple core undertones, Geophys. J.R. astr. Soc., 60, 129-161.

Greenspan, H. P., 1969. The Theory of Rotating Fluids, Cambridge University Press, Cambridge, 328 pp.

Hopkins, W., 1839. Researches in physical geology, Phil. Trans. R. Soc. Lond., 129, 381-423.

Hough, S. S., 1895. The oscillations of a rotating ellipsoidal shell containing fluid, Phil. Trans. R. Soc. Lond., A, 186, 469-506.

Ince, E. L., 1926. Ordinary Differential Equations, Longmans, Green and Company, London, 558 pp.

Johnson, I. M. and Smylie, D. E., 1977. A variational approach to whole-Earth dynamics, Geophys. J.R. astr. Soc., 50, 35-54.

Needler, G. T. and LeBlond, P.H., 1973. On the influence of the horizontal component of the Earth's rotation on long period waves. Geophys.

Fluid Dyn., 5, 23-45.
Peters, G. and Wilkinson, J. H., 1969. Eigenvalues of Ax = λ Bx with band symmetric A and B, Computer J., 12, 398-404.
Poincaré, H., 1885. Sur l'équilibre d'une masse fluide animée d'un mouvement de rotation, Acta Math., 7, 259-380.
Roberts, P. H. and Stix, M., 1972. α-Effect dynamos by the Bullard-Gellman formalism, Astron. and Astrophys., 18, 453-466.
Rochester, M. G. and Smylie, D. E., 1985. Long period core dynamics and solutions of the subseismic wave equation, in preparation.
Shen, P. Y. and Mansinha, L., 1976. Oscillation, nutation and wobble of an elliptical rotating Earth with liquid outer core, Geophys. J.R. astr. Soc., 46, 467-496.
Schultz, M. H., 1972. Spline analysis, Prentice-Hall, New Jersey.
Sludskii, F., 1896. De la rotation de la Terre supposée fluide à son intérieur, Bull. Soc. Natur. Moscou, 9, 285-318.
Smith, M. L., 1974. The scalar equation of infinitesimal elastic-gravitational motion of a rotating, slightly elliptical Earth, Geophys. J.R. astr. Soc., 37, 491-526.
Smith, M. L., 1977. Wobble and nutation of the Earth, Geophys. J. R. astr. Soc., 50, 103-140.
Smylie, D. E., 1974. Dynamics of the outer core, Veröff. Zentralinst Phys. Erde, Akad. Wiss. D.D.R., 30, 91-104.
Smylie, D. E. and Rochester, M. G., 1981. Compressibility, core dynamics and the subseismic wave equation, Phys. Earth Planet. Int., 11, 43-60.
Smylie, D. E., Szeto, A.M.K. and Rochester, M. G., 1984. The dynamics of the Earth's inner and outer cores, Rep. Prog. Phys., 47, 855-906.
Smylie, D. E. and Rochester, M. G. 1985. On a variational principle for the subseismic wave equation, submitted to Geophys. J. R. astr. Soc.
Stewartson, K., and Rickard, J. A., 1969. Pathological oscillations of a rotating fluid, J. Fluid Mech., 35, 759-773.
Stewartson, K., 1978. Homogeneous fluids in rotation. Section B: Waves, in Rotating Fluids in Geophysics, eds. P. H. Roberts and A. M. Soward, Academic Press, New York, 67-103.
Wahr, J. M., Sasao, T. and Smith, M. L. 1981. Geophys. J.R. astr. Soc. 64, 635-765.
Wahr, J. M., 1981. A normal mode expansion for the forced response of a rotating Earth, Geophys. J. R. astr. Soc., 64, 651-675.

Recommendations adopted by the Participants
of the Advanced Research Workshop
EARTH ROTATION: SOLVED AND UNSOLVED PROBLEMS

This workshop, meeting at Chateau de Bonas, Gers, France, 11-13 June 1985,

recognizing:

- that the precision and accuracy of the Earth Rotation Parameters (ERP -- including both Earth rotation and polar motion) have been dramatically improved by the various techniques such as very-long-baseline-interometry, lunar and satellite laser ranging;

- that the intercomparison of various techniques made during the recent MERIT campaign (September 1983--October 1984) has demonstrated that the accuracy of Earth rotation and polar motion at the few tenths of a millisecond and at a few milliarcsecond level respectively has been achieved and is obtainable on a regular routine basis; (MERIT, an acronym for Measurement of Earth Rotation and Intercomparison of Techniques and Analysis, is a joint International Astronomical Union and International Union of Geodesy and Geophysics/International Association of Geodesy working group on Earth rotation.)

- that the progress in ERP data collection on all time scales, from a few days to many centuries, is reaching new accuracies that permit new additional insights into geophysical phenomena;

- that the relation between variations of Earth rotation and geophysical qualities such as the geomagnetic index and atmospheric angular momenta have been studied, but are not yet fully understood;

recommends:

- that ERP data be acquired by the various techniques at the best level possible;

- that the various techniques strive to improve and document their measurements, modeling and analysis;

- that the atmospheric angular momenta data be made available from a world center, in a standard format and deduced from an adopted international reference model;

- that studies and data collection for a better understanding of the ocean-atmosphere coupling and its relation with variations of the Earth rotation be initiated;

A. Cazenave (ed.), Earth Rotation: Solved and Unsolved Problems, 325–326.

. that the collection of geophysical data such as the geomagnetic index be continued and that these data be made available to the general community via a world center.

We congratulate the MERIT working group and all organizations and individuals who have contributed to the development and implementation of the MERIT program. Finally, we thank most sincerely the NATO Science Committee and the Council of Europe for their support, and the Local Organizing Committee for the provision of excellent facilities and generous hospitality.

SUBJECT INDEX